Feature Extraction
and
Image Processing

D1323610

Dedication
We would like to dedicate this book to our parents:
To Gloria and Joaquin Aguado, and to Brenda and the late Ian Nixon.

Feature Extraction
and
Image Processing

Second edition

Mark S. Nixon

Alberto S. Aguado

AMSTERDAM • BOSTON • HEIDELBERG • LONDON • NEW YORK • OXFORD
PARIS • SAN DIEGO • SAN FRANCISCO • SINGAPORE • SYDNEY • TOKYO

Academic Press is an imprint of Elsevier

Academic Press is an imprint of Elsevier
84 Theobald's Road, London WC1X 8RR, UK
Radarweg 29, PO Box 211, 1000 AE Amsterdam, The Netherlands
30 Corporate Drive, Suite 400, Burlington, MA 01803, USA
525 B Street, Suite 1900, San Diego, CA 92101-4495, USA

First edition 2002
Reprinted 2004, 2005
Second edition 2008
Reprinted 2008

British Library Cataloguing in Publication Data
A catalogue record for this book is available from the British Library

Library of Congress Cataloging-in-Publication Data
A catalog record for this book is available from the Library of Congress

ISBN: 978-0-1237-2538-7

For information on all Academic Press publications
visit our website at www.elsevierdirect.com

Transferred to Digital Printing in 2009

Working together to grow
libraries in developing countries

www.elsevier.com | www.bookaid.org | www.sabre.org

ELSEVIER   BOOK AID
            International   Sabre Foundation

Contents

Preface   xi

1 Introduction   **1**

  1.1 Overview   1
  1.2 Human and computer vision   1
  1.3 The human vision system   3
    1.3.1 The eye   4
    1.3.2 The neural system   6
    1.3.3 Processing   7
  1.4 Computer vision systems   9
    1.4.1 Cameras   10
    1.4.2 Computer interfaces   12
    1.4.3 Processing an image   14
  1.5 Mathematical systems   15
    1.5.1 Mathematical tools   16
    1.5.2 Hello Mathcad, hello images!   16
    1.5.3 Hello Matlab!   21
  1.6 Associated literature   24
    1.6.1 Journals and magazines   24
    1.6.2 Textbooks   25
    1.6.3 The web   28
  1.7 Conclusions   29
  1.8 References   29

2 Images, sampling and frequency domain processing   **33**

  2.1 Overview   33
  2.2 Image formation   34
  2.3 The Fourier transform   37
  2.4 The sampling criterion   43
  2.5 The discrete Fourier transform   47
    2.5.1 One-dimensional transform   47
    2.5.2 Two-dimensional transform   49
  2.6 Other properties of the Fourier transform   54
    2.6.1 Shift invariance   54
    2.6.2 Rotation   56
    2.6.3 Frequency scaling   56
    2.6.4 Superposition (linearity)   57

2.7	Transforms other than Fourier	58
	2.7.1 Discrete cosine transform	58
	2.7.2 Discrete Hartley transform	59
	2.7.3 Introductory wavelets: the Gabor wavelet	61
	2.7.4 Other transforms	63
2.8	Applications using frequency domain properties	64
2.9	Further reading	65
2.10	References	66

3 Basic image processing operations   **69**

3.1	Overview	69
3.2	Histograms	70
3.3	Point operators	71
	3.3.1 Basic point operations	71
	3.3.2 Histogram normalization	74
	3.3.3 Histogram equalization	75
	3.3.4 Thresholding	77
3.4	Group operations	81
	3.4.1 Template convolution	81
	3.4.2 Averaging operator	84
	3.4.3 On different template size	87
	3.4.4 Gaussian averaging operator	88
3.5	Other statistical operators	90
	3.5.1 More on averaging	90
	3.5.2 Median filter	91
	3.5.3 Mode filter	94
	3.5.4 Anisotropic diffusion	96
	3.5.5 Force field transform	101
	3.5.6 Comparison of statistical operators	102
3.6	Mathematical morphology	103
	3.6.1 Morphological operators	104
	3.6.2 Grey-level morphology	107
	3.6.3 Grey-level erosion and dilation	108
	3.6.4 Minkowski operators	109
3.7	Further reading	112
3.8	References	113

4 Low-level feature extraction (including edge detection)   **115**

4.1	Overview	115
4.2	First order edge detection operators	117
	4.2.1 Basic operators	117
	4.2.2 Analysis of the basic operators	119
	4.2.3 Prewitt edge detection operator	121
	4.2.4 Sobel edge detection operator	123
	4.2.5 Canny edge detection operator	129

4.3	Second order edge detection operators		137
	4.3.1	Motivation	137
	4.3.2	Basic operators: the Laplacian	137
	4.3.3	Marr–Hildreth operator	139
4.4	Other edge detection operators		144
4.5	Comparison of edge detection operators		145
4.6	Further reading on edge detection		146
4.7	Phase congruency		147
4.8	Localized feature extraction		152
	4.8.1	Detecting image curvature (corner extraction)	153
		4.8.1.1 Definition of curvature	153
		4.8.1.2 Computing differences in edge direction	154
		4.8.1.3 Measuring curvature by changes in intensity (differentiation)	156
		4.8.1.4 Moravec and Harris detectors	159
		4.8.1.5 Further reading on curvature	163
	4.8.2	Modern approaches: region/patch analysis	163
		4.8.2.1 Scale invariant feature transform	163
		4.8.2.2 Saliency	166
		4.8.2.3 Other techniques and performance issues	167
4.9	Describing image motion		167
	4.9.1	Area-based approach	168
	4.9.2	Differential approach	171
	4.9.3	Further reading on optical flow	177
4.10	Conclusions		178
4.11	References		178
5	**Feature extraction by shape matching**		**183**
5.1	Overview		183
5.2	Thresholding and subtraction		184
5.3	Template matching		186
	5.3.1	Definition	186
	5.3.2	Fourier transform implementation	193
	5.3.3	Discussion of template matching	196
5.4	Hough transform		196
	5.4.1	Overview	196
	5.4.2	Lines	197
	5.4.3	Hough transform for circles	203
	5.4.4	Hough transform for ellipses	207
	5.4.5	Parameter space decomposition	210
		5.4.5.1 Parameter space reduction for lines	210
		5.4.5.2 Parameter space reduction for circles	212
		5.4.5.3 Parameter space reduction for ellipses	217
5.5	Generalized Hough transform		221
	5.5.1	Formal definition of the GHT	221
	5.5.2	Polar definition	223

	5.5.3	The GHT technique	224
	5.5.4	Invariant GHT	228
5.6	Other extensions to the Hough transform		235
5.7	Further reading		236
5.8	References		237

6   Flexible shape extraction (snakes and other techniques)   **241**

6.1	Overview		241
6.2	Deformable templates		242
6.3	Active contours (snakes)		244
	6.3.1	Basics	244
	6.3.2	The greedy algorithm for snakes	246
	6.3.3	Complete (Kass) snake implementation	252
	6.3.4	Other snake approaches	257
	6.3.5	Further snake developments	257
	6.3.6	Geometric active contours	261
6.4	Shape skeletonization		266
	6.4.1	Distance transforms	266
	6.4.2	Symmetry	268
6.5	Flexible shape models: active shape and active appearance		272
6.6	Further reading		275
6.7	References		276

7   Object description   **281**

7.1	Overview		281
7.2	Boundary descriptions		282
	7.2.1	Boundary and region	282
	7.2.2	Chain codes	283
	7.2.3	Fourier descriptors	285
		7.2.3.1 Basis of Fourier descriptors	286
		7.2.3.2 Fourier expansion	287
		7.2.3.3 Shift invariance	289
		7.2.3.4 Discrete computation	290
		7.2.3.5 Cumulative angular function	292
		7.2.3.6 Elliptic Fourier descriptors	301
		7.2.3.7 Invariance	305
7.3	Region descriptors		311
	7.3.1	Basic region descriptors	311
	7.3.2	Moments	315
		7.3.2.1 Basic properties	315
		7.3.2.2 Invariant moments	318
		7.3.2.3 Zernike moments	320
		7.3.2.4 Other moments	324
7.4	Further reading		325
7.5	References		326

8 Introduction to texture description, segmentation and classification   **329**

8.1   Overview   329
8.2   What is texture?   330
8.3   Texture description   332
    8.3.1   Performance requirements   332
    8.3.2   Structural approaches   332
    8.3.3   Statistical approaches   335
    8.3.4   Combination approaches   337
8.4   Classification   339
    8.4.1   The k-nearest neighbour rule   339
    8.4.2   Other classification approaches   343
8.5   Segmentation   343
8.6   Further reading   345
8.7   References   346

9 Appendix 1: Example worksheets   **349**

9.1   Example Mathcad worksheet for Chapter 3   349
9.2   Example Matlab worksheet for Chapter 4   352

10 Appendix 2: Camera geometry fundamentals   **355**

10.1   Image geometry   355
10.2   Perspective camera   355
10.3   Perspective camera model   357
    10.3.1   Homogeneous coordinates and projective geometry   357
        10.3.1.1   Representation of a line and duality   358
        10.3.1.2   Ideal points   358
        10.3.1.3   Transformations in the projective space   359
    10.3.2   Perspective camera model analysis   360
    10.3.3   Parameters of the perspective camera model   363
10.4   Affine camera   364
    10.4.1   Affine camera model   365
    10.4.2   Affine camera model and the perspective projection   366
    10.4.3   Parameters of the affine camera model   368
10.5   Weak perspective model   369
10.6   Example of camera models   371
10.7   Discussion   379
10.8   References   380

11 Appendix 3: Least squares analysis   **381**

11.1   The least squares criterion   381
11.2   Curve fitting by least squares   382

12   Appendix 4: Principal components analysis   385

12.1   Introduction   385
12.2   Data   385
12.3   Covariance   386
12.4   Covariance matrix   388
12.5   Data transformation   389
12.6   Inverse transformation   390
12.7   Eigenproblem   391
12.8   Solving the eigenproblem   392
12.9   PCA method summary   392
12.10 Example   393
12.11 References   398

Index   399

Preface

Why did we write this book?

We will no doubt be asked many times: why on earth write a new book on computer vision? Fair question: there are already many good books on computer vision in the bookshops, as you will find referenced later, so why add to them? Part of the answer is that any textbook is a snapshot of material that exists before it. Computer vision, the art of processing images stored within a computer, has seen a considerable amount of research by highly qualified people and the volume of research would appear even to have increased in recent years. This means that a lot of new techniques have been developed, and many of the more recent approaches have yet to migrate to textbooks.

But it is not just the new research: part of the speedy advance in computer vision technique has left some areas covered only in scanty detail. By the nature of research, one cannot publish material on technique that is seen more to fill historical gaps, rather than to advance knowledge. This is again where a new text can contribute.

Finally, the technology itself continues to advance. This means that there is new hardware, and there are new programming languages and new programming environments. In particular for computer vision, the advance of technology means that computing power and memory are now relatively cheap. It is certainly considerably cheaper than when computer vision was starting as a research field. One of the authors here notes that the laptop that his portion of the book was written on has more memory, is faster, and has bigger disk space and better graphics than the computer that served the entire university of his student days. And he is not that old! One of the more advantageous recent changes brought about by progress has been the development of mathematical programming systems. These allow us to concentrate on mathematical technique itself, rather than on implementation detail. There are several sophisticated flavours, of which Mathcad and Matlab, the chosen vehicles here, are among the most popular. We have been using these techniques in research and teaching, and we would argue that they have been of considerable benefit there. In research, they help us to develop technique more quickly and to evaluate its final implementation. For teaching, the power of a modern laptop and a mathematical system combines to show students, in lectures and in study, not only how techniques are implemented, but also how and why they work with an explicit relation to conventional teaching material.

We wrote this book for these reasons. There is a host of material that we could have included but chose to omit. Our apologies to other academics if it was your own, or your favourite, technique. By virtue of the enormous breadth of the subject of computer vision, we restricted the focus to feature extraction and image processing in computer vision, for this not only has been the focus of our research, but is also where the attention of established textbooks, with some exceptions, can be rather scanty. It is, however, one of the prime targets of applied computer vision, so would benefit from better attention. We have aimed to clarify some of its origins and development, while also exposing implementation using mathematical systems. As such, we have written this text with our original aims in mind.

Why did we produce another edition?

There are many reasons why we have updated the book to provide this new edition. First, despite its electronic submission, some of the first edition was retyped before production. This introduced errors that we have now corrected. Next, the field continues to move forward: we now include some techniques which were gaining appreciation when we first wrote the book, or have been developed since. Some areas move more rapidly than others, and this is reflected in the changes made. Also, there has been interim scholarship, especially in the form of new texts, and we include these new ones as much as we can. Matlab and Mathcad are still the computational media here, and there is a new demonstration site which uses Java. Finally, we have maintained the original format. It is always tempting to change the format, in this case even to reformat the text, but we have chosen not to do so. Apart from corrections and clarifications, the main changes from the previous edition are:

- Chapter 1: updating of eye operation, camera technology and software, updating and extension of web material and literature
- Chapter 2: very little (this is standard material), except for an excellent example of aliasing
- Chapter 3: inclusion of anisotropic diffusion for image smoothing, the force field operator and mathematical morphology
- Chapter 4: extension of frequency domain concepts and differentiation operators; inclusion of phase congruency, modern curvature operators and the scale invariant feature transform (SIFT)
- Chapter 5: emphasis of the practical attributes of feature extraction in occlusion and noise, and some moving-feature techniques
- Chapter 6: inclusion of geometric active contours and level set methods, inclusion of skeletonization, extension of active shape models
- Chapter 7: extension of the material on moments, particularly Zernike moments, including reconstruction from moments
- Chapter 8: clarification of some of the detail in feature-based recognition
- Appendices: these have been extended to cover camera models in greater detail, and principal components analysis.

As already mentioned, there is a new JAVA-based demonstration site, at http://www.ecs.soton. ac.uk/~msn/book/new_demo/, which has some of the techniques described herein and some examples of how computer vision-based biometrics work. This webpage will continue to be updated.

The book and its support

Each chapter of the book presents a particular package of information concerning feature extraction in image processing and computer vision. Each package is developed from its origins and later referenced to more recent material. Naturally, there is often theoretical development before implementation (in Mathcad or Matlab). We have provided working implementations of most of the major techniques we describe, and applied them to process a selection of imagery. Although the focus of our work has been more in analysing medical imagery or in biometrics (the science of recognizing people by behavioural or physiological characteristics, like face recognition), the techniques are general and can migrate to other application domains.

You will find a host of further supporting information at the book's website (http://www.ecs.soton.ac.uk/~msn/book/). First, you will find the *worksheets* (the Matlab and Mathcad implementations that support the text) so that you can study the techniques described herein. There are also *lecturing versions* that have been arranged for display via an overhead projector, with enlarged text and more interactive demonstration. The example questions (and, eventually, their answers) are also there. The *demonstration* site is there too. The website will be kept as up to date as possible, for it also contains links to other material such as websites devoted to techniques and to applications, as well as to available software and online literature. Finally, any errata will be reported there. It is our regret and our responsibility that these will exist, but our inducement for their reporting concerns a pint of beer. If you find an error that we do not know about (not typos such as spelling, grammar and layout) then use the mailto on the website and we shall send you a pint of good English *beer*, free!

There is a certain amount of mathematics in this book. The target audience is third or fourth year students in BSc/BEng/MEng courses in electrical or electronic engineering, software engineering and computer science, or in mathematics or physics, and this is the level of mathematical analysis here. Computer vision can be thought of as a branch of applied mathematics, although this does not really apply to some areas within its remit, but certainly applies to the material herein. The mathematics essentially concerns mainly calculus and geometry, although some of it is rather more detailed than the constraints of a conventional lecture course might allow. Certainly, not all of the material here is covered in detail in undergraduate courses at Southampton.

The book starts with an overview of computer vision hardware, software and established material, with reference to the most sophisticated vision system yet 'developed': the *human vision* system. Although the precise details of the nature of processing that allows us to see have yet to be determined, there is a considerable range of *hardware* and *software* that allow us to give a computer system the capability to acquire, process and reason with imagery, the function of 'sight'. The first chapter also provides a comprehensive *bibliography* of material on the subject, including not only textbooks, but also available software and other material. As this will no doubt be subject to change, it might well be worth consulting the website for more up-to-date information. The preferred journal references are those that are likely to be found in local university libraries or on the web, *IEEE Transactions* in particular. These are often subscribed to as they are relatively low cost, and are often of very high quality.

The next chapter concerns the basics of signal processing theory for use in computer vision. It introduces the Fourier transform, which allows you to look at a signal in a new way, in terms of its frequency content. It also allows us to work out the minimum size of a picture to conserve information and to analyse the content in terms of frequency, and even helps to speed up some of the later vision algorithms. Unfortunately, it does involve a few equations, but it is a new way of looking at data and signals, and proves to be a rewarding topic of study in its own right.

We then start to look at *basic* image-processing techniques, where image points are mapped into a new value first by considering a single point in an original image, and then by considering groups of points. We see not only common operations to make a picture's appearance better, especially for human vision, but also how to reduce the effects of different types of commonly encountered image noise. This is where the techniques are implemented as algorithms in Mathcad and Matlab to show precisely how the equations work. We shall see some of the modern ways to remove noise and thus clean images, and we shall also look at techniques which process an image using notions of shape, rather than mapping processes.

The following chapter concerns *low-level features*, which are the techniques that describe the content of an image, at the level of a whole image rather than in distinct regions of it. One of the most important processes is *edge detection*. Essentially, this reduces an image to a form of a caricaturist's sketch, but without a caricaturist's exaggerations. The major techniques are presented in detail, together with descriptions of their implementation. Other image properties we can derive include measures of *curvature* and measures of *movement*. These also are covered in this chapter.

These edges, the curvature or the motion need to be grouped in some way so that we can find shapes in an image. Our first approach to *shape extraction* concerns analysing the *match* of low-level information to a known template of a target shape. As this can be computationally very cumbersome, we then progress to a technique that improves computational performance, while maintaining an optimal performance. The technique is known as the *Hough transform*, and it has long been a popular target for researchers in computer vision who have sought to clarify its basis, improve its speed, and increase its accuracy and robustness. Essentially, by the Hough transform we estimate the parameters that govern a shape's appearance, where the shapes range from *lines* to *ellipses* and even to *unknown shapes*.

Some applications of shape extraction require the determination of rather more than the parameters that control appearance, but require the ability to *deform* or *flex* to match the image template. For this reason, the chapter on shape extraction by matching is followed by one on *flexible shape* analysis. This is a topic that has shown considerable progress of late, especially with the introduction of *snakes* (*active contours*). The newer material is the formulation by level set methods, and brings new power to shape-extraction techniques. These seek to match a shape to an image by analysing local properties. Further, we shall see how we can describe a shape by its *skeleton*, although with practical difficulty which can be alleviated by *symmetry* (though this can be slow), and also how global constraints concerning the *statistics* of a shape's appearance can be used to guide final extraction.

Up to this point, we have not considered techniques that can be used to describe the shape found in an image. We shall find that the two major approaches concern techniques that describe a shape's perimeter and those that describe its area. Some of the *perimeter description* techniques, the Fourier descriptors, are even couched using Fourier transform theory, which allows analysis of their frequency content. One of the major approaches to *area description*, statistical moments, also has a form of access to frequency components, but is of a very different nature to the Fourier analysis. One advantage is that insight into descriptive ability can be achieved by *reconstruction*, which should get back to the original shape.

The final chapter describes *texture* analysis, before some introductory material on *pattern classification*. Texture describes patterns with no known analytical description and has been the target of considerable research in computer vision and image processing. It is used here more as a vehicle for material that precedes it, such as the Fourier transform and area descriptions, although references are provided for access to other generic material. There is also introductory material on how to classify these patterns against known data, but again this is a window on a much larger area, to which appropriate pointers are given.

The *appendices* include a printout of abbreviated versions of the Mathcad and Matlab *worksheets*. The other appendices include material that is germane to the text, such as *camera models* and *coordinate geometry*, the method of *least squares* and a topic known as *principal components analysis*. These are aimed to be short introductions, and are appendices since they are germane to much of the material. Other related, especially online, material is referenced throughout the text.

In this way, the text covers all major areas of feature extraction in image processing and computer vision. There is considerably more material on the subject than is presented here; for example, there is an enormous volume of material on 3D computer vision and 2D signal processing which is only alluded to here. Topics that are specifically not included are colour, 3D processing and image coding. But to include all that would lead to a monstrous book that no one could afford, or even pick up! So we admit that we give a snapshot, but we hope that it is considered to open another window on a fascinating and rewarding subject.

In gratitude

We are immensely grateful to the input of our colleagues, in particular to Prof. Steve Gunn and to Dr John Carter. The family who put up with it are Maria Eugenia and Caz and the nippers. We are also very grateful to past and present researchers in computer vision at the Information, Signals, Images, Systems (ISIS) Research Group under (or who have survived?) Mark's supervision at the School of Electronics and Computer Science, University of Southampton. As well as Alberto and Steve, these include Dr Hani Muammar, Prof. Xiaoguang Jia, Prof. Yan Qiu Chen, Dr Adrian Evans, Dr Colin Davies, Dr David Cunado, Dr Jason Nash, Dr Ping Huang, Dr Liang Ng, Dr Hugh Lewis, Dr David Benn, Dr Douglas Bradshaw, Dr David Hurley, Dr John Manslow, Dr Mike Grant, Bob Roddis, Dr Andrew Tatem, Dr Karl Sharman, Dr Jamie Shutler, Dr Jun Chen, Dr Andy Tatem, Dr Chew Yam, Dr James Hayfron-Acquah, Dr Yalin Zheng, Dr Jeff Foster, Dr Peter Myerscough, Dr David Wagg, Dr Ahmad Al-Mazeed, Dr Jang-Hee Yoo, Dr Nick Spencer, Stuart Mowbray, Dr Stuart Prismall, Dr Peter Gething, Dr Mike Jewell, Dr David Wagg, Dr Alex Bazin, Hidayah Rahmalan, Xin Liu, Imed Bouchrika, Banafshe Arbab-Zavar, Dan Thorpe, Cem Direkoglu (the latter two especially for the new active contour material), John Evans (for the great hippo photo) and to Jamie Hutton, Ben Dowling and Sina Samangooei (for the Java demonstrations site). We are also very grateful to other past Southampton students of BEng and MEng Electronic Engineering, MEng Information Engineering, BEng and MEng Computer Engineering, MEng Software Engineering and BSc Computer Science who have pointed our earlier mistakes, noted areas for clarification and in some cases volunteered some of the material herein. Beyond Southampton, we remain grateful to the reviewers of the two editions and to Prof. Daniel Cremers, Dr Timor Kadir and Prof. Tim Cootes for observations on and improvements to the text and for permission to use images. To all of you, our very grateful thanks.

Final message

We ourselves have already benefited much by writing this book, and this second edition. As we already know, previous students have also benefited, and contributed to it as well. But it remains our hope that it will inspire people to join in this fascinating and rewarding subject that has proved to be such a source of pleasure and inspiration to its many workers.

Mark S. Nixon   Alberto S. Aguado
University of Southampton   University of Surrey

1

Introduction

1.1 Overview

This is where we start, by looking at the human visual system to investigate what is meant by vision, then on to how a computer can be made to sense pictorial data and then how we can process an image. The overview of this chapter is shown in Table 1.1; you will find a similar overview at the start of each chapter. There are no references (citations) in the overview, citations are made in the text and are collected at the end of each chapter.

Table 1.1 Overview of Chapter 1

Main topic	Sub topics	Main points
Human vision system	How the *eye* works, how visual *information* is *processed* and how it can *fail*.	Sight, lens, retina, image, colour, monochrome, processing, brain, visual illusions.
Computer vision systems	How electronic *images* are formed, how *video* is fed into a *computer* and how we can *process* the information using a computer.	Picture elements, pixels, video standard, camera technologies, pixel technology, performance effects, specialist cameras, video conversion, computer languages, processing packages. Demonstrations of working techniques.
Mathematical systems	How we can process images using *mathematical packages*; introduction to the *Matlab* and *Mathcad* systems.	Ease, consistency, support, visualization of results, availability, introductory use, example worksheets.
Literature	Other *textbooks* and other places to find *information* on image processing, computer vision and feature extraction.	Magazines, textbooks, websites and this book's website.

1.2 Human and computer vision

A computer vision system processes images acquired from an electronic camera, which is like the human vision system where the brain processes images derived from the eyes. Computer vision is a rich and rewarding topic for study and research for electronic engineers, computer scientists and many others. Increasingly, it has a commercial future. There are now many vision systems in routine industrial use: cameras inspect mechanical parts to check size, food is inspected for quality, and images used in astronomy benefit from computer vision techniques. Forensic studies and biometrics (ways to recognize people) using computer vision include automatic face recognition and recognizing people by the 'texture' of their irises. These studies are paralleled by biologists and psychologists who continue to study how our human vision system works, and how we see and recognize objects (and people).

A selection of (computer) images is given in Figure 1.1; these images comprise a set of points or *picture elements* (usually concatenated to *pixels*) stored as an *array of numbers* in a *computer*. To recognize faces, based on an image such as Figure 1.1(a), we need to be able to analyse constituent shapes, such as the shape of the nose, the eyes and the eyebrows, to make some measurements to describe, and then recognize, a face. (Figure 1.1a is perhaps one of the most famous images in image processing. It is called the Lena image, and is derived from a picture of Lena Sjööblom in *Playboy* in 1972.) Figure 1.1(b) is an ultrasound image of the carotid artery (which is near the side of the neck and supplies blood to the brain and the face), taken as a cross-section through it. The top region of the image is near the skin; the bottom is inside the neck. The image arises from combinations of the reflections of the ultrasound radiation by tissue. This image comes from a study that aimed to produce three-dimensional (3D) models of arteries, to aid vascular surgery. Note that the image is very *noisy*, and this obscures the shape of the (elliptical) artery. Remotely sensed images are often analysed by their *texture* content. The perceived texture is different between the road junction and the different types of foliage seen in Figure 1.1(c). Finally, Figure 1.1(d) is a magnetic resonance imaging (MRI) image of a cross-section near the middle of a human body. The chest is at the top of the image, the lungs and blood vessels are the dark areas, and the internal organs and the fat appear grey. Nowadays, MRI images are in routine medical use, owing to their ability to provide high-quality images.

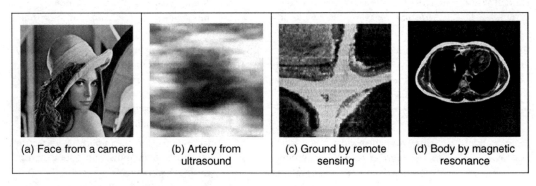

| (a) Face from a camera | (b) Artery from ultrasound | (c) Ground by remote sensing | (d) Body by magnetic resonance |

Figure 1.1   Real images from different sources

There are many different image sources. In medical studies, MRI is good for imaging soft tissue, but does not reveal the bone structure (the spine cannot be seen in Figure 1.1d); this can be achieved by using computed tomography (CT), which is better at imaging bone, as opposed to soft tissue. Remotely sensed images can be derived from infrared (thermal) sensors or synthetic-aperture radar, rather than by cameras, as in Figure 1.1(c). Spatial information can be provided by two-dimensional arrays of sensors, including sonar arrays. There are perhaps more varieties of sources of spatial data in medical studies than in any other area. But computer vision techniques are used to analyse any form of data, not just the images from cameras.

Synthesized images are good for *evaluating* techniques and finding out how they work, and some of the bounds on *performance*. Two synthetic images are shown in Figure 1.2. Figure 1.2(a) is an image of circles that were specified *mathematically*. The image is an ideal case: the circles are perfectly defined and the brightness levels have been specified to be constant. This type of synthetic image is good for evaluating techniques which find the borders of the shape (its edges) and the shape itself, and even for making a description of the shape. Figure 1.2(b) is a synthetic image made up of sections of real image data. The borders between the regions of image data are exact, again specified by a program. The image data comes from a well-known texture database, the Brodatz album of textures. This was scanned and stored as a computer image. This image can be used to analyse how well computer vision algorithms can identify regions of differing texture.

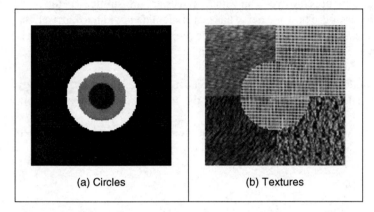

(a) Circles   (b) Textures

Figure 1.2 Examples of synthesized images

This chapter will show you how basic computer vision systems work, in the context of the human vision system. It covers the main elements of human vision, showing you how your eyes work (and how they can be deceived). For computer vision, this chapter covers the hardware and the software used for image analysis, giving an introduction to Mathcad and Matlab, the software tools used throughout this text to implement computer vision algorithms. Finally, a selection of pointers to other material is provided, especially those for more detail on the topics covered in this chapter.

1.3 The human vision system

Human vision is a sophisticated system that senses and acts on *visual stimuli*. It has evolved for millions of years, primarily for defence or survival. Intuitively, computer and human vision appear to have the same function. The purpose of both systems is to interpret *spatial* data, data that is indexed by more than one dimension. Even though computer and human vision are functionally similar, you cannot expect a computer vision system to replicate exactly the function of the human eye. This is partly because we do not understand fully how the vision system of the eye and brain works, as we shall see in this section. Accordingly, we cannot design a system to replicate its function exactly. In fact, some of the properties of the human eye are useful when developing computer vision techniques, whereas others are actually undesirable in a computer vision system. But we shall see computer vision techniques which can, to some extent, replicate, and in some cases even improve upon, the human vision system.

You might ponder this, so put one of the fingers from each of your hands in front of your face and try to estimate the distance between them. This is difficult, and I am sure you would agree that your measurement would not be very accurate. Now put your fingers very close together. You can still tell that they are apart even when the distance between them is tiny. So human vision can distinguish *relative* distance well, but is poor for *absolute* distance. Computer vision is the other way around: it is good for estimating absolute difference, but with relatively poor resolution for relative difference. The number of pixels in the image imposes the accuracy of the computer vision system, but that does not come until the next chapter. Let us start at the beginning, by seeing how the human vision system works.

In human vision, the sensing element is the eye from which images are transmitted via the optic nerve to the brain, for further processing. The optic nerve has insufficient bandwidth to carry all the information sensed by the eye. Accordingly, there must be some preprocessing before the image is transmitted down the optic nerve. The human vision system can be modelled in three parts:

- the eye: this is a physical model since much of its function can be determined by pathology
- a processing system: this is an experimental model since the function can be modelled, but not determined precisely
- analysis by the brain: this is a psychological model since we cannot access or model such processing directly, but only determine behaviour by experiment and inference.

1.3.1 The eye

The function of the eye is to form an image; a cross-section of the eye is illustrated in Figure 1.3. Vision requires an ability to focus selectively on objects of interest. This is achieved by the *ciliary muscles* that hold the *lens*. In old age, it is these muscles which become slack and the eye loses its ability to focus at short distance. The *iris*, or pupil, is like an *aperture* on a camera and controls the amount of light entering the eye. It is a delicate system and needs protection, which is provided by the cornea (sclera). This is outside the *choroid*, which has blood vessels

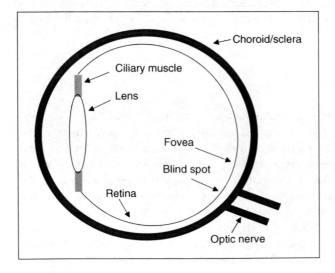

Figure 1.3 Human eye

that supply nutrition and is *opaque* to cut down the amount of light. The *retina* is on the inside of the eye, which is where light falls to form an image. By this system, muscles rotate the eye, and shape the lens, to form an image on the *fovea* (focal point), where the majority of sensors are situated. The *blind spot* is where the optic nerve starts; there are no sensors there.

Focusing involves *shaping* the lens, rather than positioning it as in a camera. The lens is shaped to refract close images greatly, and distant objects little, essentially by 'stretching' it. The distance of the focal centre of the lens varies from approximately 14 mm to around 17 mm, depending on the lens shape. This implies that a world scene is translated into an area of about $2 \, mm^2$. Good vision has high *acuity* (sharpness), which implies that there must be very many sensors in the area where the image is formed.

There are nearly 100 million sensors dispersed around the retina. Light falls on these sensors to stimulate photochemical transmissions, which results in nerve impulses that are collected to form the signal transmitted by the eye. There are two types of sensor: first, the *rods*, which are used for *black and white* (*scotopic*) vision; and secondly, the *cones*, which are used for *colour* (*photopic*) vision. There are approximately 10 million cones and nearly all are found within 5° of the fovea. The remaining 100 million rods are distributed around the retina, with the majority between 20° and 5° of the fovea. Acuity is expressed in terms of spatial resolution (sharpness) and brightness/colour resolution and is greatest within 1° of the fovea.

There is only one type of rod, but there are three types of cone. These are:

- S (short wavelength): these sense light towards the blue end of the visual spectrum
- M (medium wavelength): these sense light around green
- L (long wavelength): these sense light towards the red region of the spectrum.

The total response of the cones arises from summing the response of these three types of cone, which gives a response covering the whole of the visual spectrum. The rods are sensitive to light within the entire visual spectrum, giving the monochrome capability of scotopic vision. Accordingly, when the light level is low, images are formed away from the fovea, to use the superior sensitivity of the rods, but without the colour vision of the cones. Note that there are very few of the bluish cones, and there are many more of the others. But we can still see a lot of blue (especially given ubiquitous denim!). So, somehow, the human vision system compensates for the lack of blue sensors, to enable us to perceive it. The world would be a funny place with red water! The vision response is logarithmic and depends on brightness adaptation from dark conditions where the image is formed on the rods, to brighter conditions where images are formed on the cones.

One inherent property of the eye, known as *Mach bands*, affects the way we perceive images. These are illustrated in Figure 1.4 and are the bands that appear to be where two stripes of constant shade join. By assigning values to the image brightness levels, the cross-section of plotted brightness is shown in Figure 1.4(a). This shows that the picture is formed from stripes of constant brightness. Human vision *perceives* an image for which the cross-section is as plotted in Figure 1.4(c). These Mach bands do not really exist, but are introduced by your eye. The bands arise from overshoot in the eyes' response at boundaries of regions of different intensity (this aids us to differentiate between objects in our field of view). The *real* cross-section is illustrated in Figure 1.4(b). Note also that a human eye can distinguish only relatively few grey levels. It has a capability to discriminate between 32 levels (equivalent to five bits), whereas the image of Figure 1.4(a) could have many more brightness levels. This is why your perception finds it more difficult to discriminate between the low-intensity bands on the left of Figure 1.4(a). (Note that Mach bands cannot be seen in the earlier image of circles, Figure 1.2a, owing to the arrangement

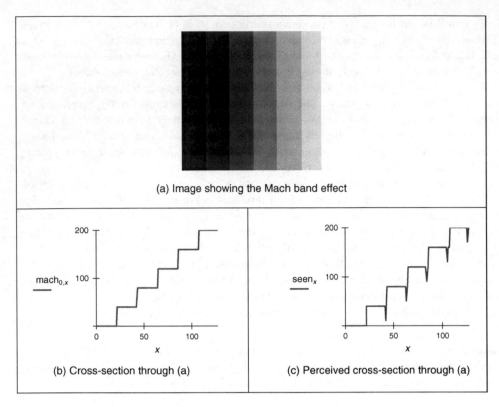

(a) Image showing the Mach band effect

(b) Cross-section through (a)

(c) Perceived cross-section through (a)

Figure 1.4 Illustrating the Mach band effect

of grey levels.) This is the limit of our studies of the first level of human vision; for those who are interested, Cornsweet (1970) provides many more details concerning visual perception.

So we have already identified two properties associated with the eye that it would be difficult to include, and would often be unwanted, in a computer vision system: Mach bands and sensitivity to unsensed phenomena. These properties are integral to human vision. At present, human vision is far more sophisticated than we can hope to achieve with a computer vision system. Infrared guided-missile vision systems can have difficulty in distinguishing between a bird at 100 m and a plane at 10 km. Poor birds! (Lucky plane?) Human vision can handle this with ease.

1.3.2 The neural system

Neural signals provided by the eye are essentially the transformed response of the wavelength-dependent receptors, the cones and the rods. One *model* is to combine these transformed signals by addition, as illustrated in Figure 1.5. The response is transformed by a logarithmic function, mirroring the known response of the eye. This is then multiplied by a weighting factor that controls the contribution of a particular sensor. This can be arranged to allow combination of responses from a particular region. The weighting factors can be chosen to afford particular filtering properties. For example, in *lateral inhibition*, the weights for the centre sensors are much greater than the weights for those at the extreme. This allows the response of the centre sensors to dominate the combined response given by addition. If the weights in one half are chosen to

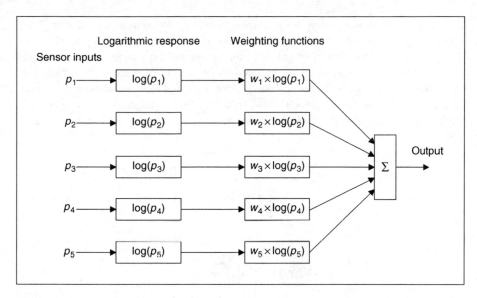

Figure 1.5 Neural processing

be negative, while those in the other half are positive, then the output will show detection of contrast (change in brightness), given by the differencing action of the weighting functions.

The signals from the cones can be combined in a manner that reflects *chrominance (colour)* and *luminance (brightness)*. This can be achieved by subtraction of logarithmic functions, which is then equivalent to taking the logarithm of their ratio. This allows measures of chrominance to be obtained. In this manner, the signals derived from the sensors are combined before transmission through the optic nerve. This is an experimental model, since there are many ways possible to combine the different signals together.

Visual information is then sent back to arrive at the *lateral geniculate nucleus* (LGN), which is in the thalamus and is the primary processor of visual information. This is a layered structure containing different types of cells, with differing functions. The axons from the LGN pass information on to the visual cortex. The function of the LGN is largely unknown, although it has been shown to play a part in coding the signals that are transmitted. It is also considered to help the visual system to focus its attention, such as on sources of sound. For further information on retinal neural networks, see Ratliff (1965); an alternative study of neural processing can be found in Overington (1992).

1.3.3 Processing

The neural signals are then transmitted to two areas of the brain for further processing. These areas are the *associative cortex*, where *links* between objects are made, and the *occipital cortex*, where *patterns* are processed. It is naturally difficult to determine precisely what happens in this region of the brain. To date, there have been no volunteers for detailed study of their brain's function (although progress with new imaging modalities such as positive emission tomography or electrical impedance tomography will doubtless help). For this reason, there are only psychological models to suggest how this region of the brain operates.

It is well known that one function of the human vision system is to use edges, or boundaries, of objects. We can easily read the word in Figure 1.6(a); this is achieved by filling in the missing boundaries in the knowledge that the pattern is likely to represent a printed word. But we can infer more about this image; there is a suggestion of illumination, causing shadows to appear in unlit areas. If the light source is bright, then the image will be washed out, causing the disappearance of the boundaries which are interpolated by our eyes. So there is more than just physical response, there is also knowledge, including prior knowledge of solid geometry. This situation is illustrated in Figure 1.6(b), which could represent three 'pacmen' about to collide, or a white triangle placed on top of three black circles. Either situation is possible.

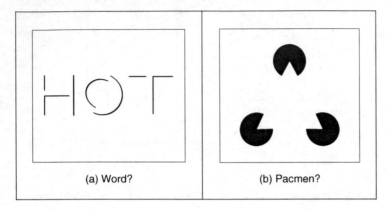

| (a) Word? | (b) Pacmen? |

Figure 1.6 How human vision uses edges

It is also possible to deceive human vision, primarily by imposing a scene that it has not been trained to handle. In the famous *Zollner illusion* (Figure 1.7a), the bars appear to be *slanted*, whereas in reality they are *vertical* (check this by placing a pen between the lines): the small cross-bars mislead your eye into perceiving the vertical bars as slanting. In the *Ebbinghaus illusion* (Figure 1.7b), the inner circle appears to be *larger* when surrounded by *small* circles than it is when surrounded by larger circles.

There are dynamic illusions too: you can always impress children with the 'see my wobbly pencil' trick. Just hold the pencil loosely between your fingers then, to whoops of childish glee, when the pencil is shaken up and down, the solid pencil will appear to bend. *Benham's disk*

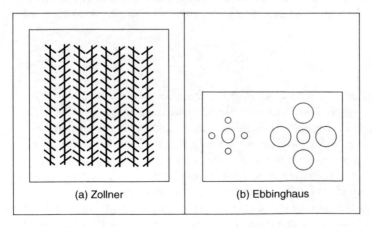

| (a) Zollner | (b) Ebbinghaus |

Figure 1.7 Static illusions

(Figure 1.8) shows how hard it is to model vision accurately. If you make up a version of this disk into a spinner (push a matchstick through the centre) and spin it anticlockwise, you do not see three dark rings, you will see three *coloured* ones. The outside one will appear to be *red*, the middle one a sort of *green* and the inner one deep *blue*. (This can depend greatly on lighting, and contrast between the black and white on the disk. If the colours are not clear, try it in a different place, with different lighting.) You can appear to explain this when you notice that the red colours are associated with the long lines and the blue with short lines. But this is from physics, not psychology. Now spin the disk clockwise. The order of the colours reverses: *red* is associated with the *short* lines (inside) and *blue* with the *long* lines (outside). So the argument from physics is clearly incorrect, since red is now associated with short lines, not long ones, revealing the need for psychological explanation of the eyes' function. This is not colour perception; see Armstrong (1991) for an interesting (and interactive) study of colour theory and perception.

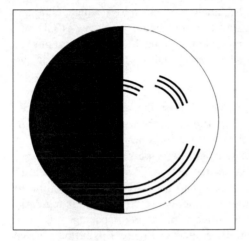

Figure 1.8 Benham's disk

There are many texts on human vision. One popular text on human visual perception is by Schwartz (2004) and there is an online book, *The Joy of Vision* (http://www.yorku. ca/eye/thejoy.htm): useful, despite its title! Marr's seminal text (Marr, 1982) is a computational investigation into human vision and visual perception, investigating it from a computer vision viewpoint. For further details on pattern processing in human vision, see Bruce and Green (1990); for more illusions see Rosenfeld and Kak (1982). Many of the properties of human vision are hard to include in a computer vision system, but let us now look at the basic components that are used to make computers see.

1.4 Computer vision systems

Given the progress in computer technology, computer vision hardware is now relatively inexpensive; a basic computer vision system requires a camera, a camera interface and a computer. These days, some personal computers offer the capability for a basic vision system, by including a camera and its interface within the system. There are specialized systems for vision, offering high performance in more than one aspect. These can be expensive, as any specialist system is.

1.4.1 Cameras

A *camera* is the *basic sensing element*. In simple terms, most cameras rely on the property of light to cause hole/electron pairs (the charge carriers in electronics) in a conducting material. When a potential is applied (to attract the charge carriers), this charge can be sensed as current. By Ohm's law, the voltage across a resistance is proportional to the current through it, so the current can be turned in to a voltage by passing it through a resistor. The number of hole/electron pairs is proportional to the amount of incident light. Accordingly, greater charge (and hence greater voltage and current) is caused by an increase in brightness. In this manner cameras can provide as output, a voltage that is proportional to the brightness of the points imaged by the camera. Cameras are usually arranged to supply video according to a specified standard. Most will aim to satisfy the *CCIR standard* that exists for closed circuit television (CCTV) systems.

There are three main types of camera: *vidicons*, *charge coupled devices* (CCDs) and, more recently, *CMOS* cameras (complementary metal oxide silicon, now the dominant technology for logic circuit implementation). Vidicons are the *older* (analogue) technology which, although cheap (mainly by virtue of longevity in production), are being replaced by the *newer* CCD and CMOS *digital* technologies. The digital technologies now dominate much of the camera market because they are *lightweight* and *cheap* (with other advantages) and are therefore used in the domestic video market.

Vidicons operate in a manner akin to a television in reverse. The image is formed on a screen, and then sensed by an electron beam that is scanned across the screen. This produces an output which is continuous; the output *voltage* is proportional to the *brightness* of points in the scanned line, and is a continuous signal, a voltage which varies continuously with time. In contrast, CCDs and CMOS cameras use an array of sensors; these are regions where *charge* is collected which is proportional to the *light* incident on that region. This is then available in discrete, or *sampled*, form as opposed to the continuous sensing of a vidicon. This is similar to human vision with its array of cones and rods, but digital cameras use a *rectangular* regularly spaced lattice, whereas human vision uses a *hexagonal* lattice with irregular spacing.

Two main types of semiconductor pixel sensor are illustrated in Figure 1.9. In the *passive sensor*, the charge generated by incident light is presented to a bus through a pass transistor.

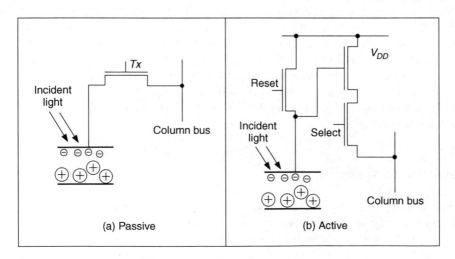

Figure 1.9 Pixel sensors

When the signal Tx is activated, the pass transistor is enabled and the sensor provides a capacitance to the bus, one that is proportional to the incident light. An *active pixel* includes an amplifier circuit that can compensate for limited fill factor of the photodiode. The select signal again controls presentation of the sensor's information to the bus. A further reset signal allows the charge site to be cleared when the image is rescanned.

The basis of a CCD sensor is illustrated in Figure 1.10. The number of charge sites gives the resolution of the CCD sensor; the contents of the charge sites (or buckets) need to be converted to an output (voltage) signal. In simple terms, the contents of the buckets are emptied into vertical transport registers which are shift registers moving information towards the horizontal transport registers. This is the column bus supplied by the pixel sensors. The horizontal transport registers empty the information row by row (point by point) into a signal conditioning unit, which transforms the sensed charge into a voltage which is proportional to the charge in a bucket, and hence proportional to the brightness of the corresponding point in the scene imaged by the camera. The CMOS cameras are like a form of memory: the charge incident on a particular site in a two-dimensional lattice is proportional to the brightness at a point. The charge is then read like computer memory. (In fact, a computer RAM chip can act as a rudimentary form of camera when the circuit, the one buried in the chip, is exposed to light.)

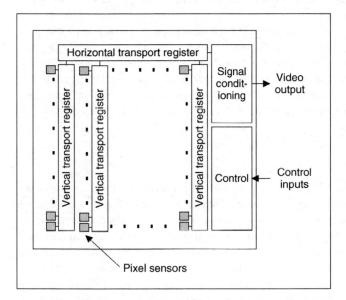

Figure 1.10 CCD sensing element

There are many more varieties of vidicon (Chalnicon, etc.) than there are of CCD technology (charge injection device, etc.), perhaps owing to the greater age of basic vidicon technology. Vidicons are cheap but have a number of intrinsic performance problems. The scanning process essentially relies on moving parts. As such, the camera performance will change with time, as parts *wear*; this is known as *ageing*. Also, it is possible to *burn* an image into the scanned screen by using high incident light levels; vidicons can also suffer *lag*, that is, a delay in response to moving objects in a scene. The digital technologies are dependent on the physical arrangement of charge sites and as such do not suffer from ageing, but can suffer from irregularity in the charge sites' (silicon) material. The underlying technology also makes CCD and CMOS cameras less sensitive to lag and burn, but the signals associated with the CCD transport registers can give rise to *readout effects*. Charge coupled devices only came to dominate camera technology when technological difficulty associated with *quantum efficiency* (the magnitude of response to incident light) for the shorter, blue, wavelengths was solved. One of the major problems in CCD cameras is *blooming*, where bright (incident) light causes a bright spot to grow and disperse in the image (this used to happen in the analogue technologies too). This happens much less in CMOS cameras because the charge sites can be much better defined and reading their data is equivalent to reading memory sites as opposed to shuffling charge between sites. Also, CMOS cameras have now overcome the problem of *fixed pattern noise* that plagued earlier MOS cameras. The CMOS cameras are actually much more recent than CCDs. This begs a question as to which is better: CMOS or CCD? Given that they will be both be subject to much continued development, CMOS is a cheaper technology and it lends itself directly to intelligent cameras with on-board processing. This is mainly because the feature size of points (pixels) in a CCD sensor is limited to be about $4\,\mu$m so that enough light is collected. In contrast, the feature size in CMOS technology is considerably smaller, currently at around $0.1\,\mu$m. Accordingly, it is now possible to integrate signal processing within the camera chip and thus it is perhaps possible that CMOS cameras will eventually replace CCD technologies for many applications. However, the more modern CCDs also have on-board circuitry, and their process technology is more mature, so the debate will continue.

Finally, there are specialist cameras, which include *high-resolution* devices, which can give pictures with a great number of points, *low-light level* cameras, which can operate in very dark conditions (this is where vidicon technology is still found), and *infrared* cameras, which sense heat to provide thermal images. Increasingly, *hyperspectral* cameras are available, which have more sensing bands. For more detail concerning modern camera practicalities and imaging systems, see Nakamura (2005). For more detail on sensor development, particularly CMOS, Fossum (1997) is well worth a look.

1.4.2 Computer interfaces

This technology is in a rapid state of change, owing to the emergence of digital cameras. Essentially, the image sensor converts light into a signal which is expressed either as a continuous signal or in sampled (digital) form. Some (older) systems expressed the camera signal as an analogue continuous signal, according to a standard, often the CCIR standard, and this was converted at the computer (and still is in some cases). Modern digital systems convert the sensor information into digital information with on-chip circuitry and then provide the digital information according to a specified standard. The older systems, such as surveillance systems, supplied (or supply) video, whereas the newer systems are digital. Video implies delivering the moving image as a sequence of *frames* and these can be in analogue (continuous) or discrete (sampled) form, of which one format is digital video (DV).

An interface that converts an *analogue* signal into a set of digital numbers is called a *framegrabber*, since it grabs frames of data from a *video sequence*, and is illustrated in Figure 1.11. Note that cameras that provide digital information do not need this particular interface (it is inside the camera). However, an analogue camera signal is *continuous* and is transformed into *digital* (discrete) format using an analogue-to-digital (A/D) converter. *Flash converters* are usually used owing to the high speed required for conversion, say 11 MHz, which cannot be met by any other conversion technology. Usually, 8 bit A/D converters are used; at 6 dB/bit, this gives 48 dB, which just satisfies the CCIR stated *bandwidth* of approximately 45 dB. The output of the A/D converter is often fed to *look-up tables* (LUTs), which implement

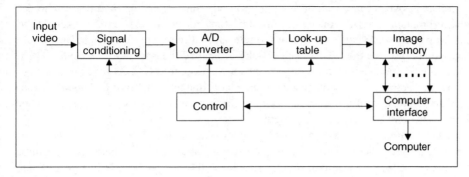

Figure 1.11 A computer interface: a framegrabber

designated conversion of the input data, but in hardware rather than in software, and this is very fast. The outputs of the A/D converter are then stored. Note that there are aspects of the sampling process that are of considerable interest in computer vision; these are covered in Chapter 2.

In digital camera systems this processing is usually performed on the camera chip, and the camera eventually supplies digital information, often in coded form. IEEE 1394 (or *firewire*) is a way of connecting devices external to a computer and is often used for digital video cameras as it supports high-speed digital communication and can provide power; this is similar to universal serial bus (USB), which can be used for still cameras. Firewire needs a connection system and software to operate it, and these can be easily acquired. One important aspect of Firewire is its support of isochronous transfer operation which guarantees timely delivery of data, which is of importance in video-based systems.

There are many different ways to design framegrabber units, especially for specialist systems. Note that the control circuitry has to determine exactly when image data is to be sampled. This is controlled by synchronization pulses that are supplied within the video signal: the sync signals, which control the way video information is constructed. Television pictures are constructed from a set of *lines*, those lines scanned by a camera. To reduce requirements on transmission (and for viewing), the 625 lines in the *PAL* system (*NTSC* is of lower resolution) are transmitted in two *fields*, each of 312.5 lines, as illustrated in Figure 1.12. (Currently, in high-definition television,

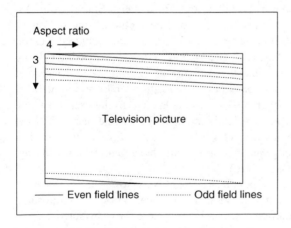

Figure 1.12 Interlacing in television pictures

there is some debate between the corporations who do not want interlacing, and those who do, e.g. the television broadcasters.) If you look at a television, but not directly, the flicker due to *interlacing* can be perceived. When you look at the television directly, persistence in the human eye ensures that you do not see the *flicker*. These fields are called the *odd* and *even* fields. There is also an *aspect ratio* in picture transmission: pictures are arranged to be 1.33 times longer than they are high. These factors are chosen to make television images attractive to *human* vision, and can complicate the design of a framegrabber unit. Some conversion systems accept PAL or NTSC video and convert it to the firewire system.

Nowadays, *digital video cameras* can provide digital output, in *progressive scan* (without interlacing), delivering sequences of images that are readily processed. Or there are *webcams*, or just *digital camera systems* that deliver images straight to the computer. Life just gets easier!

This completes the material we need to cover for basic computer vision systems. For more detail concerning the practicalities of computer vision systems see, for example, Davies (2005) (especially for product inspection) or Umbaugh (2005) (and both offer much more than this).

1.4.3 Processing an image

Most image processing and computer vision techniques are implemented in computer *software*. Often, only the simplest techniques migrate to hardware, although coding techniques to maximize efficiency in image transmission are of sufficient commercial interest that they have warranted extensive, and very sophisticated, hardware development. The systems include the Joint Photographic Expert Group (JPEG) and the Moving Picture Expert Group (MPEG) image coding formats. C, C++ and Java™ are by now the most popular languages for vision system implementation, because of strengths in integrating high- and low-level functions, and the availability of good compilers. As systems become more complex, C++ and Java become more attractive when encapsulation and polymorphism may be exploited. Many people use Java as a development language partly because of platform independence, but also because of ease in implementation (although some claim that speed and efficiency are not as good as in C/C++). There is considerable implementation advantage associated with use of the Java Advanced Imaging API (application programming interface). There is an online demonstration site, for educational purposes only, associated with this book, to be found of the book's website at http://www.ecs.soton.ac.uk/~msn/book/new_demo/. This is based around Java, so that the site can be used over the web (as long as Java is installed and up to date). Some textbooks offer image processing systems implemented in these languages. Many commercial packages are available, although these are often limited to basic techniques, and do not include the more sophisticated shape extraction techniques. The Visiquest (was Khoros) image processing system has attracted much interest; this is a schematic (data-flow) image processing system where a user links together chosen modules. This allows for better visualization of information flow during processing. However, the underlying mathematics is not made clear to the user, as it can be when a mathematical system is used. There is a textbook, and a very readable one at that, by Efford (2000), which is based entirely on Java and includes, via a CD, the classes necessary for image processing software development. Other popular texts include those that present working algorithms, such as Seul et al. (2001) and Parker (1996).

In terms of *software packages*, one of the most popular is OpenCV, whose philosophy is to 'aid commercial uses of computer vision in human–computer interface, robotics, monitoring, biometrics and security by providing a free and open infrastructure where the distributed efforts of the vision community can be consolidated and performance optimized'. This contains a wealth of technique and (optimized) implementation; there is even a Wikipedia entry and a discussion website supporting it. Then there are the VXL libraries (the Vision-*something*-Libraries, groan). This is 'a collection of C++ libraries designed for computer vision research and implementation'. Finally, there is Adobe's Generic Image Library (GIL), which aims to ease difficulties with writing imaging-related code that is both generic and efficient. Note that these are open source, but there are licences and conditions on use and exploitation.

A set of web links is shown in Table 1.2 for established freeware and commercial software image processing systems. Perhaps the best selection can be found at the general site, from the computer vision homepage software site at Carnegie Mellon (repeated later in Table 1.5).

Table 1.2 Software websites

Packages (freeware or student version indicated by *)

General Site	Carnegie Mellon	http://www.cs.cmu.edu/afs/cs/project/cil/ftp/html/v-source.html (large popular index including links to research code, image processing toolkits, and display tools)
Visiquest (Khoros)	Accusoft	http://www.accusoft.com/
	Hannover University	http://www.tnt.uni-hannover.de/soft/imgproc/khoros/
AdOculos* (+ Textbook)	The Imaging Source	http://www.theimagingsource.com/
CVIPtools*	Southern Illinois University	http://www.ee.siue.edu/CVIPtools/
LaboImage*	Geneva University	http://cuiwww.unige.ch/~vision/LaboImage/labo.html
TN-Image*	Thomas J. Nelson	http://brneurosci.org/tnimage.html (scientific image analysis)
OpenCV	Intel	http://www.intel.com/technology/computing/opencv/index.htm and http://sourceforge.net/projects/opencvlibrary/
VXL	Many international contributors	http://vxl.sourceforge.net/
GIL	Adobe	http://opensource.adobe.com/gil/

1.5 Mathematical systems

Several *mathematical systems* have been developed. These offer what is virtually a word-processing system for mathematicians. Many are screen based, using a Windows system. The advantage of these systems is that you can transpose mathematics pretty well directly from textbooks, and see how it works. Code functionality is not obscured by the use of data structures, although this can make the code appear cumbersome. A major advantage is that the system provides low-level functionality and data visualization schemes, allowing the user to concentrate on techniques alone. Accordingly, these systems afford an excellent route to understand, and appreciate, mathematical systems before the development of application code, and to check that the final code works correctly.

1.5.1 Mathematical tools

Mathcad, Mathematica, Maple and Matlab are among the most popular of current tools. There have been surveys that compare their efficacy, but it is difficult to ensure precise comparison owing to the impressive speed of development of techniques. Most systems have their protagonists and detractors, as in any commercial system. Many books use these packages for particular subjects, and there are often handbooks as addenda to the packages. We shall use both Matlab and Mathcad throughout this text as they two very popular mathematical systems. We shall describe Matlab later, as it is different from Mathcad, although the aim is the same. The website links for the main mathematical packages are given in Table 1.3.

Table 1.3 Mathematical package websites

General		
Guide to available mathematical software	NIST	http://gams.nist.gov/
Vendors		
Mathcad	MathSoft	http://www.mathcad.com/
Mathematica	Wolfram Research	http://www.wolfram.com/
Matlab	Mathworks	http://www.mathworks.com/
Maple	Maplesoft	http://www.maplesoft.com/

1.5.2 Hello Mathcad, hello images!

Mathcad uses *worksheets* to implement mathematical analysis. The flow of calculation is very similar to using a piece of paper: calculation starts at the top of a document, and flows left to right and downwards. Data is available to *later* calculation (and to calculation to the right), but is not available to prior calculation, much as is this case when calculation is written manually on paper. Mathcad uses the Maple mathematical library to extend its functionality. To ensure that equations can migrate easily from a textbook to application, Mathcad uses a WYSIWYG (what you see is what you get) notation [its equation editor is not dissimilar to the Microsoft Equation (Word) editor]. Mathcad offers a compromise between many performance factors, and is available at low cost. There used to be a free worksheet viewer called Mathcad Explorer which operated in read-only mode, which is an advantage lost. An image processing handbook is available with Mathcad, but it does not include many of the more sophisticated feature extraction techniques.

Images are actually spatial data, data which is indexed by two spatial coordinates. The camera senses the *brightness* at a point with coordinates x,y. Usually, x and y refer to the horizontal and vertical axes, respectively. Throughout this text we shall work in *orthographic projection*, ignoring *perspective*, where real-world coordinates map directly to x and y coordinates in an image. The *homogeneous coordinate system* is a popular and proven method for handling three-dimensional coordinate systems (x, y and z, where z is depth). Since it is not used directly in the text, it is included as Appendix 2 (Section 10.3). The brightness sensed by the camera is transformed to a signal which is then fed to the A/D converter and stored as a value within the computer, referenced to the coordinates x,y in the image. Accordingly, a computer image is a matrix of points. For a greyscale image, the value of each point is proportional to the brightness of the corresponding point in the scene viewed, and imaged, by the camera. These points are the picture elements, or *pixels*.

Consider, for example, the matrix of pixel values in Figure 1.13(a). This can be viewed as a surface (or function) in Figure 1.13(b), or as an image in Figure 1.13(c). In Figure 1.13(c) the brightness of each point is proportional to the value of its pixel. This gives the synthesized image of a bright square on a dark background. The square is bright where the pixels have a value around 40 brightness levels; the background is dark, and these pixels have a value near 0 brightness levels. This image is first given a label, `pic`, and then `pic` is allocated, `:=`, to the matrix defined by using the matrix dialogue box in Mathcad, specifying a matrix with eight rows and eight columns. The pixel values are then entered one by one until the matrix is complete (alternatively, the matrix can be specified by using a subroutine, but that comes later). Note that neither the background nor the square has a constant brightness. This is because noise has been added to the image. If we want to evaluate the performance of a computer vision technique on an image, but without the noise, we can simply remove it (one of the advantages of using synthetic images). The matrix becomes an image when it is viewed as a picture, in Figure 1.13(c). This is done either by presenting it as a surface plot, rotated by zero^{degrees} and viewed from above, or by using Mathcad's picture facility. As a surface plot, Mathcad allows the user to select a greyscale image, and the patch plot option allows an image to be presented as point values.

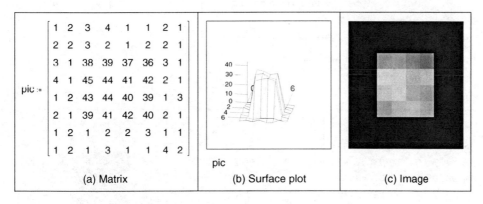

Figure 1.13 Synthesized image of a square

Mathcad stores matrices in row–column format. The coordinate system used throughout this text has x as the horizontal axis and y as the vertical axis (as conventional). Accordingly, x is the column count and y is the row count, so a point (in Mathcad) at coordinates x,y is actually accessed as $pic_{y,x}$. The origin is at coordinates $x = 0$ and $y = 0$, so $pic_{0,0}$ is the magnitude of the point at the origin and $pic_{2,2}$ is the point at the third row and third column and $pic_{3,2}$ is the point at the third column and fourth row, as shown in Code 1.1 (the points can be seen

```
pic₂,₂=38  pic₃,₂=45
rows(pic)=8   cols(pic)=8
```

Code 1.1 Accessing an image in Mathcad

in Figure 1.13a). Since the origin is at (0,0) the bottom right-hand point, at the last column and row, has coordinates (7,7). The number of rows and the number of columns in a matrix, the dimensions of an image, can be obtained by using the Mathcad rows and cols functions, respectively, and again in Code 1.1.

This synthetic image can be processed using the Mathcad programming language, which can be invoked by selecting the appropriate dialogue box. This allows for conventional for, while and if statements, and the earlier assignment operator which is := in non-code sections, is replaced by [back-arrow] in sections of code. A subroutine that inverts the brightness level at each point, by subtracting it from the maximum brightness level in the original image, is illustrated in Code 1.2. This uses for loops to index the rows and the columns, and then calculates a new pixel value by subtracting the value at that point from the maximum obtained by Mathcad's max function. When the whole image has been processed, the new picture is returned to be assigned to the label newpic. The resulting matrix is shown in Figure 1.14(a). When this is viewed as a surface (Figure 1.14b), the inverted brightness levels mean that the square appears dark and its surroundings appear white, as in Figure 1.14(c).

```
new_pic:=   for x∈0..cols(pic)-1
               for y∈0..rows(pic)-1
                  newpicture_{y,x}←max(pic)-pic_{y,x}
            newpicture
```

Code 1.2   Processing image points in Mathcad

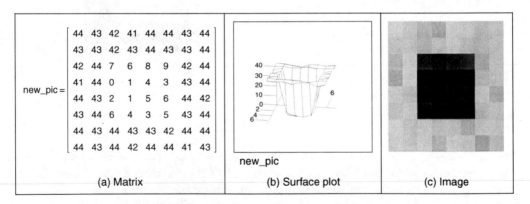

$$new\_pic = \begin{bmatrix} 44 & 43 & 42 & 41 & 44 & 44 & 43 & 44 \\ 43 & 43 & 42 & 43 & 44 & 43 & 43 & 44 \\ 42 & 44 & 7 & 6 & 8 & 9 & 42 & 44 \\ 41 & 44 & 0 & 1 & 4 & 3 & 43 & 44 \\ 44 & 43 & 2 & 1 & 5 & 6 & 44 & 42 \\ 43 & 44 & 6 & 4 & 3 & 5 & 43 & 44 \\ 44 & 43 & 44 & 43 & 43 & 42 & 44 & 44 \\ 44 & 43 & 44 & 42 & 44 & 44 & 41 & 43 \end{bmatrix}$$

(a) Matrix   (b) Surface plot   (c) Image

Figure 1.14   Image of a square after division

Routines can be formulated as *functions*, so they can be invoked to process a chosen picture, rather than restricted to a specific image. Mathcad functions are conventional; we simply add two arguments (one is the image to be processed, the other is the brightness to be added) and use the arguments as local variables, to give the add function illustrated in Code 1.3. To add a value, we simply call the function and supply an image and the chosen brightness level as the arguments.

```add_value(inpic,value):=```	```for x∈0..cols(inpic)-1```
	``` for y∈0..rows(inpic)-1```
	``` newpicture_{y,x}←inpic_{y,x}+value```
	```newpicture```

```
add_value(inpic,value):= for x∈0..cols(inpic)-1
                             for y∈0..rows(inpic)-1
                                 newpicture_{y,x}←inpic_{y,x}+value
                         newpicture
```

Code 1.3 Function to add a value to an image in Mathcad

Mathematically, for an image which is a matrix of $N \times N$ points, the brightness of the pixels in a new picture (matrix), **N**, are the result of adding b brightness values to the pixels in the old picture, **O**, given by:

$$\mathbf{N}_{x,y} = \mathbf{O}_{x,y} + b \qquad \forall x, y \in 1, N \qquad\qquad (1.1)$$

Real images have many points. Unfortunately, the Mathcad matrix dialogue box only allows matrices that are 10 rows and 10 columns at most, i.e. a 10×10 matrix. Real images can be 512×512, but are often 256×256 or 128×128; this implies a storage requirement for 262×144, 65×536 and 16×384 pixels, respectively. Since Mathcad stores all points as high-precision, complex floating point numbers, 512×512 images require too much storage, but 256×256 and 128×128 images can be handled with ease. Since this cannot be achieved by the dialogue box, Mathcad has to be 'tricked' into accepting an image of this size. Figure 1.15 shows an image

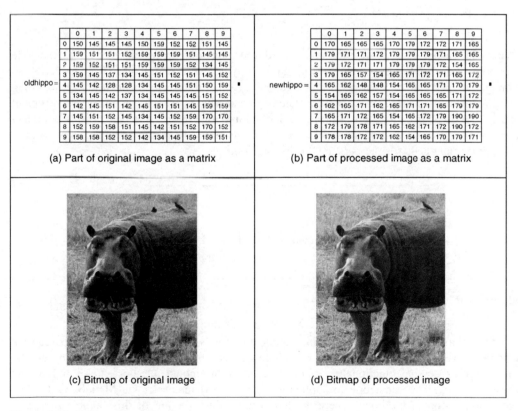

oldhippo =

	0	1	2	3	4	5	6	7	8	9
0	150	145	145	145	150	159	152	152	151	145
1	159	151	151	152	159	159	159	151	145	145
2	159	152	151	151	159	159	159	152	134	145
3	159	145	137	134	145	151	152	151	145	152
4	145	142	128	128	134	145	145	151	150	159
5	134	145	142	137	134	145	145	145	151	152
6	142	145	151	142	145	151	151	145	159	159
7	145	151	152	145	134	145	152	159	170	170
8	152	159	158	151	145	142	151	152	170	152
9	158	158	152	152	142	134	145	159	159	151

(a) Part of original image as a matrix

newhippo =

	0	1	2	3	4	5	6	7	8	9
0	170	165	165	165	170	179	172	172	171	165
1	179	171	171	172	179	179	179	171	165	165
2	179	172	171	171	179	179	179	172	154	165
3	179	165	157	154	165	171	172	171	165	172
4	165	162	148	148	154	165	165	171	170	179
5	154	165	162	157	154	165	165	165	171	172
6	162	165	171	162	165	171	171	165	179	179
7	165	171	172	165	154	165	172	179	190	190
8	172	179	178	171	165	162	171	172	190	172
9	178	178	172	172	162	154	165	179	179	171

(b) Part of processed image as a matrix

(c) Bitmap of original image

(d) Bitmap of processed image

Figure 1.15 Processing an image

captured by a camera. This image has been stored in Windows bitmap (.BMP) format. This can be read into a Mathcad worksheet using the READBMP command (in capitals – Mathcad cannot handle `readbmp`), and is assigned to a variable. It is inadvisable to attempt to display this using the Mathcad surface plot facility, as it can be slow for images and requires a lot of memory. It is best to view an image using Mathcad's picture facility or to store it using the WRITEBMP command, and then look at it using a bitmap viewer.

So, we can make an image brighter, by addition, by the routine in Code 1.3, via the code in Code 1.4. The result is shown in Figure 1.15. The matrix listings in Figure 1.15(a) and (b) show that 20 has been added to each point (these only show the top left-hand section of the image where the bright points relate to the grass; the darker points on, say, the ear cannot be seen). The effect will be to make each point appear brighter, as seen by comparison of the (darker) original image (Figure 1.15c) with the (brighter) result of addition (Figure 1.15d). In Chapter 3 we will investigate techniques that can be used to manipulate the image brightness to show the face in a much better way. For the moment, we are just seeing how Mathcad can be used, in a simple way, to process pictures.

```
oldhippo:=READBMP("hippo_orig")
newhippo:=add_value(hippo,20)
WRITEBMP("hippo_brighter.bmp"):=newhippo
```

Code 1.4 Processing an image

Mathcad was used to generate the image used to demonstrate the *Mach band* effect; the code is given in Code 1.5. First, an image is defined by copying the `hippo` image (from Code 1.4) to an image labelled `mach`. Then, the `floor` function (which returns the nearest integer less than its argument) is used to create the bands, scaled by an amount appropriate to introduce sufficient contrast (the division by 21.5 gives six bands in the image of Figure 1.4a). The cross-section and the perceived cross-section of the image were both generated by Mathcad's X–Y plot facility, using appropriate code for the perceived cross-section.

$$
\text{mach}:= \quad
\begin{array}{l}
\text{for } x \in 0\,..\,\text{cols}(\text{mach})-1 \\
\quad \text{for } y \in 0\,..\,\text{rows}(\text{mach})-1 \\
\qquad \text{mach}_{y,x} \leftarrow \text{brightness}\cdot\text{floor}\left(\dfrac{x}{\text{bar\_width}}\right) \\
\text{mach}
\end{array}
$$

Code 1.5 Creating the image of Figure 4(a)

The translation of the Mathcad code into application can be rather prolix when compared with the Mathcad version by the necessity to include low-level functions. Since these can obscure the basic image processing functionality, Mathcad is used throughout this book to show how the techniques work. The translation to application code is perhaps easier via Matlab, as it offers direct compilation of the code. There is also an electronic version of this book, which is a collection of worksheets to help you to learn the subject; and an example Mathcad worksheet is given in Appendix 1 (Section 9.1 for Mathcad; 9.2 for Matlab). You can download these worksheets from this book's website (http://www.ecs.soton.ac.uk/~msn/book/) and there is a link to the old Mathcad Explorer there too. You can then use the algorithms as a basis for developing your own application code. This provides a good way to verify that your code actually works: you can compare the results of your final application code with those of the original mathematical description. If your final application code and the Mathcad implementation are both correct, the results should be the same. Your application code will be much faster than in Mathcad, and will benefit from the graphical user interface (GUI) that you have developed.

1.5.3 Hello Matlab!

Matlab is rather different from Mathcad. It is not a WYSIWYG system, but instead it is more screen based. It was originally developed for matrix functions, hence the 'Mat' in the name. Like Mathcad, it offers a set of mathematical tools and visualization capabilities in a manner arranged to be very similar to conventional computer programs. In some users' views, a WYSIWYG system like Mathcad is easier to start with, but there are a number of advantages to Matlab, not least the potential speed advantage in computation and the facility for debugging, together with a considerable amount of established support. Again, there is an image processing toolkit supporting Matlab, but it is rather limited compared with the range of techniques exposed in this text. Its popularity is reflected in a book dedicated to use of Matlab for image processing (Gonzalez et al., 2003), by perhaps one of the subject's most popular authors.

Essentially, Matlab is the set of instructions that process the data stored in a workspace, which can be extended by user-written commands. The workspace stores the different lists of data and these data can be stored in a MAT file; the user-written commands are functions that are stored in M-files (files with extension .M). The procedure operates by instructions at the command line to process the workspace data using either one of Matlab's commands or your own commands. The results can be visualized as graphs, surfaces or images, as in Mathcad.

Matlab provides powerful matrix manipulations to develop and test complex implementations. In this book, we avoid matrix implementations in favour of a more C++ algorithmic form. Thus, matrix expressions are transformed into loop sequences. This helps students without experience in matrix algebra to understand and implement the techniques without dependency on matrix manipulation software libraries. Implementations in this book only serve to gain understanding of the techniques' performance and correctness, and favour clarity rather than speed.

The system runs on Unix/Linux or Windows and on Macintosh systems. A student version is available at low cost. There is no viewer available for Matlab; you have to have access to a system for which it is installed. As the system is not based around worksheets, we shall use a script which is the simplest type of M-file, as illustrated in Code 1.6. To start the Matlab system, type MATLAB at the command line. At the Matlab prompt ($>>$) type chapter1 to load and run the script (given that the file chapter1.m is saved in the directory you are working in). Here, we can see that there are no text boxes and so comments are preceded by %. The first command is one that allocates data to our variable pic. There is a more sophisticated way to input this in the Matlab system, but that is not available here. The points are addressed in row–column format and the origin is at coordinates $y = 1$ and $x = 1$. So we access these points $pic_{3,3}$ as the third column of the third row and $pic_{4,3}$ as the point in the third column of the fourth row. Having set the display facility to black and white, we can view the array pic as a surface. When the surface (Figure 1.16a), is plotted, then Matlab has been made to pause until you press Return before moving on. Here, when you press Return, you will next see the image of the array (Figure 1.16b).

```
%Chapter 1 Introduction (Hello Matlab) CHAPTER1.M
%Written by: Mark S. Nixon

disp('Welcome to the Chapter1 script')
disp('This worksheet is the companion to Chapter 1 and is an
introduction.')
disp('It is the source of Section 1.4.3 Hello Matlab.')
disp('The worksheet follows the text directly and allows you to
process basic images.')

disp('Let us define a matrix, a synthetic computer image called
pic.')

pic =[1   2   3   4   1   1   2   1;
      2   2   3   2   1   2   2   1;
      3   1  38  39  37  36   3   1;
      4   1  45  44  41  42   2   1;
      1   2  43  44  40  39   1   3;
      2   1  39  41  42  40   2   1;
      1   2   1   2   2   3   1   1;
      1   2   1   3   1   1   4   2]

%Pixels are addressed in row-column format.
%Using x for the horizontal axis (a column count), and y for the
%vertical axis (a row count) then picture points are addressed as
%pic(y,x). The origin is at coordinates (1,1), so the point
%pic(3,3)  is on the third row and third column; the point pic(4,3)
%is on the fourth row, at the third column. Let's print them:
disp ('The element pic(3,3) is')
pic(3,3)
disp('The element pic(4,3)is')
pic(4,3)

%We'll set the output display to black and white
colormap(gray);
%We can view the matrix as a surface plot
disp ('We shall now view it as a surface plot (play with the
controls to see it in relief)')
disp('When you are ready to move on, press RETURN')
surface(pic);
%Let's hold awhile so we can view it
pause;
%Or view it as an image
disp ('We shall now view the array as an image')
disp('When you are ready to move on, press RETURN')
imagesc(pic);
%Let's hold awhile so we can view it
pause;
```

```
%Let's look at the array's dimensions
disp('The dimensions of the array are')
size(pic)

%now let's invoke a routine that inverts the image
inverted_pic=invert(pic);
%Let's print it out to check it
disp('When we invert it by subtracting each point from the
maximum, we get')
inverted_pic
%And view it
disp('And when viewed as an image, we see')
disp('When you are ready to move on, press RETURN')
imagesc(inverted_pic);
%Let's hold awhile so we can view it
pause;
disp('We shall now read in a bitmap image, and view it')
disp('When you are ready to move on, press RETURN')
face=imread('rhdark.bmp','bmp');
imagesc(face);
pause;
%Change from unsigned integer(uint8) to double precision so we can
process it
face=double(face);
disp('Now we shall invert it, and view the inverted image')
inverted_face=invert(face);
imagesc(inverted_face);
disp('So we now know how to process images in Matlab. We shall be
using this later!')
```

Code 1.6 Matlab script for Chapter 1

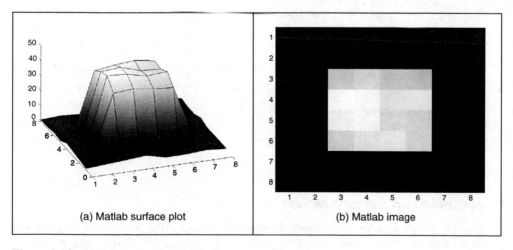

(a) Matlab surface plot

(b) Matlab image

Figure 1.16 Matlab image visualization

We can use Matlab's own command to interrogate the data; these commands find use in the M-files that store subroutines. An example routine is called after this. This subroutine is stored in a file called `invert.m` and is a function that inverts brightness by subtracting the value of each point from the array's maximum value. The code is illustrated in Code 1.7. Note that this code uses `for` loops, which are best avoided to improve speed, using Matlab's vectorized operations (as in Mathcad). The whole procedure can be implemented by the command `inverted=max(max(pic))-pic`. One of Matlab's assets is a 'profiler' which allows you to determine exactly how much time is spent on different parts of your programs. There is facility for importing graphics files, which is actually rather more extensive (i.e. it accepts a wider range of file formats) than available in Mathcad. When images are used, this reveals that unlike Mathcad, which stores all variables as full precision real numbers, Matlab has a range of datatypes. We must move from the unsigned integer datatype, used for images, to the double precision datatype to allow processing as a set of real numbers. In these ways Matlab can and will be used to process images throughout this book. As with the Mathcad worksheets, there are Matlab scripts available at the website for online tutorial support of the material in this book; an abbreviated example worksheet is given in Appendix 1 (Section 9.2).

```
function inverted=invert(image)
%Subtract image point brightness from maximum
%
%Usage:[new image]=invert(image)
%
%Parameters: image-array of points
%
%Author: Mark S.Nixon
%get dimensions
[rows,cols]=size(image);

%find the maximum
maxi=max(max(image));

%subtract image points from maximum
for x=1:cols %address all columns
   for y=1:rows %address all rows
      inverted(y,x)=maxi-image(y,x);
   end
end
```

Code 1.7   Matlab function (invert.m) to invert an image

1.6   Associated literature

1.6.1   Journals and magazines

As in any academic subject, there are many sources of literature. The *professional magazines* include those that are more systems orientated, like *Vision Systems Design* and *Advanced Imaging*. These provide more general articles, and are often a good source of information about new computer vision products. For example, they survey available equipment, such as cameras and monitors, and provide listings of those available, including some of the factors by which you might choose to purchase them.

There is a wide selection of *research journals*, probably more than you can find in your nearest library unless it is particularly well stocked. These journals have different merits: some are targeted at short papers only, whereas some have short and long papers; some are more dedicated to the development of new theory, whereas others are more pragmatic and focus more on practical, working, image processing systems. But it is rather naive to classify journals in this way, since all journals welcome good research, with new ideas, which has been demonstrated to satisfy promising objectives.

The main research journals include: *IEEE Transactions on: Pattern Analysis and Machine Intelligence* (in later references this will be abbreviated to *IEEE Trans. on PAMI*); *Image Processing (IP); Systems, Man and Cybernetics (SMC)*; and *Medical Imaging* (there are many more IEEE transactions, some of which sometimes publish papers of interest in image processing and computer vision). The *IEEE Transactions* are usually found in (university) libraries since they are available at comparatively low cost; they are online to subscribers at the IEEE Explore site (http://ieeexplore.ieee.org/) and this includes conferences and IEE Proceedings (described soon). *Computer Vision and Image Understanding* and *Graphical Models and Image Processing* arose from the splitting of one of the subject's earlier journals, *Computer Vision, Graphics and Image Processing (CVGIP)*, into two parts. Do not confuse *Pattern Recognition (Pattern Recog.)* with *Pattern Recognition Letters (Pattern Recog. Lett.)*, published under the aegis of the Pattern Recognition Society and the International Association of Pattern Recognition, respectively, since the latter contains shorter papers only. The *International Journal of Computer Vision* is a more recent journal, whereas *Image and Vision Computing* was established in the early 1980s. Finally, do not miss out on the *IEE Proceedings – Vision, Image and Signal Processing* (now called *IET Computer Vision*).

Most journals are now online but usually to subscribers only; some go back a long way. Academic Press titles include *Computer Vision and Image Understanding, Graphical Models and Image Processing* and *Real-Time Imaging* (which will reappear as Springer's *Real-Time Image Processing*).

There are plenty of conferences too: the Proceedings of the IEEE conferences are held on the IEEE Explore site; *Lecture Notes in Computer Science* are hosted by Springer (http://www.springer.com/). Some conferences such as the *British Machine Vision Conference* series maintain their own site (http://www.bmva.ac.uk). The excellent Computer Vision Conferences page in Table 1.5 is brought to us by Keith Price and lists conferences in computer vision, image processing and pattern recognition.

1.6.2 Textbooks

There are many textbooks in this area. Increasingly, there are web versions, or web support, as summarized in Table 1.4. The difficulty is one of access, as you need a subscription to be able to access the online book (and sometimes even to see that it is available online). For example, this book is available online to those subscribing to Referex in Engineering Village (http://www.engineeringvillage.org). The site given in Table 1.4, as this book is the support site which includes demonstrations, worksheets, errata and other information. The site given next, at Edinburgh University UK, is part of the excellent CVOnline site (many thanks to Bob Fisher there) and it lists current books as well pdfs, some of which are more dated, but still excellent

Table 1.4 Web textbooks and homepages

This book's homepage	Southampton University	http://www.ecs.soton.ac.uk/~msn/book/
CVOnline: online book compendium	Edinburgh University	http://homepages.inf.ed.ac.uk/rbf/CVonline/books.htm
Image Processing Fundamentals	Delft University	http://www.ph.tn.tudelft.nl/Courses/FIP/noframes/fip.html
World of Mathematics	Wolfram Research	http://mathworld.wolfram.com
Numerical Recipes	Cambridge University Press	http://www.nr.com/
Digital Signal Processing	Steven W. Smith	http://www.dspguide.com/
The Joy of Visual Perception	York University	http://www.yorku.ca/research/vision/eye/thejoy.htm

(e.g. Ballard and Brown, 1982). There is also continuing debate on appropriate education in image processing and computer vision, for which review material is available (Bebis et al., 2003).

The *CVOnline* site also describes a great deal of technique. *Image Processing Fundamentals* is an online textbook for image processing. The *World of Mathematics* comes from Wolfram research (the distributors of Mathematica) and gives an excellent web-based reference for mathematics. *Numerical Recipes* (Press et al., 2002) is one of the best established texts in signal processing. It is beautifully written, with examples and implementation, and is on the web too. *Digital Signal Processing* is an online site with focus on the more theoretical aspects which will be covered in Chapter 2. As previously mentioned, *The Joy of Visual Perception* is an online site on how the human vision system works.

By way of context, for comparison with other textbooks, this text aims to start at the foundation of computer vision, and ends very close to a research level. Its content specifically addresses techniques for image analysis, considering shape analysis in particular. Mathcad and Matlab are used as a vehicle to demonstrate implementation, which is not always considered in other texts. But there are other texts, and these can help you to develop your interest in other areas of computer vision.

This section includes only a selection of some of the texts. There are more than these, some of which will be referred to in later chapters; each offers a particular view or insight into computer vision and image processing. Some of the *main textbooks* include: Marr, *Vision* (1982), which concerns vision and visual perception (as mentioned previously); Jain, *Fundamentals of Computer Vision* (1989), which is stacked with theory and technique, but omits implementation and some image analysis, and *Robot Vision* (Horn, 1986); Sonka et al., *Image Processing, Analysis and Computer Vision* (1998), offers more modern coverage of computer vision including many more recent techniques, together with pseudocode implementation, but omitting some image processing theory (the support site http://css.engineering.uiowa.edu/%7Edip/LECTURE/lecture.html has teaching material too); Jain et al., *Machine Vision* (1995), offers concise and modern coverage of 3D and motion (there is an online website at http://vision.cse.psu.edu/with code and images, together with corrections); Gonzalez and Wintz, *Digital Image Processing* (1987), has more tutorial element than many of the basically theoretical texts and has a fantastic reputation for introducing people to the field; Rosenfeld and Kak, *Digital Picture Processing* (1982) is very dated now, but is a well-proven text for much of the basic material; and Pratt, *Digital Image Processing* (2001), which was originally one of the earliest books on image processing and, like Rosenfeld and Kak, is a well-proven text for much of the basic material, particularly image transforms. Despite its name, the recent text called *Active Contours* (Blake and Isard, 1998) concentrates rather more on models of motion and deformation and probabilistic treatment of shape and motion, than on the active contours which we shall find here. As such, it is more a research text, reviewing many of the advanced techniques to describe shapes and their motion. *Image Processing – The Fundamentals* (Petrou and Bosdogianni, 1999) surveys the subject (as its title implies) from an image processing viewpoint, covering not only image transforms, but also restoration and enhancement before edge detection. The latter of these is most appropriate for one of the major contributors to that subject. A newer text (Shapiro and Stockman, 2001) includes chapters on image databases and on virtual and augmented reality. Umbaugh's *Computer Imaging: Digital Image Analysis and Processing* (2005) reflects recent interest in implementation by giving many programming examples. One of the most modern books is Forsyth and Ponce's *Computer Vision: A Modern Approach* (2002), which offers much new, and needed, insight into this continually developing subject (two chapters that did not make the final text, on probability and on tracking, are available at the book's website http://www.cs.berkeley.edu/%7Edaf/book.html)

Kasturi and Jain's *Computer Vision: Principles* (1991) and *Computer Vision: Advances and Applications* (1991) present a collection of seminal papers in computer vision, many of which are cited in their original form (rather than in this volume) in later chapters. There are other interesting edited collections (Chellappa, 1992), and one edition (Bowyer and Ahuja, 1996) honours Azriel Rosenfeld's many contributions.

Books that include a *software implementation* include Lindley, *Practical Image Processing in C* (1991), and Pitas, *Digital Image Processing Algorithms* (1993), which both cover basic image processing and computer vision algorithms. Parker, *Practical Computer Vision Using C* (1994), offers an excellent description and implementation of low-level image processing tasks within a well-developed framework, but again does not extend to some of the more recent and higher level processes in computer vision and includes little theory, although there is more in his later text *Image Processing and Computer Vision* (Parker, 1996). There is excellent coverage of practicality in Seul et al. (2000) and the book's support site is at http://www.mlmsoftwaregroup.com/. As mentioned previously, a recent text, *Computer Vision and Image Processing* (Umbaugh, 2005), takes an applications-orientated approach to computer vision and image processing, offering a variety of techniques in an engineering format. One recent text concentrates on Java only, *Image Processing in Java* (Lyon, 1999), and concentrates more on image processing systems implementation than on feature extraction (giving basic methods only). As already mentioned, a newer textbook (Efford, 2000) offers Java implementation, although it omits much of the mathematical detail, making it a lighter (more enjoyable?) read. Masters, *Signal and Image Processing with Neural Networks – A C++ Sourcebook* (1994), offers good guidance in combining image processing technique with neural networks and gives code for basic image processing technique, such as frequency domain transformation.

Other textbooks include Russ, *The Image Processing Handbook* (2002), which contains much basic technique with excellent visual support, but without any supporting theory, and has many practical details concerning image processing systems; Davies, *Machine Vision: Theory, Algorithms and Practicalities* (2005), which is targeted primarily at (industrial) machine vision systems, but covers much basic technique, with pseudo-code to describe their implementation; and the *Handbook of Pattern Recognition and Computer Vision* (Cheng, 2005). Last but by no means least, there is even an illustrated dictionary (Fisher, 2005) to guide you through the terms that are used.

1.6.3 The web

The web entries continue to proliferate. A list of web pages is given in Table 1.5 and these give you a starting point from which to build up your own list of favourite bookmarks. All these links, and more, are available at this book's homepage (http://www.ecs.soton.ac.uk/~msn/book/). This will be checked regularly and kept up to date. The web entries in Table 1.5 start with the Carnegie Mellon homepage (called the computer vision homepage). The Computer Vision Online CVOnline homepage has been brought to us by Bob Fisher from the University of Edinburgh. There is a host of material there, including its description. Their group also proves the Hypermedia Image Processing Website and, in their words: 'HIPR2 is a free www-based set of tutorial materials for the 50 most commonly used image processing operators. It contains tutorial text, sample results and JAVA demonstrations of individual operators and collections'. It covers a lot of basic material and shows you the results of various processing options. A big list of active groups can be found at the Computer Vision Homepage. If your University has access to the web-based indexes of published papers, the ISI index gives you journal papers (and allows for citation search), but unfortunately including medicine and science (where you can

Table 1.5 Computer vision and image processing websites

Name/scope	Host	Address
Vision and its applications		
The Computer Vision Homepage	Carnegie Mellon University	http://www.cs.cmu.edu/afs/cs/project/cil/ftp/html/vision.html
Computer Vision Online	Edinburgh University	http://www.dai.ed.ac.uk/CVonline/
Hypermedia Image Processing Reference 2	Edinburgh University	http://www.dai.ed.ac.uk/HIPR2
Image Processing Archive	PEIPA	http://peipa.essex.ac.uk/
3D Reconstruction	Stanford University	http://biocomp.stanford.edu/3dreconstruction/index.html
Face Recognition	Zagreb University	http://www.face-rec.org/
Conferences		
Computer Vision (and image processing)	Keith Price, USC	http://iris.usc.edu/Information/Iris-Conferences.html
Newsgroups		
Computer Vision Image Processing	Vision List	comp.ai.vision (http://www.vislist.com/) sci.image.processing

get papers with over 30 authors). Alternatively, Compendex and INSPEC include papers more related to engineering, together with papers in conferences, and hence vision, but without the ability to search citations. More recently, many turn to Citeseer and Google Scholar with direct ability to retrieve the papers, as well as to see where they have been used. Two newsgroups can be found at the addresses given in Table 1.5 to provide what is perhaps the most up-to-date information.

1.7 Conclusions

This chapter has covered most of the prerequisites for feature extraction in image processing and computer vision. We need to know how we see, in some form, where we can find information and how to process data. More importantly, we need an image, or some form of spatial data. This is to be stored in a computer and processed by our new techniques. As it consists of data points stored in a computer, this data is sampled or discrete. Extra material on image formation, camera models and image geometry is to be found in Appendix 2, but we shall be considering images as a planar array of points from here on. We need to know some of the bounds on the sampling process, on how the image is formed. These are the subjects of the next chapter, which also introduces a new way of looking at the data, and how it is interpreted (and processed) in terms of frequency.

1.8 References

Armstrong, T., *Colour Perception – A Practical Approach to Colour Theory*, Tarquin Publications, Diss, 1991

Ballard, D. H. and Brown, C. M., *Computer Vision*, Prentice-Hall, New Jersey, 1982

Bebis, G., Egbert, D. and Shah, M., Review of Computer Vision Education, *IEEE Trans. Educ.*, **46**(1), pp. 2–21, 2003

Blake, A. and Isard, M., *Active Contours*, Springer, London, 1998

Bowyer, K. and Ahuja, N. (Eds), *Advances in Image Understanding, A Festschrift for Azriel Rosenfeld*, IEEE Computer Society Press, Los Alamitos, CA, 1996

Bruce, V. and Green, P., *Visual Perception: Physiology, Psychology and Ecology*, 2nd edn, Lawrence Erlbaum Associates, Hove, 1990

Chellappa, R., *Digital Image Processing*, 2nd edn, IEEE Computer Society Press, Los Alamitos, CA, 1992

Cheng, C. H. and Wang, P. S. P., *Handbook of Pattern Recognition and Computer Vision*, 3rd edn, World Scientific, Singapore, 2005

Cornsweet, T. N., *Visual Perception*, Academic Press, New York, 1970

Davies, E. R., *Machine Vision: Theory, Algorithms and Practicalities*, 3rd edn, Morgan Kaufmann (Elsevier), Amsterdam, 2005

Efford, N., *Digital Image Processing – A Practical Introduction Using JAVA*, Pearson Education, Harlow, 2000

Fisher, R. B., Dawson-Howe, K., Fitzgibbon, A. and Robertson, C., *Dictionary of Computer Vision and Image Processing*, John Wiley & Sons, New York, 2005

Forsyth, D. and Ponce, J., *Computer Vision: A Modern Approach*, Prentice Hall, New Jersey, 2002

Fossum, E. R., CMOS Image Sensors: Electronic Camera-On-A-Chip, *IEEE Trans. Electron. Devices*, **44**(10), pp. 1689–1698, 1997

Gonzalez, R. C. and Wintz, P., *Digital Image Processing*, 2nd edn, Addison Wesley, Reading, MA, 1987

Gonzalez, R. C., Woods, R. E. and Eddins, S., *Digital Image Processing using MATLAB*, 1st edn, Prentice Hall, New Jersey, 2003

Horn, B. K. P., *Robot Vision*, MIT Press, Boston, MA, 1986

Jain, A. K., *Fundamentals of Computer Vision*, Prentice Hall International (UK), Hemel Hempstead, 1989

Jain, R. C., Kasturi, R. and Schunk, B. G., *Machine Vision*, McGraw-Hill Book Co., Singapore, 1995

Kasturi, R. and Jain, R. C., *Computer Vision: Principles*, IEEE Computer Society Press, Los Alamitos, CA, 1991

Kasturi, R. and Jain, R. C., *Computer Vision: Advances and Applications*, IEEE Computer Society Press, Los Alamitos, CA, 1991

Lindley, C. A., *Practical Image Processing in C*, Wiley & Sons, New York, 1991

Lyon, D. A., *Image Processing in Java*, Prentice Hall, New Jersey, 1999

Maple, Waterloo Maple Software, Ontario, Canada

Marr, D., *Vision*, W. H. Freeman and Co., New York, 1982

Masters, T. *Signal and Image Processing with Neural Networks – A C++ Sourcebook*, Wiley and Sons, New York, 1994

MATLAB, The MathWorks, 24 Prime Way Park, Natick, MA, USA

Mathcad, Mathsoft, 101 Main St., Cambridge, MA, USA

Mathematica, Wolfram Research, 100 Trade Center Drive, Champaign, IL, USA

Nakamura, J., *Image Sensors and Signal Processing for Digital Still Cameras*, CRC Press, Boca Raton, FL, 2005

Overington, I., *Computer Vision – A Unified, Biologically-Inspired Approach*, Elsevier Science Press, Amsterdam, 1992

Parker, J. R., *Practical Computer Vision Using C*, Wiley & Sons, New York, 1994

Parker, J. R., *Algorithms for Image Processing and Computer Vision*, Wiley & Sons, New York, 1996

Petrou, M. and Bosdogianni, P., *Image Processing – The Fundamentals*, John Wiley & Sons, London, 1999

Pitas, I., *Digital Image Processing Algorithms*, Prentice Hall International (UK), Hemel Hempstead, 1993

Pratt, W. K., *Digital Image Processing: PIKS Inside*, 3rd edn, Wiley, New York, 2001

Press, W. H., Teukolsky, S. A., Vetterling, W. T. and Flannery, B. P., *Numerical Recipes in C++: The Art of Scientific Computing*, 2nd edn, Cambridge University Press, Cambridge, 2002

Ratliff, F., *Mach Bands: Quantitative Studies on Neural Networks in the Retina*, Holden-Day Inc., San Francisco, USA, 1965

Rosenfeld, A. and Kak A. C., *Digital Picture Processing*, 2nd edn, Vols 1 and 2, Academic Press, Orlando, FL, 1982

Russ, J. C., *The Image Processing Handbook*, 4th edn, CRC Press (IEEE Press), Boca Raton, FL, 2002

Schwarz, S. H., *Visual Perception*, 3rd edn, McGraw-Hill, New York, 2004

Seul, M., O'Gorman, L. and Sammon, M. J., *Practical Algorithms for Image Analysis: Descriptions, Examples, and Code*, Cambridge University Press, Cambridge, 2000

Shapiro, L. G. and Stockman, G. C., *Computer Vision*, Prentice Hall, New Jersey, USA, 2001

Sonka, M., Hllavac, V. and Boyle, R, *Image Processing, Analysis and Computer Vision*, 2nd edn, Chapman Hall, London, 1998

Umbaugh, S. E., *Computer Imaging: Digital Image Analysis and Processing*, CRC Press, Boca Raton, FL, 2005

2

Images, sampling and frequency domain processing

2.1 Overview

In this chapter, we shall look at the basic theory which underlies image formation and processing. We shall start by investigating what makes up a picture and look at the consequences of having a different number of points in the image. We shall also look at images in a different representation, known as the frequency domain. In this, as the name implies, we consider an image as a collection of frequency components. We can operate on images in the frequency domain and we shall also consider different transformation processes. These allow us different insights into images and image processing which will be used in later chapters not only as a means to develop techniques, but also to give faster (computer) processing.

Table 2.1 Overview of Chapter 2

Main topic	Sub topics	Main points
Images	Effects of differing *numbers* of points and of number *range* for those points.	Greyscale, colour, resolution, dynamic range, storage.
Fourier transform theory	What is meant by the *frequency domain*, how it applies to *discrete* (sampled) images, how it allows us to *interpret* images and the *sampling resolution* (number of points).	Continuous Fourier transform and properties, sampling criterion, discrete Fourier transform and properties, image transformation, transform duals. Inverse Fourier transform.
Consequences of transform approach	Basic *properties* of Fourier transforms, *other transforms*, frequency domain *operations*.	Translation (shift), rotation and scaling. Principle of superposition and linearity. Walsh, Hartley, discrete cosine and wavelet transforms. Filtering and other operations.

2.2 Image formation

A computer image is a matrix (a two-dimensional array) of *pixels*. The value of each pixel is proportional to the *brightness* of the corresponding point in the scene; its value is usually derived from the output of an analogue-to-digital (A/D) converter. The matrix of pixels, the image, is usually square and an image may be described as $N \times N$ m-bit pixels, where N is the number of points and m controls the number of brightness values. Using m bits gives a range of 2^m values, ranging from 0 to $2^m - 1$. If m is 8 this gives brightness levels ranging between 0 and 255, which are usually displayed as black and white, respectively, with shades of grey in between, as they are for the *greyscale image* of a walking man in Figure 2.1(a). Smaller values of m give fewer available levels, reducing the available contrast in an image.

The ideal value of m is related to the signal-to-noise ratio (bandwidth) of the camera. This is stated as approximately 45 dB for an analogue camera, and since there are 6 dB per bit, 8 bits will cover the available range. Choosing 8 bit pixels has further advantages in that it is very convenient to store pixel values as *bytes*, and 8 bit A/D converters are cheaper than those with a higher resolution. For these reasons images are nearly always stored as 8 bit bytes, although some applications use a different range. The relative influence of the 8 bits is shown in the image of the walking subject in Figure 2.1. Here, the least significant bit, bit 0 (Figure 2.1b), carries the least information (it changes most rapidly). As the order of the bits increases, they change less rapidly and carry more information. The most information is carried by the most significant bit, bit 7 (Figure 2.1i). Clearly, the fact that there is a walker in the original image can be recognized much more reliably from the high-order bits than it can from the other bits (notice too the odd effects in the bits, which would appear to come from lighting at the top of the image).

Colour images follow a similar storage strategy to specify pixels' intensities. However, instead of using just one image plane, colour images are represented by three intensity components. These components generally correspond to red, green and blue (the RGB model), although there are other colour schemes. For example, the CMYK colour model is defined by the components cyan, magenta, yellow and black. In any colour mode, the pixel's colour can be specified in two main ways. First, you can associate an integer value with each pixel, which can be used as an index to a table that stores the intensity of each colour component. The index is used to recover the actual colour from the table when the pixel is going to be displayed, or processed. In this scheme, the table is known as the image's *palette* and the display is said to be performed by *colour mapping*. The main reason for using this colour representation is to reduce memory requirements. That is, we only store a single image plane (i.e. the indices) and the palette. This is less than storing the red, green and blue components separately and so makes the hardware cheaper, and it can have other advantages, for example when the image is transmitted. The main disadvantage is that the quality of the image is reduced since only a reduced collection of colours is actually used. An alternative to represent colour is to use several image planes to store the colour components of each pixel. This scheme is known as *true colour* and it represents an image more accurately, essentially by considering more colours. The most common format uses 8 bits for each of the three RGB components. These images are known as *24 bit* true colour and they can contain 16 777 216 different colours simultaneously. In spite of requiring significantly more memory, the image quality and the continuing reduction in cost of computer memory make this format a good alternative, even for storing the image frames from a video sequence. A good compression algorithm is always helpful in these cases, particularly if images

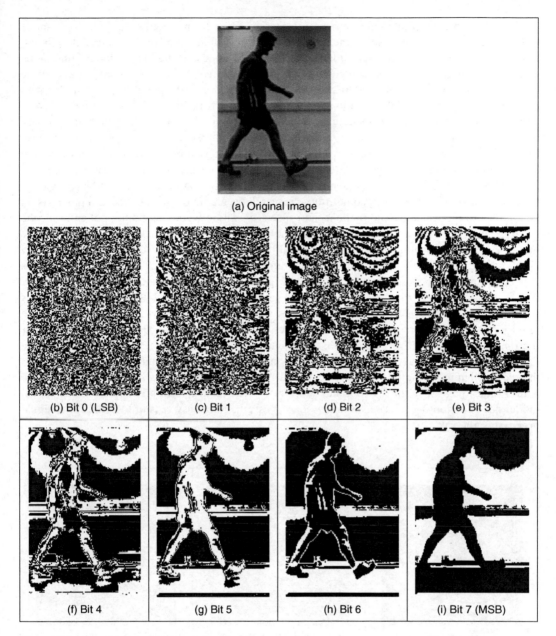

(a) Original image

(b) Bit 0 (LSB) (c) Bit 1 (d) Bit 2 (e) Bit 3

(f) Bit 4 (g) Bit 5 (h) Bit 6 (i) Bit 7 (MSB)

Figure 2.1 Decomposing an image into its bits

need to be transmitted on a network. Here we will consider the processing of grey-level images only, since they contain enough information to perform feature extraction and image analysis. Should the image be originally colour, we will consider processing its luminance only, often computed in a standard way. In any case, the amount of memory used is always related to the image size.

Choosing an appropriate value for the image size, N, is far more complicated. We want N to be sufficiently large to resolve the required level of spatial detail in the image. If N is too *small*, the image will be coarsely quantized: lines will appear to be very 'blocky' and some of the detail will be *lost*. Larger values of N give more *detail*, but need more storage space and the images will take longer to process, since there are more pixels. For example, with reference to the image of the walking subject in Figure 2.1(a), Figure 2.2 shows the effect of taking the image at different resolutions. Figure 2.2(a) is a 64×64 image, which shows only the broad structure. It is impossible to see any detail in the subject's face, or anywhere else. Figure 2.2(b) is a 128×128 image, which is starting to show more of the detail, but it would be hard to determine the subject's identity. The original image, repeated in Figure 2.2(c), is a 256×256 image which shows a much greater level of detail, and the subject can be recognized from the image. (These images come from a research programme aimed to use computer vision techniques to recognize people by their gait; face recognition would have little potential for the low-resolution image, which is often the sort of image that security cameras provide.) If the image were a pure photographic image, some of the much finer detail like the hair would show up in much greater detail. This is because the grains in film are very much smaller than the pixels in a computer image. Note that the images in Figure 2.2 have been scaled to be the same size. As such, the pixels in Figure 2.2(a) are much larger than in Figure 2.2(c), which emphasizes its blocky structure. The most common choices are for 256×256 images or 512×512. These require 64 and 256 kbytes of storage, respectively. If we take a sequence of, say, 20 images for motion analysis, we will need more than 1 Mbyte to store the 20 256×256 images, and more than 5 Mbytes if the images were 512×512. Even though memory continues to become cheaper, this can still impose high cost. But it is not just cost which motivates an investigation of the appropriate image size, the appropriate value for N. The main question is: are there theoretical guidelines for choosing it? The short answer is 'yes'; the long answer is to look at digital signal processing theory.

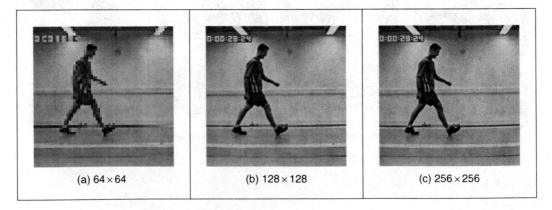

(a) 64×64 (b) 128×128 (c) 256×256

Figure 2.2 Effects of differing image resolution

The choice of sampling frequency is dictated by the *sampling criterion*. Presenting the sampling criterion requires understanding of how we interpret signals in the *frequency domain*. The way in is to look at the Fourier transform. This is a highly theoretical topic, but do not let that put you off (it leads to image coding, like the JPEG format, so it is very useful indeed).

The Fourier transform has found many uses in image processing and understanding; it might appear to be a complex topic (that's actually a horrible pun!), but it is a very rewarding one to study. The particular concern is the appropriate sampling frequency of (essentially, the value for N), or the rate at which pixel values are taken from, a camera's video signal.

2.3 The Fourier transform

The *Fourier transform* is a way of mapping a signal into its component frequencies. *Frequency* measures in Hertz (Hz) the rate of repetition with *time*, measured in seconds (s); time is the *reciprocal* of frequency and vice versa (Hertz = 1/second; s = 1/Hz).

Consider a music centre: the sound comes from a CD player (or a tape, etc.) and is played on the speakers after it has been processed by the amplifier. On the amplifier, you can change the bass or the treble (or the loudness, which is a combination of bass and treble). *Bass* covers the *low*-frequency components and *treble* covers the *high*-frequency ones. The Fourier transform is a way of mapping the signal from the CD player, which is a signal varying continuously with time, into its frequency components. When we have transformed the signal, we know which frequencies made up the original sound.

So why do we do this? We have not changed the signal, only its representation. We can now visualize it in terms of its frequencies, rather than as a voltage which changes with time. But we can now change the frequencies (because we can see them clearly) and this will change the sound. If, say, there is hiss on the original signal then since hiss is a high-frequency component, it will show up as a high-frequency component in the Fourier transform. So we can see how to remove it by looking at the Fourier transform. If you have ever used a graphic equalizer, you have done this before. The graphic equalizer is a way of changing a signal by interpreting its frequency domain representation; you can selectively control the frequency content by changing the positions of the controls of the graphic equalizer. The equation that defines the *Fourier transform*, *Fp*, of a signal *p*, is given by a complex integral:

$$\Im(p(\omega)) = Fp(\omega) = \int_{-\infty}^{\infty} p(t)e^{-j\omega t}\mathrm{d}t \tag{2.1}$$

where $Fp(\omega)$ is the Fourier transform; ω is the *angular* frequency, $\omega = 2\pi f$, measured in *radians/s* (where the frequency f is the reciprocal of time t, $f = 1/t$); j is the complex variable (electronic engineers prefer j to i since they cannot confuse it with the symbol for current – perhaps they don't want to be mistaken for mathematicians!); $p(t)$ is a *continuous* signal (varying continuously with time); and $e^{-j\omega t} = \cos(\omega t) - j\sin(\omega t)$ gives the frequency components in $p(t)$.

We can derive the Fourier transform by applying Equation 2.1 to the signal of interest. We can see how it works by constraining our analysis to simple signals. (We can then say that complicated signals are just made up by adding up lots of simple signals.) If we take a pulse which is of amplitude (size) A between when it starts at time $t = -T/2$ and when it ends at $t = T/2$, and is zero elsewhere, the pulse is:

$$p(t) = \begin{cases} A & \text{if} -T/2 \leq t \leq T/2 \\ 0 & \text{otherwise} \end{cases} \tag{2.2}$$

To obtain the Fourier transform, we substitute for $p(t)$ in Equation 2.1. $p(t) = A$ only for a specified time, so we choose the limits on the integral to be the start and end points of our pulse (it is zero elsewhere) and set $p(t) = A$, its value in this time interval. The Fourier transform of this pulse is the result of computing:

$$Fp(\omega) = \int_{-T/2}^{T/2} Ae^{-j\omega t}dt \tag{2.3}$$

When we solve this we obtain an expression for $Fp(\omega)$:

$$Fp(\omega) = -\frac{Ae^{-j\omega T/2} - Ae^{j\omega T/2}}{j\omega} \tag{2.4}$$

By simplification, using the relation $\sin(\theta) = (e^{j\theta} - e^{-j\theta})/2j$, the Fourier transform of the pulse is:

$$Fp(\omega) = \begin{vmatrix} \dfrac{2A}{\omega} \sin\left(\dfrac{\omega T}{2}\right) & \text{if } \omega \neq 0 \\ \\ AT & \text{if } \omega = 0 \end{vmatrix} \tag{2.5}$$

This is a version of the *sinc* function, $\text{sinc}(x) = \sin(x)/x$. The original pulse, and its transform are illustrated in Figure 2.3. Equation 2.5 (as plotted in Figure 2.3b) suggests that a pulse is made up of a lot of low frequencies (the main body of the pulse) and a few higher frequencies (which give the edges of the pulse). (The range of frequencies is symmetrical around zero frequency; negative frequency is a necessary mathematical abstraction.) The plot of the Fourier transform is called the *spectrum* of the signal, which can be considered akin to the spectrum of light.

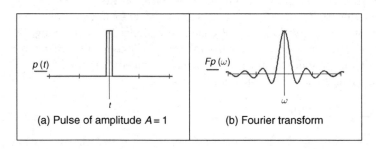

(a) Pulse of amplitude $A = 1$ (b) Fourier transform

Figure 2.3   A pulse and its Fourier transform

So what actually is this Fourier transform? It tells us what frequencies make up a time domain signal. The magnitude of the transform at a particular frequency is the amount of that frequency in the original signal. If we collect together sinusoidal signals in amounts specified by the Fourier transform, we should obtain the originally transformed signal. This process is illustrated in Figure 2.4 for the signal and transform illustrated in Figure 2.3. Note that since the Fourier transform is a *complex* number it has real and imaginary parts, and we only plot the *real* part here. A low frequency, that for $\omega = 1$, in Figure 2.4(a), contributes a large component of the original signal; a higher frequency, that for $\omega = 2$, contributes less, as in Figure 2.4(b). This is because the transform coefficient is less for $\omega = 2$ than it is for $\omega = 1$. There is a very small contribution for $\omega = 3$ (Figure 2.4c), although there is more for $\omega = 4$ (Figure 2.4d). This is because there are frequencies for which there is no contribution, where the transform is zero. When these signals are integrated together, we achieve a signal that looks similar to the original pulse (Figure 2.4e). Here we have only considered frequencies from $\omega = -6$ to $\omega = 6$. If the

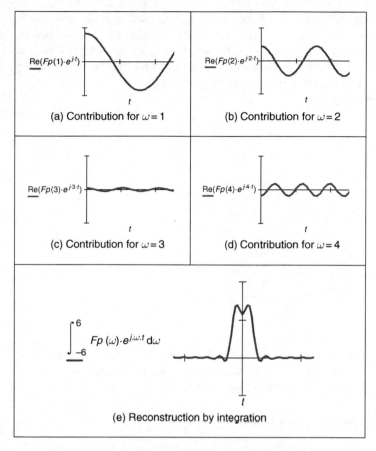

Figure 2.4 Reconstructing a signal from its transform

frequency range in integration were larger, more high frequencies would be included, leading to a more faithful reconstruction of the original pulse.

The result of the Fourier transform is a *complex* number. As such, it is usually represented in terms of its *magnitude* (or size, or modulus) and *phase* (or argument). The transform can be represented as:

$$\int_{-\infty}^{\infty} p(t)e^{-j\omega t}dt = \text{Re}[Fp(\omega)] + j\ \text{Im}[Fp(\omega)] \tag{2.6}$$

where $\text{Re}(\omega)$ and $\text{Im}(\omega)$ are the real and imaginary parts of the transform, respectively. The *magnitude* of the transform is then:

$$\left| \int_{-\infty}^{\infty} p(t)e^{-j\omega t}dt \right| = \sqrt{\text{Re}[Fp(\omega)]^2 + \text{Im}[Fp(\omega)]^2} \tag{2.7}$$

and the *phase* is:

$$\left\langle \int_{-\infty}^{\infty} p(t)e^{-j\omega t}dt = \tan^{-1} \frac{\text{Im}[Fp(\omega)]}{\text{Re}[Fp(\omega)]} \tag{2.8}$$

Images, sampling and frequency domain processing 39

where the signs of the real and the imaginary components can be used to determine which quadrant the phase is in (since the phase can vary from 0 to 2π radians). The *magnitude* describes the *amount* of each frequency component, while the *phase* describes *timing*, when the frequency components occur. The magnitude and phase of the transform of a pulse are shown in Figure 2.5, where the magnitude returns a positive transform, and the phase is either 0 or 2π radians (consistent with the sine function).

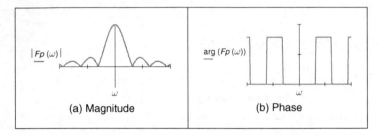

(a) Magnitude   (b) Phase

Figure 2.5 Magnitude and phase of the Fourier transform of a pulse

To return to the time domain signal, from the frequency domain signal, we require the *inverse Fourier transform*. This is the process by which we reconstructed the pulse from its transform components. The inverse FT calculates $p(t)$ from $Fp(\omega)$ according to:

$$p(t) = \Im^{-1}(Fp(\omega)) = \frac{1}{2\pi}\int_{-\infty}^{\infty} Fp(\omega)e^{-j\omega t}d\omega \qquad (2.9)$$

Together, Equations 2.1 and 2.9 form a relationship known as a *transform pair* that allows us to transform into the frequency domain, and back again. By this process, we can perform operations in the frequency domain or in the time domain, since we have a way of changing between them. One important process is known as *convolution*. The convolution of one signal $p_1(t)$ with another signal $p_2(t)$, where the convolution process denoted by $*$, is given by the integral

$$p_1(t) * p_2(t) = \int_{-\infty}^{\infty} p_1(\tau)p_2(t - \tau)d\tau \qquad (2.10)$$

This is the basis of systems theory, where the output of a system is the convolution of a stimulus, say p_1, and a system's *response*, p_2, By inverting the time axis of the system response, to give $p_2(t - \tau)$, we obtain a *memory* function. The convolution process then sums the effect of a stimulus multiplied by the memory function: the current output of the system is the cumulative response to a stimulus. By taking the Fourier transform of Equation 2.10, where the Fourier transformation is denoted by $\Im$, the Fourier transform of the convolution of two signals is

$$\Im[p_1(t) * p_2(t)] = \int_{-\infty}^{\infty} \left\{\int_{-\infty}^{\infty} p_1(\tau)p_2(t - \tau)d\tau\right\} e^{-j\omega t}dt$$

$$= \int_{-\infty}^{\infty} \left\{\int_{-\infty}^{\infty} p_2(t - \tau)e^{-j\omega t}dt\right\} p_1(\tau)d\tau \qquad (2.11)$$

Now, since $\Im[p_2(t - \tau)] = e^{-j\omega t}Fp_2(\omega)$ (to be considered later in Section 2.6.1),

$$\Im[p_1(t) * p_2(t)] = \int_{-\infty}^{\infty} Fp_2(\omega)p_1(\tau)e^{-j\omega t}d\tau$$

$$= Fp_2(\omega) \int_{-\infty}^{\infty} p_1(\tau)e^{-j\omega t}d\tau \qquad (2.12)$$

$$= Fp_2(\omega) \times Fp_1(\omega)$$

As such, the frequency domain dual of convolution is multiplication; the *convolution integral* can be performed by *inverse* Fourier transformation of the *product* of the transforms of the two signals. A frequency domain representation essentially presents signals in a different way, but it also provides a different way of processing signals. Later, we shall use the duality of convolution to speed up the computation of vision algorithms considerably.

Further, *correlation* is defined to be

$$p_1(t) \otimes p_2(t) = \int_{-\infty}^{\infty} p_1(\tau)p_2(t + \tau)d\tau \qquad (2.13)$$

where $\otimes$ denotes correlation ($\odot$ is another symbol which is used sometimes, but there is not much consensus on this symbol). Correlation gives a measure of the *match* between the two signals $p_2(\omega)$ and $p_1(\omega)$. When $p_2(\omega) = p_1(\omega)$ we are correlating a signal with itself and the process is known as *autocorrelation*. We shall be using correlation later, to *find* things in images.

Before proceeding further, we also need to define the *delta function*, which can be considered to be a function occurring at a particular time interval:

$$delta(t - \tau) = \begin{vmatrix} 1 & \text{if } t = \tau \\ 0 & \text{otherwise} \end{vmatrix} \qquad (2.14)$$

The relationship between a signal's time domain representation and its frequency domain version is also known as a *transform pair*: the transform of a pulse (in the time domain) is a sinc function in the frequency domain. Since the transform is symmetrical, the Fourier transform of a sinc function is a pulse.

There are other Fourier transform pairs, as illustrated in Figure 2.6. First, Figure 2.6(a) and (b) show that the Fourier transform of a cosine function is two points in the frequency domain (at the same value for positive and negative frequency); we expect this since there is only one frequency in the cosine function, the frequency shown by its transform. Figure 2.6(c) and (d) show that the transform of the *Gaussian function* is another Gaussian function; this illustrates linearity (for linear systems it is Gaussian in, Gaussian out, which is another version of GIGO). Figure 2.6(e) is a single point (the delta function) which has a transform that is an infinite set of frequencies (Figure 2.6f); an alternative interpretation is that a delta function contains an equal amount of all frequencies. This can be explained by using Equation 2.5, where if the pulse is of shorter duration (T tends to zero), the sinc function is wider; as the pulse becomes infinitely thin, the spectrum becomes infinitely flat.

Finally, Figure 2.6(g) and (h) show that the transform of a set of uniformly spaced delta functions is another set of uniformly spaced delta functions, but with a different spacing. The spacing in the frequency domain is the reciprocal of the spacing in the time domain. By way of

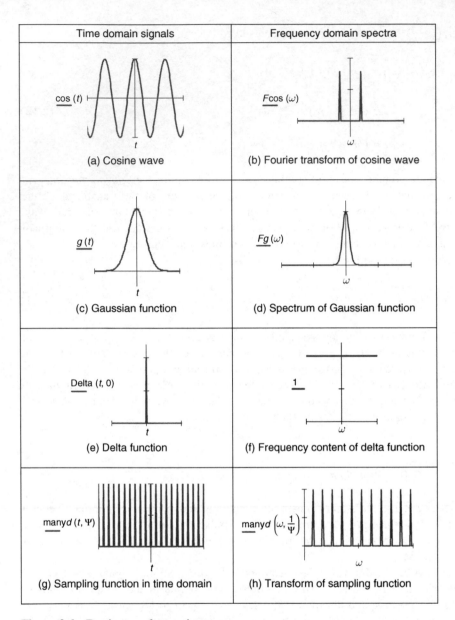

Time domain signals	Frequency domain spectra
$\underline{\cos}(t)$ (a) Cosine wave	$\underline{F\cos}(\omega)$ (b) Fourier transform of cosine wave
$\underline{g}(t)$ (c) Gaussian function	$\underline{Fg}(\omega)$ (d) Spectrum of Gaussian function
Delta $(t, 0)$ (e) Delta function	$\underline{1}$ (f) Frequency content of delta function
$\underline{manyd}(t, \Psi)$ (g) Sampling function in time domain	$\underline{manyd}\left(\omega, \dfrac{1}{\Psi}\right)$ (h) Transform of sampling function

Figure 2.6 Fourier transform pairs

a (non-mathematical) explanation, let us consider that the Gaussian function in Figure 2.6(c) is actually made up by summing a set of closely spaced (and very thin) Gaussian functions. Then, since the spectrum for a delta function is infinite, as the Gaussian function is stretched in the time domain (eventually to be a set of pulses of uniform height) we obtain a set of pulses in the frequency domain, but spaced by the reciprocal of the time domain spacing. This transform pair is the basis of sampling theory (which we aim to use to find a criterion that guides us to an appropriate choice for the image size).

2.4 The sampling criterion

The *sampling criterion* specifies the condition for the *correct choice* of sampling frequency. *Sampling* concerns taking *instantaneous* values of a continuous signal; physically, these are the outputs of an A/D converter sampling a camera signal. The samples are the values of the signal at sampling instants. This is illustrated in Figure 2.7, where Figure 2.7(a) concerns taking samples at a *high* frequency (the spacing between samples is low), compared with the amount of change seen in the signal of which the samples are taken. Here, the samples are taken sufficiently fast to notice the slight dip in the sampled signal. Figure 2.7(b) concerns taking samples at a *low* frequency, compared with the rate of change of (the maximum frequency in) the sampled signal. Here, the slight dip in the sampled signal is *not* seen in the samples taken from it.

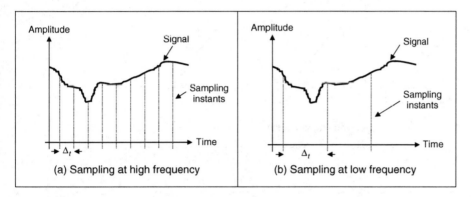

Figure 2.7 Sampling at different frequencies

We can understand the process better in the frequency domain. Let us consider a time-variant signal which has a range of frequencies between $-f_{max}$ and f_{max} as illustrated in Figure 2.8(b). This range of frequencies is shown by the Fourier transform, where the signal's *spectrum* exists only between these frequencies. This function is sampled every Δ_t s: this is a sampling function of spikes occurring every Δ_t s. The Fourier transform of the sampling function is a series of spikes separated by $f_{sample} = 1/\Delta_t$ Hz. The Fourier pair of this transform was illustrated earlier (Figure 2.6g and h).

The sampled signal is the result of multiplying the time-variant signal by the sequence of spikes; this gives samples that occur every Δ_t s, and the sampled signal is shown in Figure 2.8(a). These are the outputs of the A/D converter at sampling instants. The frequency domain analogue of this sampling process is to *convolve* the spectrum of the time-variant signal with the spectrum of the sampling function. Convolving the signals, the convolution process, implies that we take the spectrum of one, *flip* it along the horizontal axis and then *slide* it across the other. Taking the spectrum of the time-variant signal and sliding it over the spectrum of the spikes results in a spectrum where the spectrum of the original signal is *repeated* every $1/\Delta_t$ Hz, f_{sample} in Figure 2.8(b–d). If the spacing between samples is Δ_t, the repetitions of the time-variant signal's spectrum are spaced at intervals of $1/\Delta_t$, as in Figure 2.8(b). If the sample spacing is large, then the time-variant signal's spectrum is replicated close together and the spectra *collide*, or interfere, as in Figure 2.8(d). The spectra just *touch* when the sampling frequency is *twice* the maximum frequency in the signal. If the frequency domain spacing, f_{sample}, is *more* than twice the maximum frequency, f_{max}, the spectra do *not* collide or interfere, as in Figure 2.8(c). If the

Images, sampling and frequency domain processing   43

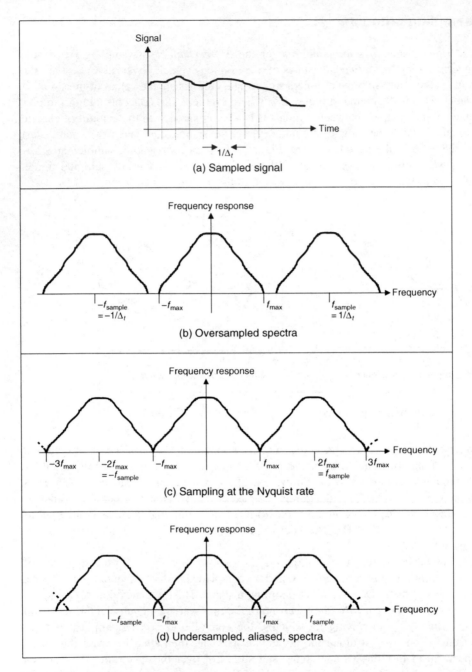

Figure 2.8 Sampled spectra

sampling frequency exceeds twice the maximum frequency then the spectra cannot collide. This is the Nyquist sampling criterion:

> In order to reconstruct a signal from its samples, the sampling frequency must be at least twice the highest frequency of the sampled signal.

If we do not obey Nyquist's sampling theorem the spectra collide. When we inspect the sampled signal, whose spectrum is within $-f_{max}$ to f_{max}, wherein the spectra collided, the corrupt spectrum implies that by virtue of sampling, we have *ruined* some of the information. If we were to attempt to reconstruct a signal by inverse Fourier transformation of the sampled signal's spectrum, processing Figure 2.8(d) would lead to the wrong signal, whereas inverse Fourier transformation of the frequencies between $-f_{max}$ and f_{max} in Figure 2.8(b) and (c) would lead back to the original signal. This can be seen in computer images as illustrated in Figure 2.9, which show an image of a group of people (the computer vision research team at Southampton) displayed at different spatial resolutions (the contrast has been increased to the same level in each subimage, so that the effect we want to demonstrate should definitely show up in the print copy). Essentially, the people become less distinct in the lower resolution image (Figure 2.9b). Now, look closely at the two sets of window blinds behind the people. At higher resolution, in Figure 2.9(a), these appear as normal window blinds. In Figure 2.9(b), which is sampled at a much lower resolution, a new pattern appears: the pattern appears to be curved, and if you consider the blinds' relative size the shapes actually appear to be much larger than normal window blinds. So by reducing the resolution, we are seeing something different, an *alias* of the true information: something that is not actually there at all, but appears to be there as a result of sampling. This is the result of sampling at too low a frequency: if we sample at high frequency, the interpolated result matches the original signal; if we sample at *too low* a frequency we can get the *wrong* signal. (For these reasons people on television tend not to wear chequered clothes, or should not!)

(a) High resolution   (b) Low resolution – aliased

Figure 2.9   Aliasing in sampled imagery

Obtaining the wrong signal is called *aliasing*: our interpolated signal is an alias of its proper form. Clearly, we want to *avoid* aliasing, so according to the sampling theorem we must sample at twice the maximum frequency of the signal coming out of the camera. The maximum frequency is defined to be 5.5 MHz, so we must sample the camera signal at 11 MHz. (For information, when using a computer to analyse speech we must sample the speech at a minimum frequency of 12 kHz, since the maximum speech frequency is 6 kHz.) Given the timing of a video signal, sampling at 11 MHz implies a minimum image resolution of 576×576 pixels.

This is unfortunate: 576 is not an integer power of two, which has poor implications for storage and processing. Accordingly, since many image processing systems have a maximum resolution of 512×512, they must anticipate aliasing. This is mitigated somewhat by the observations that:

- globally, the *lower* frequencies carry *more* information, whereas *locally* the *higher* frequencies contain more information, so the corruption of high-frequency information is of less importance
- there is *limited* depth of focus in imaging systems (reducing high frequency content).

But aliasing can, and does, occur and we must remember this when interpreting images. A different form of this argument applies to the images derived from digital cameras. The basic argument that the precision of the estimates of the high-order frequency components is dictated by the relationship between the effective sampling frequency (the number of image points) and the imaged structure, still applies.

The effects of sampling can often be seen in films, especially in the rotating wheels of cars, as illustrated in Figure 2.10. This shows a wheel with a single spoke, for simplicity. The film is a sequence of frames starting on the left. The sequence of frames plotted Figure 2.10(a) is for a wheel which rotates by 20° between frames, as illustrated in Figure 2.10(b). If the wheel is rotating much faster, by 340° between frames, as in Figure 2.10(c) and (d), the wheel will appear to rotate the other way. If the wheel rotates by 360° between frames, it will appear to be stationary. To perceive the wheel as rotating forwards, then the rotation between frames must be 180° at most. This is consistent with sampling at least twice the maximum frequency. Our eye can resolve this in films (when watching a film, I bet you haven't thrown a wobbly because the car's going forwards whereas the wheels say it's going the other way) since we know that the direction of the car must be consistent with the motion of its wheels, and we expect to see the wheels appear to go the wrong way, sometimes.

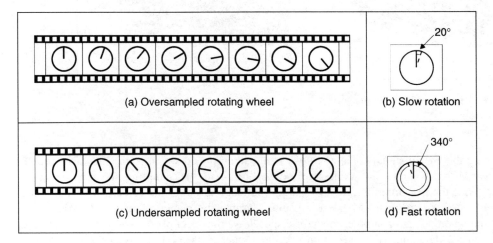

Figure 2.10 Correct and incorrect apparent wheel motion

2.5 The discrete Fourier transform

2.5.1 One-dimensional transform

Given that image processing concerns sampled data, we require a version of the Fourier transform which handles this. This is known as the *discrete Fourier transform* (DFT). The DFT of a set of N points $\mathbf{p}_x$ (sampled at a frequency which at least equals the Nyquist sampling rate) into sampled frequencies $\mathbf{Fp}_u$ is:

$$\mathbf{Fp}_u = \frac{1}{\sqrt{N}} \sum_{x=0}^{N-1} \mathbf{p}_x e^{-j\left(\frac{2\pi}{N}\right)xu} \tag{2.15}$$

This is a discrete analogue of the continuous Fourier transform: the continuous signal is replaced by a set of samples, the continuous frequencies by sampled ones, and the integral is replaced by a summation. If the DFT is applied to samples of a pulse in a window from sample 0 to sample $N/2 - 1$ (when the pulse ceases), the equation becomes:

$$\mathbf{Fp}_u = \frac{1}{\sqrt{N}} \sum_{x=0}^{\frac{N}{2}-1} A e^{-j\left(\frac{2\pi}{N}\right)xu} \tag{2.16}$$

Since the sum of a geometric progression can be evaluated according to:

$$\sum_{k=0}^{n} a_0 r^k = \frac{a_0(1 - r^{n+1})}{1-r} \tag{2.17}$$

the discrete Fourier transform of a sampled pulse is given by:

$$\mathbf{Fp}_u = \frac{A}{\sqrt{N}} \left(\frac{1 - e^{-j\left(\frac{2\pi}{N}\right)\left(\frac{N}{2}\right)u}}{1 - e^{-j\left(\frac{2\pi}{N}\right)u}} \right) \tag{2.18}$$

By rearrangement, we obtain:

$$\mathbf{Fp}_u = \frac{A}{\sqrt{N}} e^{-j\left(\frac{\pi u}{2}\right)\left(1 - \frac{2}{N}\right)} \frac{\sin(\pi u/2)}{\sin(\pi u/N)} \tag{2.19}$$

The modulus of the transform is:

$$|\mathbf{Fp}_u| = \frac{A}{\sqrt{N}} \left| \frac{\sin(\pi u/2)}{\sin(\pi u/N)} \right| \tag{2.20}$$

since the magnitude of the exponential function is 1. The original pulse is plotted Figure 2.11(a) and the magnitude of the Fourier transform plotted against frequency is given in Figure 2.11(b).

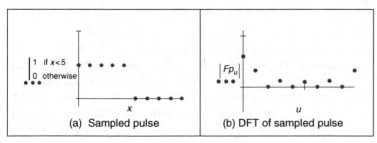

(a) Sampled pulse   (b) DFT of sampled pulse

Figure 2.11   Transform pair for sampled pulse

This is clearly comparable with the result of the continuous Fourier transform of a pulse (Figure 2.3), since the transform involves a similar, sinusoidal, signal. The spectrum is equivalent to a set of sampled frequencies; we can build up the sampled pulse by adding up the frequencies according to the Fourier description. Consider a signal such as that shown in Figure 2.12(a). This has no explicit analytic definition, and as such it does not have a closed Fourier transform; the Fourier transform is generated by direct application of Equation 2.15. The result is a set of samples of frequency (Figure 2.12b).

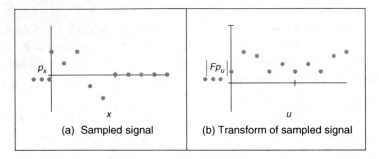

(a) Sampled signal	(b) Transform of sampled signal

Figure 2.12 A sampled signal and its discrete transform

The Fourier transform in Figure 2.12(b) can be used to reconstruct the original signal in Figure 2.12(a), as illustrated in Figure 2.13. Essentially, the coefficients of the Fourier transform tell us how much there is of each of a set of sinewaves (at different frequencies) in the original signal. The lowest frequency component $\mathbf{Fp}_0$, for zero frequency, is called the *d.c. component* (it is constant and equivalent to a sinewave with no frequency) and it represents the *average* value of the samples. Adding the contribution of the first coefficient $\mathbf{Fp}_0$ (Figure 2.13b) to the contribution of the second coefficient $\mathbf{Fp}_1$ (Figure 2.13c), is shown in Figure 2.13(d). This shows how addition of the first two frequency components approaches the original sampled pulse. The approximation improves when the contribution due to the fourth component, $\mathbf{Fp}_3$, is included, as shown in Figure 2.13(e). Finally, adding up all six frequency components gives a close approximation to the original signal, as shown in Figure 2.13(f).

This process is the *inverse DFT*. This can be used to reconstruct a sampled signal from its frequency components by:

$$\mathbf{p}_x = \sum_{u=0}^{N-1} \mathbf{Fp}_u e^{j\left(\frac{2\pi}{N}\right)ux} \tag{2.21}$$

Note that there are several assumptions made before application of the DFT. The first is that the sampling criterion has been satisfied. The second is that the sampled function replicates to infinity. When generating the transform of a pulse, Fourier theory assumes that the pulse repeats outside the window of interest. (There are window operators that are designed specifically to handle difficulty at the ends of the sampling window.) Finally, the maximum frequency corresponds to half the sampling period. This is consistent with the assumption that the sampling criterion has not been violated, otherwise the high frequency spectral estimates will be corrupt.

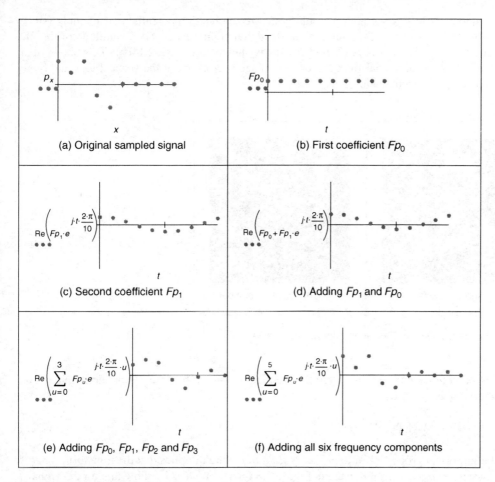

Figure 2.13 Signal reconstruction from its transform components

2.5.2 Two-dimensional transform

Equation 2.15 gives the DFT of a one-dimensional (1D) signal. We need to generate Fourier transforms of images, so we need a *two-dimensional* (2D) *discrete Fourier transform*. This is a transform of pixels (sampled picture points) with a 2D spatial location indexed by coordinates x and y. This implies that we have two dimensions of frequency, u and v, which are the horizontal and vertical spatial frequencies, respectively. Given an image of a set of vertical lines, the Fourier transform will show only horizontal spatial frequency. The vertical spatial frequencies are zero since there is no vertical variation along the y-axis. The 2D Fourier transform evaluates the frequency data, $\mathbf{FP}_{u,v}$, from the $N \times N$ pixels $\mathbf{P}_{x,y}$ as:

$$\mathbf{FP}_{u,v} = \frac{1}{N}\sum_{x=0}^{N-1}\sum_{y=0}^{N-1}\mathbf{P}_{x,y}e^{-j\left(\frac{2\pi}{N}\right)(ux+vy)} \tag{2.22}$$

The Fourier transform of an image can be obtained *optically* by transmitting a laser through a photographic slide and forming an image using a lens. The Fourier transform of the image of the slide is formed in the front focal plane of the lens. This is still restricted to transmissive systems, whereas reflective formation would widen its application potential considerably (since optical computation is just slightly faster than its digital counterpart). The magnitude of the 2D DFT to an image of vertical bars (Figure 2.14a) is shown in Figure 2.14(b). This shows that there are only horizontal spatial frequencies; the image is constant in the vertical axis and there are no vertical spatial frequencies.

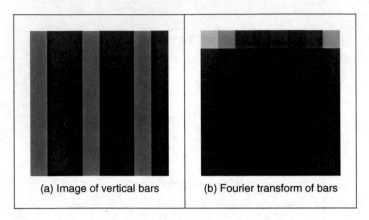

(a) Image of vertical bars   (b) Fourier transform of bars

Figure 2.14   Applying the 2D discrete Fourier transform

The *2D inverse DFT* transforms from the frequency domain back to the image domain. The 2D inverse DFT is given by:

$$\mathbf{P}_{x,y} = \sum_{u=0}^{N-1} \sum_{v=0}^{N-1} \mathbf{FP}_{u,v} e^{j\left(\frac{2\pi}{N}\right)(ux+vy)} \qquad (2.23)$$

One of the important properties of the FT is *replication*, which implies that the transform *repeats* in frequency up to *infinity*, as indicated in Figure 2.8 for 1D signals. To show this for 2D signals, we need to investigate the Fourier transform, originally given by $\mathbf{FP}_{u,v}$, at integer multiples of the number of sampled points $\mathbf{FP}_{u+mM,\ v+nN}$ (where m and n are integers). The Fourier transform $\mathbf{FP}_{u+mM,\ v+nN}$ is, by substitution in Equation 2.22:

$$\mathbf{FP}_{u+mN,v+nN} = \frac{1}{N} \sum_{x=0}^{N-1} \sum_{y=0}^{N-1} \mathbf{P}_{x,y} e^{-j\left(\frac{2\pi}{N}\right)((u+mN)x+(v+nN)y)} \qquad (2.24)$$

so,

$$\mathbf{FP}_{u+mN,v+nN} = \frac{1}{N} \sum_{x=0}^{N-1} \sum_{y=0}^{N-1} \mathbf{P}_{x,y} e^{-j\left(\frac{2\pi}{N}\right)(ux+vy)} \times e^{-j2\pi(mx+ny)} \qquad (2.25)$$

and since $e^{-j2\pi(mx+ny)} = 1$ (since the term in brackets is always an integer and then the exponent is always an integer multiple of 2π), then

$$\mathbf{FP}_{u+mN,v+nN} = \mathbf{FP}_{u,v} \qquad (2.26)$$

which shows that the replication property does hold for the Fourier transform. However, Equations 2.22 and 2.23 are very slow for large image sizes. They are usually implemented by using the *fast Fourier transform* (FFT), which is a splendid rearrangement of the Fourier transform's computation that improves speed dramatically. The FFT algorithm is beyond the scope of this text, but is also a rewarding topic of study (particularly for computer scientists or software engineers). The FFT can only be applied to square images whose size is an integer power of 2 (without special effort). Calculation involves the *separability* property of the Fourier transform. Separability means that the Fourier transform is calculated in two stages: the rows are first transformed using a 1D FFT, then this data is transformed in columns, again using a 1D FFT. This process can be achieved since the sinusoidal *basis functions* are orthogonal. Analytically, this implies that the 2D DFT can be decomposed as in Equation 2.27:

$$\frac{1}{N}\sum_{x=0}^{N-1}\sum_{y=0}^{N-1}\mathbf{P}_{x,y}e^{-j\left(\frac{2\pi}{N}\right)(ux+vy)} = \frac{1}{N}\sum_{x=0}^{N-1}\left\{\sum_{y=0}^{N-1}\mathbf{P}_{x,y}e^{-j\left(\frac{2\pi}{N}\right)(vy)}\right\}e^{-j\left(\frac{2\pi}{N}\right)(ux)} \tag{2.27}$$

showing how separability is achieved, since the inner term expresses transformation along one axis (the y-axis) and the outer term transforms this along the other (the x-axis).

Since the computational cost of a 1D FFT of N points is $O(N \log(N))$, the cost (by separability) for the 2D FFT is $O(N^2 \log(N))$, whereas the computational cost of the 2D DFT is $O(N^3)$. This implies a considerable saving since it suggests that the FFT requires much less time, particularly for large image sizes (so for a 128×128 image, if the FFT takes minutes, the DFT will take days). The 2D FFT is available in Mathcad using the `icfft` function, which gives a result equivalent to Equation 2.22. The inverse 2D FFT (Equation 2.23) can be implemented using the Mathcad `cfft` function. (The difference between many Fourier transform implementations essentially concerns the chosen scaling factor.) The Mathcad implementations of the 2D DFT and the inverse 2D DFT are given in Code 2.1(a) and (b), respectively. The implementations using the Mathcad functions using the FFT are given in Code 2.1(c) and (d), respectively.

$$FP_{u,v} := \frac{1}{\text{rows}(P)} \cdot \sum_{y=0}^{\text{rows}(P)-1}\ \sum_{x=0}^{\text{cols}(P)-1} P_{y,x} \cdot e^{\frac{-j \cdot 2 \cdot \pi \cdot (u \cdot y + v \cdot x)}{\text{rows}(P)}}$$

(a) 2D DFT, Equation 2.22

$$IFP_{y,x} := \sum_{u=0}^{\text{rows}(FP)-1}\ \sum_{v=0}^{\text{cols}(FP)-1} FP_{u,v} \cdot e^{\frac{j \cdot 2 \cdot \pi \cdot (u \cdot y + v \cdot x)}{\text{rows}(FP)}}$$

(b) Inverse 2D DFT, Equation 2.23

```
Fourier(pic):=icfft(pic)
```
(c) 2D FFT

```
inv_Fourier(trans):=cfft(trans)
```
(d) inverse 2D FFT

Code 2.1 Implementing Fourier transforms

For reasons of speed, the 2D FFT is the algorithm commonly used in application. One (unfortunate) difficulty is that the nature of the Fourier transform produces an image which, at first, is difficult to interpret. The Fourier transform of an image gives the frequency components. The position of each component reflects its frequency: *low*-frequency components are *near* the origin and *high*-frequency components are further *away*. As before, the lowest frequency component, for zero frequency, the d.c. component, represents the *average* value of the samples. Unfortunately, the arrangement of the 2D Fourier transform places the low-frequency components at the *corners* of the transform. The image of the square in Figure 2.15(a) shows this in its transform (Figure 2.15b). A spatial transform is easier to visualize if the d.c. (zero frequency) component is in the *centre*, with frequency increasing towards the edge of the image. This can be arranged either by rotating each of the four quadrants in the Fourier transform by 180°, or by *reordering* the original image to give a transform which shifts the transform to the centre. Both operations result in the image in Figure 2.15(c), wherein the transform is much more easily seen. Note that this is aimed to improve visualization and does not change any of the frequency domain information, only the way it is displayed.

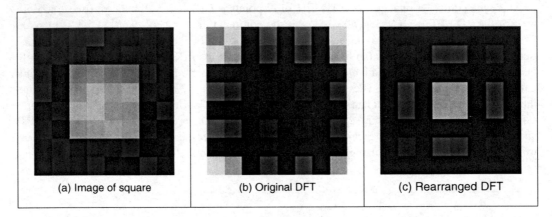

| (a) Image of square | (b) Original DFT | (c) Rearranged DFT |

Figure 2.15 Rearranging the 2D DFT for display purposes

To rearrange the image so that the d.c. component is in the centre, the frequency components need to be reordered. This can be achieved simply by multiplying each image point $\mathbf{P}_{x,y}$ by $-1^{(x+y)}$. Since $\cos(-\pi) = -1$, then $-1 = e^{-j\pi}$ (the minus sign is introduced just to keep the analysis neat), so we obtain the transform of the multiplied image as:

$$\frac{1}{N}\sum_{x=0}^{N-1}\sum_{y=0}^{N-1}\mathbf{P}_{x,y}e^{-j\left(\frac{2\pi}{N}\right)(ux+vy)} \times -1^{(x+y)} = \frac{1}{N}\sum_{x=0}^{N-1}\sum_{y=0}^{N-1}\mathbf{P}_{x,y}e^{-j\left(\frac{2\pi}{N}\right)(ux+vy)} \times e^{-j\pi(x+y)}$$

$$= \frac{1}{N}\sum_{x=0}^{N-1}\sum_{y=0}^{N-1}\mathbf{P}_{x,y}e^{-j\left(\frac{2\pi}{N}\right)\left(\left(u+\frac{N}{2}\right)x+\left(v+\frac{N}{2}\right)y\right)} \quad (2.28)$$

$$= \mathbf{FP}_{u+\frac{N}{2},v+\frac{N}{2}}$$

According to Equation 2.28, when pixel values are multiplied by $-1^{(x+y)}$, the Fourier transform becomes shifted along each axis by half the number of samples. According to the replication theorem (Equation 2.26), the transform replicates along the frequency axes. This implies that the centre of a transform image will now be the d.c. component. (Another way of interpreting this is that rather than look at the frequencies centred on where the image is, our viewpoint has been shifted so as to be centred on one of its corners, thus invoking the replication property.) The operator `rearrange`, in Code 2.2, is used before transform calculation, and results in the image of Figure 2.15(c) and all later transform images.

```
rearrange(picture):=    for y∈0..rows(picture)-1
                            for x∈0..cols(picture)-1
                                rearranged_pic_{y,x} ← picture_{y,x} · (-1)^{(y+x)}
                        rearranged_pic
```

Code 2.2 Reordering for transform calculation

The full effect of the Fourier transform is shown by application to an image of much higher resolution. Figure 2.16(a) shows the image of a face and Figure 2.16(b) shows its transform. The transform reveals that much of the information is carried in the *lower* frequencies, since this is where most of the spectral components concentrate. This is because the face image has many regions where the brightness does not change a lot, such as the cheeks and forehead. The *high*-frequency components reflect *change* in intensity. Accordingly, the higher frequency components arise from the hair (and that awful feather!) and from the borders of features of the human face, such as the nose and eyes.

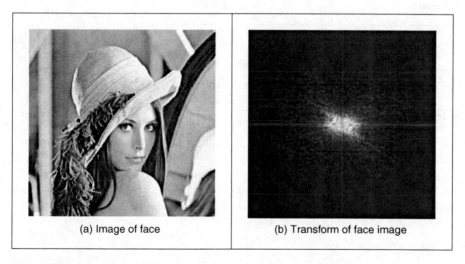

(a) Image of face   (b) Transform of face image

Figure 2.16 Applying the Fourier transform to the image of a face

As with the 1D Fourier transform, there are 2D Fourier transform pairs, illustrated in Figure 2.17. The 2D Fourier transform of a 2D pulse (Figure 2.17a) is a 2D sinc function (Figure 2.17b). The 2D Fourier transform of a Gaussian function (Figure 2.17c) is again a 2D Gaussian function in the frequency domain (Figure 2.17d).

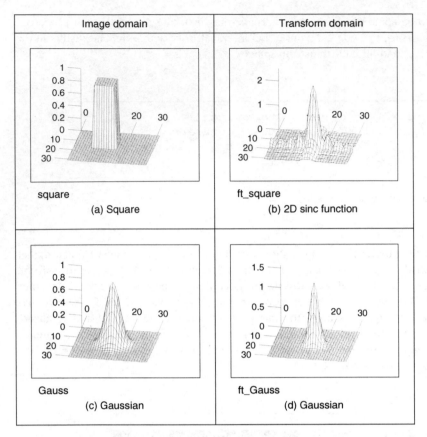

Figure 2.17 2D Fourier transform pairs

2.6 Other properties of the Fourier transform

2.6.1 Shift invariance

The decomposition into spatial frequency does not depend on the position of features within the image. If we shift all the features by a fixed amount, or acquire the image from a different position, the magnitude of its Fourier transform does not change. This property is known as *shift invariance*. By denoting the delayed version of $p(t)$ as $p(t - \tau)$, where τ is the delay, and the Fourier transform of the shifted version as $\Im[p(t - \tau)]$, we obtain the relationship between a time domain shift in the time and frequency domains as:

$$\Im[p(t - \tau)] = e^{-j\omega t} P(\omega) \tag{2.29}$$

Accordingly, the magnitude of the Fourier transform is:

$$|\Im[p(t - \tau)]| = |e^{-j\omega t} P(\omega)| = |e^{-j\omega t}||P(\omega)| = |P(\omega)| \tag{2.30}$$

and since the magnitude of the exponential function is 1.0, the magnitude of the Fourier transform of the shifted image equals that of the original (unshifted) version. We shall use this property later in Chapter 7, when we use Fourier theory to describe shapes. There, it will allow us to give the same description to different instances of the same shape, but a different description to a different shape. You do not get something for nothing: even though the magnitude of the Fourier transform remains constant, its phase does not. The phase of the shifted transform is:

$$\langle \Im[p(t-\tau)] = \langle e^{-j\omega t} P(\omega) \rangle \tag{2.31}$$

The Mathcad implementation of a `shift` operator (Code 2.3) uses the modulus operation to enforce the cyclic shift. The arguments fed to the function are: the image to be shifted (`pic`), the horizontal shift along the x-axis (`x_value`) and the vertical shift along the y-axis (`y_value`).

```
shift(pic,y_val,x_val):=   NC←cols(pic)
                           NR←rows(pic)
                           for y∈0..NR-1
                             for x∈0..NC-1
                               shifted_{y,x}←pic_{mod(y+y_val,NR),mod(x+x_val,NC)}
                           shifted
```

Code 2.3 Shifting an image

This process is illustrated in Figure 2.18. An original image (Figure 2.18a) is shifted along the x- and y-axes (Figure 2.18d). The shift is cyclical, so parts of the image wrap around; those parts at the top of the original image appear at the base of the shifted image. The Fourier transform of the original image and that of the shifted image are identical: Figure 2.18(b) appears the same as Figure 2.18(e). The phase differs: the phase of the original image (Figure 2.18c) is clearly different from the phase of the shifted image (Figure 2.18f).

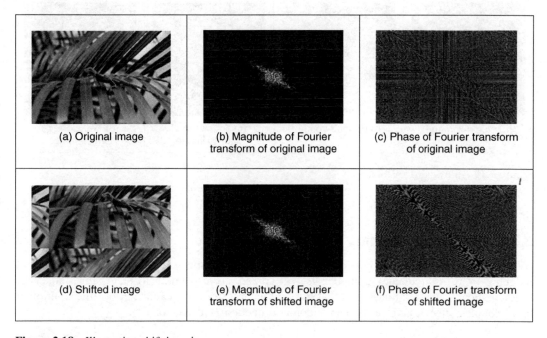

(a) Original image

(b) Magnitude of Fourier transform of original image

(c) Phase of Fourier transform of original image

(d) Shifted image

(e) Magnitude of Fourier transform of shifted image

(f) Phase of Fourier transform of shifted image

Figure 2.18 Illustrating shift invariance

The differing phase implies that, in application, the magnitude of the Fourier transform of a face, say, will be the same irrespective of the position of the face in the image (i.e. the camera or the subject can move up and down), assuming that the face is much larger than its image version. This implies that if the Fourier transform is used to analyse an image of a human face or one of cloth, to describe it by its spatial frequency, we do not need to control the position of the camera, or the object, precisely.

2.6.2 Rotation

The Fourier transform of an image *rotates* when the source image *rotates*. This is to be expected since the decomposition into spatial frequency reflects the orientation of features within the image. As such, orientation dependency is built into the Fourier transform process.

This implies that if the frequency domain properties are to be used in image analysis, via the Fourier transform, the orientation of the original image needs to be known, or fixed. It is often possible to fix orientation, or to estimate its value when a feature's orientation cannot be fixed. Alternatively, there are techniques to impose invariance to rotation, say by translation to a polar representation, although this can prove to be complex.

The effect of rotation is illustrated in Figure 2.19. An image (Figure 2.19a) is rotated by 90° to give the image in Figure 2.19(b). Comparison of the transform of the original image (Figure 2.19c) with the transform of the rotated image (Figure 2.19d) shows that the transform has been rotated by 90°, by the same amount as the image. In fact, close inspection of Figure 2.19(c) and (d) shows that the diagonal axis is consistent with the normal to the axis of the leaves (where the change mainly occurs), and this is the axis that rotates.

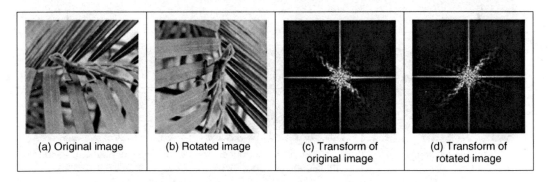

| (a) Original image | (b) Rotated image | (c) Transform of original image | (d) Transform of rotated image |

Figure 2.19 Illustrating rotation

2.6.3 Frequency scaling

By definition, time is the reciprocal of frequency. So if an image is compressed, equivalent to reducing time, its frequency components will spread, corresponding to increasing frequency. Mathematically, the relationship is that the Fourier transform of a function of time multiplied by a scalar λ, $p(\lambda t)$, gives a frequency domain function $P(\omega/\lambda)$, so:

$$\Im[p(\lambda t)] = \frac{1}{\lambda} P\left(\frac{\omega}{\lambda}\right) \tag{2.32}$$

This is illustrated in Figure 2.20, where the texture image of a chain-link fence (Figure 2.20a) is reduced in scale (Figure 2.20b), thereby increasing the spatial frequency. The DFT of the original texture image is shown in Figure 2.20(c), which reveals that the large spatial frequencies in the original image are arranged in a star-like pattern. As a consequence of scaling the original image, the spectrum will spread from the origin consistent with an increase in spatial frequency, as shown in Figure 2.20(d). This retains the star-like pattern, but with points at a greater distance from the origin.

| (a) Texture image | (b) Scaled texture image | (c) Transform of original texture | (d) Transform of scaled texture |

Figure 2.20 Illustrating frequency scaling

The implications of this property are that if we reduce the scale of an image, say by imaging at a greater distance, we will alter the frequency components. The relationship is linear: the amount of reduction, say the proximity of the camera to the target, is directly proportional to the scaling in the frequency domain.

2.6.4 Superposition (linearity)

The *principle of superposition* is very important in systems analysis. Essentially, it states that a system is linear if its response to two combined signals equals the sum of the responses to the individual signals. Given an output O which is a function of two inputs I_1 and I_2, the response to signal I_1 is $O(I_1)$, that to signal I_2 is $O(I_2)$, and the response to I_1 and I_2, when applied together, is $O(I_1 + I_2)$, the superposition principle states:

$$O(I_1 + I_2) = O(I_1) + O(I_2) \tag{2.33}$$

Any system which satisfies the principle of superposition is termed linear. The Fourier transform is a linear operation since, for two signals p_1 and p_2:

$$\Im[p_1 + p_2] = \Im[p_1] + \Im[p_2] \tag{2.34}$$

In application this suggests that we can separate images by looking at their frequency domain components. This is illustrated for 1D signals in Figure 2.21. One signal is shown in Figure 2.21(a) and a second is shown in Figure 2.21(c). The Fourier transforms of these signals are shown in Figure 2.21(b) and (d). The addition of these signals is shown in Figure 2.21(e) and its transform in Figure 2.21(f). The Fourier transform of the added signals differs little from the addition of their transforms (Figure 2.21g). This is confirmed by subtraction of the two (Figure 2.21d) (some slight differences can be seen, but these are due to numerical error).

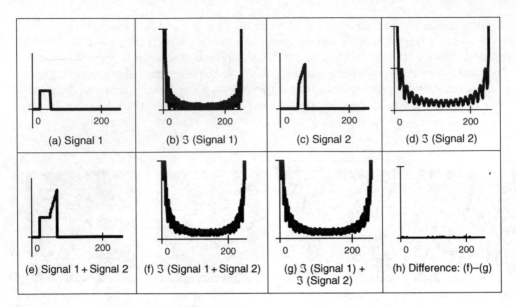

Figure 2.21 Illustrating superposition

By way of example, given the image of a fingerprint in blood on cloth it is very difficult to separate the fingerprint from the cloth by analysing the combined image. However, by translation to the frequency domain, the Fourier transform of the combined image shows strong components due to the texture (this is the spatial frequency of the cloth's pattern) and weaker, more scattered, components due to the fingerprint. If we suppress the frequency components due to the cloth's texture, and invoke the inverse Fourier transform, then the cloth will be removed from the original image. The fingerprint can now be seen in the resulting image.

2.7 Transforms other than Fourier

2.7.1 Discrete cosine transform

The *discrete cosine transform* (DCT) (Ahmed et al., 1974) is a real transform that has great advantages in *energy compaction*. Its definition for spectral components $\mathbf{DP}_{u,v}$ is:

$$\mathbf{DP}_{u,v} = \begin{vmatrix} \dfrac{1}{N} \displaystyle\sum_{x=0}^{N-1}\sum_{y=0}^{N-1} \mathbf{P}_{x,y} & \text{if } u=0 \quad \text{and} \quad v=0 \\[2ex] \dfrac{2}{N} \displaystyle\sum_{x=0}^{N-1}\sum_{y=0}^{N-1} \mathbf{P}_{x,y} \times \cos\left(\dfrac{(2x+1)u\pi}{2N}\right) \times \cos\left(\dfrac{(2y+1)v\pi}{2N}\right) & \text{otherwise} \end{vmatrix} \tag{2.35}$$

The inverse DCT is defined by

$$\mathbf{P}_{x,y} = \frac{2}{N} \sum_{u=0}^{N-1}\sum_{v=0}^{N-1} \mathbf{DP}_{u,v} \times \cos\left(\frac{(2x+1)u\pi}{2N}\right) \times \cos\left(\frac{(2y+1)v\pi}{2N}\right) \tag{2.36}$$

A fast version of the DCT is available, like the FFT, and calculation can be based on the FFT. Both implementations offer about the same speed. The Fourier transform is not actually optimal for *image coding*, since the DCT can give a higher compression rate for the same image quality. This is because the cosine basis functions can afford for high-energy compaction. This can be seen by comparison of Figure 2.22(b) with Figure 2.22(a), which reveals that the DCT components are much more concentrated around the origin, than those for the Fourier transform. This is the compaction property associated with the DCT. The DCT has been considered as optimal for image coding, and this is why it is found in the JPEG and MPEG standards for coded image transmission.

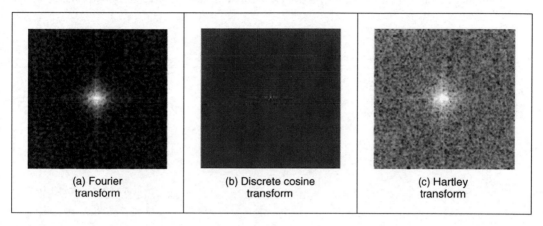

(a) Fourier	(b) Discrete cosine	(c) Hartley
transform	transform	transform

Figure 2.22 Comparing transforms of the Lena image

The DCT is actually shift variant, owing to its cosine basis functions. In other respects, its properties are very similar to the DFT, with one important exception: it has not yet proved possible to implement convolution with the DCT. It is possible to calculate the DCT via the FFT. This has been performed in Figure 2.22(b), since there is no fast DCT algorithm in Mathcad and, as shown earlier, fast implementations of transform calculation can take a fraction of the time of the conventional counterpart.

The Fourier transform essentially decomposes, or *decimates*, a signal into sine and cosine components, so the natural partner to the DCT is the discrete sine transform (DST). However, the DST transform has odd basis functions (sine) rather than the even ones in the DCT. This lends the DST transform some less desirable properties, and it finds much less application than the DCT.

2.7.2 Discrete Hartley transform

The Hartley transform (Hartley, 1942) is a form of the Fourier transform, but without complex arithmetic, with a result for the face image shown in Figure 2.22(c). Oddly, although it sounds like a very rational development, the Hartley transform was first invented in 1942, but not rediscovered and then formulated in discrete form until 1983 (Bracewell, 1983). One advantage of the Hartley transform is that the forward and inverse transform is the same operation; a disadvantage is that phase is built into the order of frequency components since it is not readily available as the argument of a complex number. The definition of the discrete Hartley transform (DHT) is that transform components $\mathbf{HP}_{u,v}$ are:

$$\mathbf{HP}_{u,v} = \frac{1}{N} \sum_{x=0}^{N-1} \sum_{y=0}^{N-1} \mathbf{P}_{x,y} \times \left(\cos\left(\frac{2\pi}{N} \times (ux+vy)\right) + \sin\left(\frac{2\pi}{N} \times (ux+vy)\right) \right) \quad (2.37)$$

The inverse Hartley transform is the same process, but applied to the transformed image.

$$\mathbf{P}_{x,y} = \frac{1}{N} \sum_{u=0}^{N-1} \sum_{v=0}^{N-1} \mathbf{HP}_{x,y} \times \left(\cos\left(\frac{2\pi}{N} \times (ux+vy)\right) + \sin\left(\frac{2\pi}{N} \times (ux+vy)\right) \right) \quad (2.38)$$

The implementation is then the same for both the forward and the inverse transforms, as given in Code 2.4.

```
Hartley(pic):=    NC←cols(pic)
                  NR←row(pic)
                  for v∈0.. NR-1
                    for u∈0.. NC-1

                                      1    NR-1 NC-1            ⎡    2·π·(u·x+v·y)⎤
                        trans_{v,u}←  ──·   Σ    Σ   pic_{y,x}·⎢cos──────────────
                                      NC   y=0  x=0            ⎣         NR

                                                               ⎡ 2·π·(u·x+v·y)⎤⎤
                                                        + sin⎢──────────────⎥⎥
                                                               ⎣      NC      ⎦⎦

                  trans
```

Code 2.4 Implementing the Hartley transform

Again, a fast implementation is available, the fast Hartley transform (Bracewell, 1984) (although some suggest that it should be called the Bracewell transform, eponymously). It is possible to calculate the DFT of a function, $F(u)$, from its Hartley transform, $H(u)$. The analysis here is based on 1D data, but only for simplicity since the argument extends readily to two dimensions. By splitting the Hartley transform into its odd and even parts, $O(u)$ and $E(u)$, respectively, we obtain:

$$H(u) = O(u) + E(u) \quad (2.39)$$

where:

$$E(u) = \frac{H(u) + H(N-u)}{2} \quad (2.40)$$

and

$$O(u) = \frac{H(u) - H(N-u)}{2} \quad (2.41)$$

The DFT can then be calculated from the DHT simply by

$$F(u) = E(u) - j \times O(u) \quad (2.42)$$

Conversely, the Hartley transform can be calculated from the Fourier transform by:

$$H(u) = \text{Re}[F(u)] - \text{Im}[F(u)] \tag{2.43}$$

where Re[ ] and Im[ ] denote the real and the imaginary parts, respectively. This emphasizes the natural relationship between the Fourier and the Hartley transform. The image of Figure 2.22(c) has been calculated via the 2D FFT using Equation 2.43. Note that the transform in Figure 2.22(c) is the complete transform, whereas the Fourier transform in Figure 2.22(a) shows magnitude only. As with the DCT, the properties of the Hartley transform mirror those of the Fourier transform. Unfortunately, the Hartley transform does not have shift invariance, but there are ways to handle this. Also, convolution requires manipulation of the odd and even parts.

2.7.3 Introductory wavelets: the Gabor wavelet

Wavelets are a comparatively recent approach to signal processing, being introduced only in 1990 (Daubechies, 1990). Their main advantage is that they allow multiresolution analysis (analysis at different scales, or resolution). Furthermore, wavelets allow decimation in space and frequency, *simultaneously*. Earlier transforms actually allow decimation in frequency, in the forward transform, and in time (or position) in the inverse. In this way, the Fourier transform gives a measure of the frequency content of the whole image: the contribution of the image to a particular frequency component. Simultaneous decimation allows us to describe an image in terms of frequency which occurs at a position, as opposed to an ability to measure frequency content across the whole image. This gives us greater descriptional power, which can be used to good effect.

First, though, we need a basis function, so that we can decompose a signal. The basis functions in the Fourier transform are sinusoidal waveforms at different frequencies. The function of the Fourier transform is to convolve these sinusoids with a signal to determine how much of each is present. The *Gabor wavelet* is well suited to introductory purposes, since it is essentially a sinewave modulated by a Gaussian envelope. The Gabor wavelet gw is given by

$$gw(t, \omega_0, t_0, a) = e^{-j\omega_0 t} e^{-\left(\frac{t - t_0}{a}\right)^2} \tag{2.44}$$

where $\omega_0 = 2\pi f_0$ is the modulating frequency, t_0 dictates position and a controls the width of the Gaussian envelope which embraces the oscillating signal. An example Gabor wavelet is shown in Figure 2.23, which shows the real and the imaginary parts (the modulus is the Gaussian envelope). Increasing the value of ω_0 increases the frequency content within the envelope,

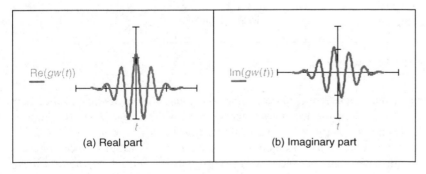

(a) Real part   (b) Imaginary part

Figure 2.23 An example Gabor wavelet

whereas increasing the value of a spreads the envelope without affecting the frequency. So why does this allow simultaneous analysis of time and frequency? Given that this function is the one convolved with the test data, then we can compare it with the Fourier transform. In fact, if we remove the term on the right-hand side of Equation 2.44, we return to the sinusoidal basis function of the Fourier transform, the exponential in Equation 2.1. Accordingly, we can return to the Fourier transform by setting a to be very large. Alternatively, setting f_0 to zero removes frequency information. Since we operate in between these extremes, we obtain position and frequency information simultaneously.

An infinite class of wavelets exists which can be used as an expansion basis in signal decimation. One approach (Daugman, 1988) has generalized the Gabor function to a 2D form aimed to be optimal in terms of spatial and spectral resolution. These 2D Gabor wavelets are given by

$$gw2D(x, y) = \frac{1}{\sigma\sqrt{\pi}} e^{-\left(\frac{(x-x_0)^2+(y-y_0)^2}{2\sigma^2}\right)} e^{-j2\pi f_0((x-x_0)\cos(\theta)+(y-y_0)\sin(\theta))} \qquad (2.45)$$

where x_0, y_0 control position, f_0 controls the frequency of modulation along either axis, and θ controls the *direction* (orientation) of the wavelet (as implicit in a 2D system). The shape of the area imposed by the 2D Gaussian function could be elliptical if different variances were allowed along the x- and y-axes (the frequency can also be modulated differently along each axis). Figure 2.24, an example of a 2D Gabor wavelet, shows that the real and imaginary parts are even and odd functions, respectively; again, different values for f_0 and σ control the frequency and envelope's spread, respectively, the extra parameter θ controls rotation.

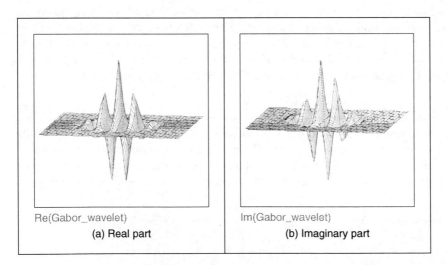

(a) Real part   (b) Imaginary part

Figure 2.24 Example 2D Gabor wavelet

The function of the wavelet transform is to determine where and how each wavelet specified by the range of values for each of the free parameters occurs in the image. Clearly, there is a wide choice which depends on application. An example transform is given in Figure 2.25. Here, the Gabor wavelet parameters have been chosen in such a way as to select face features: the eyes, nose and mouth have come out very well. These features are where there is local frequency content with orientation according to the head's inclination. These are not the only

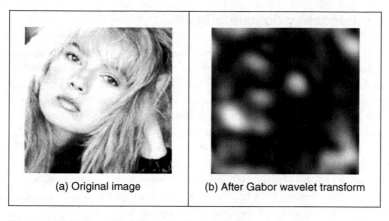

| (a) Original image | (b) After Gabor wavelet transform |

Figure 2.25 Example Gabor wavelet transform

features with these properties: the cuff of the sleeve is highlighted too! But this does show the Gabor wavelet's ability to select and analyse localized variation in image intensity.

The conditions under which a set of continuous Gabor wavelets will provide a complete representation of any image (i.e. that any image can be reconstructed) have only recently been developed. However, the theory is very powerful, since it accommodates frequency and position simultaneously, and further it facilitates multiresolution analysis. We shall find wavelets again, when processing images to find low-level features. Applications of Gabor wavelets include measurement of iris texture to give a very powerful security system (Daugman, 1993) and face feature extraction for automatic face recognition (Lades et al., 1993). Wavelets continue to develop (Daubechies, 1990) and have found applications in image texture analysis (Laine and Fan, 1993), coding (da Silva and Ghanbari, 1996) and image restoration (Banham and Katsaggelos, 1996). Unfortunately, the discrete wavelet transform is not shift invariant, although there are approaches aimed to remedy this (see, for example, Donoho, 1995). As such, we shall not study it further and just note that there is an important class of transforms that combine spatial and spectral sensitivity, and it is likely that this importance will continue to grow.

2.7.4 Other transforms

Decomposing a signal into sinusoidal components was one of the first approaches to transform calculus, and this is why the Fourier transform is so important. The sinusoidal functions are called *basis functions*; the implicit assumption is that the basis functions map well to the signal components. There is (theoretically) an infinite range of basis functions. Discrete signals can map better into collections of binary components rather than sinusoidal ones. These collections (or sequences) of binary data are called sequency components and form the basis functions of the *Walsh transform* (Walsh, 1923). This has found wide application in the interpretation of digital signals, although it is less widely used in image processing. The Karhunen–Loéve transform (Karhunen, 1947; Loéve, 1948) (also called the *Hotelling* transform from which it was derived, or more popularly *principal components analysis*; see Chapter 12, Appendix 4) is a way of analysing (statistical) data to reduce it to those data which are *informative*, discarding those which are not.

2.8 Applications using frequency domain properties

Filtering is a major use of Fourier transforms, particularly because we can understand an image, and how to process it, much better in the frequency domain. An analogy is the use of a graphic equalizer to control the way music sounds. In images, if we want to remove high-frequency information (like the hiss on sound) then we can filter, or remove, it by inspecting the Fourier transform. If we retain low-frequency components, we implement a *low-pass filter*. The low-pass filter describes the area in which we retain spectral components; the size of the area dictates the range of frequencies retained, and is known as the filter's *bandwidth*. If we retain components within a circular region centred on the d.c. component, and inverse Fourier transform the filtered transform, then the resulting image will be *blurred*. Higher spatial frequencies exist at the sharp *edges* of features, so removing them causes blurring. But the amount of fluctuation is reduced too; any high-frequency noise will be removed in the filtered image.

The implementation of a low-pass filter which retains frequency components within a circle of specified radius is the function `low_filter`, given in Code 2.5. This operator assumes that the radius and centre coordinates of the circle are specified before its use. Points within the circle remain unaltered, whereas those outside the circle are set to zero, black.

$$\text{low\_filter(pic)} := \begin{vmatrix} \text{for } y \in 0.. \text{ rows(pic)}-1 \\ \quad \text{for } x \in 0.. \text{ cols(pic)}-1 \\ \qquad \text{filtered}_{y,x} \leftarrow \begin{vmatrix} \text{pic}_{y,x} & \text{if } \left(y - \dfrac{\text{rows(pic)}}{2}\right)^2 + \left(x - \dfrac{\text{cols(pic)}}{2}\right)^2 \\ & \qquad\qquad - \text{radius}^2 \le 0 \\ 0 & \text{otherwise} \end{vmatrix} \\ \text{filtered} \end{vmatrix}$$

Code 2.5 Implementing low-pass filtering

When applied to an image we obtain a low-pass filtered version. In application to an image of a face, the low spatial frequencies are the ones which change slowly, as reflected in the resulting, blurred image (Figure 2.26a). The high-frequency components have been removed as shown in the transform (Figure 2.26b). The radius of the circle controls how much of the original

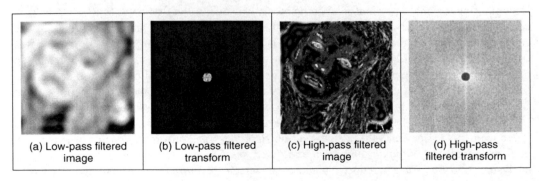

| (a) Low-pass filtered image | (b) Low-pass filtered transform | (c) High-pass filtered image | (d) High-pass filtered transform |

Figure 2.26 Illustrating low- and high-pass filtering

image is retained. In this case, the radius is 10 pixels (and the image resolution is 256×256). If a larger circle were to be used, more of the high-frequency detail would be retained (and the image would look more like its original version); if the circle was very small, an even more blurred image would result, since only the lowest spatial frequencies would be retained. This differs from the earlier Gabor wavelet approach, which allows for *localized* spatial frequency analysis. Here, the analysis is *global*: we are filtering the frequency across the *whole* image.

Alternatively, we can retain high-frequency components and remove low-frequency ones. This is a *high-pass filter*. If we remove components near the d.c. component and retain all the others, the result of applying the inverse Fourier transform to the filtered image will be to emphasize the features that were removed in low-pass filtering. This can lead to a popular application of the high-pass filter: to 'crispen' an image by emphasizing its high-frequency components. An implementation using a circular region merely requires selection of the set of points outside the circle, rather than inside as for the low-pass operator. The effect of high-pass filtering can be observed in Figure 2.26(c), which shows removal of the low-frequency components: this emphasizes the hair and the borders of a face's features, since these are where brightness varies rapidly. The retained components are those which were removed in low-pass filtering, as illustrated in the transform, Figure 2.26(d).

It is also possible to retain a specified range of frequencies. This is known as *band-pass filtering*. It can be implemented by retaining frequency components within an annulus centred on the d.c. component. The width of the annulus represents the bandwidth of the band-pass filter.

This leads to digital signal processing theory. There are many considerations to be made in the way you select, and the manner in which frequency components are retained or excluded. This is beyond a text on computer vision. For further study in this area, Rabiner and Gold (1975), or Oppenheim et al. (1999), although published (in their original form) a long time ago now, remain as popular introductions to digital signal processing theory and applications.

It is possible to recognize the object within the low pass filtered image. Intuitively, this implies that we could store only the frequency components selected from the transform data, rather than all the image points. In this manner a fraction of the information would be stored, and still provide a recognizable image, albeit slightly blurred. This concerns *image coding*, which is a popular target for image processing techniques. For further information see Clarke (1985) or a newer text, such as Woods (2006).

2.9 Further reading

We shall meet the frequency domain throughout this book, since it allows for an alternative interpretation of operation, in the frequency domain as opposed to the time domain. This will occur in low- and high-level feature extraction, and in shape description. Further, it allows for some of the operations we shall cover. Because of the availability of the FFT, it is also used to speed up algorithms.

Given these advantages, it is well worth looking more deeply. For introductory study, there is *Who is Fourier* (Lex, 1995), which offers a lighthearted and completely digestible overview of the Fourier transform, and is simply excellent for a starter view of the topic. For further study (and entertaining study too!) of the Fourier transform, try *The Fourier Transform and its Applications* by Bracewell (1986). Some of the standard image processing texts include much coverage of transform calculus, such as Jain (1989), Gonzalez and Wintz (1987) and Pratt (2007). For more coverage of the DCT try Jain (1989); for excellent coverage of the Walsh transform try Beauchamp's superb text (Beauchamp, 1975). For wavelets, the book by Wornell (1996) introduces wavelets from a signal processing standpoint, or there is Mallat's classic text (Mallat, 1999). For general signal processing theory there are introductory texts (e.g. Meade and Dillon, 1986) or Ifeachor's excellent book (Ifeachor and Jervis, 2002), for more complete coverage try Rabiner and Gold (1975) or Oppenheim and Schafer (1999) (as mentioned earlier). Finally, on the implementation side of the FFT (and for many other signal processing algorithms), *Numerical Recipes in C* (Press et al., 2002) is an excellent book. It is extremely readable, full of practical detail and well worth a look. *Numerical Recipes* is on the web too, together with other signal processing sites, as listed in Table 1.4.

2.10 References

Ahmed, N., Natarajan, T. and Rao, K. R., Discrete Cosine Transform, *IEEE Trans. Comput.*, pp. 90–93, 1974

Banham, M. R. and Katsaggelos, K., Spatially Adaptive Wavelet-Based Multiscale Image Restoration, *IEEE Trans. Image Process.*, **5**(4), pp. 619–634, 1996

Beauchamp, K. G., *Walsh Functions and Their Applications*, Academic Press, London, 1975

Bracewell, R. N., The Discrete Hartley Transform, *J. Opt. Soc. Am.*, **73**(12), pp. 1832–1835, 1983

Bracewell, R. N., The Fast Hartley Transform, *Proc. IEEE*, **72**(8), pp. 1010–1018, 1984

Bracewell, R. N., *The Fourier Transform and its Applications*, Revised 2nd edn, McGraw-Hill Book Co., Singapore, 1986

Clarke, R. J., *Transform Coding of Images*, Addison Wesley, Reading, MA, 1985

Daubechies, I., The Wavelet Transform, Time Frequency Localization and Signal Analysis, *IEEE Trans. Inform. Theory*, **36**(5), pp. 961–1004, 1990

Daugman, J. G., Complete Discrete 2D Gabor Transforms by Neural Networks for Image Analysis and Compression, *IEEE Trans. Acoust. Speech Signal Process.*, **36**(7), pp. 1169–1179, 1988

Daugman, J. G., High Confidence Visual Recognition of Persons by a Test of Statistical Independence, *IEEE Trans. PAMI*, **15**(11), pp. 1148–1161, 1993

Donoho, D. L., Denoizing by Soft Thresholding, *IEEE Trans. Inform. Theory*, **41**(3), pp. 613–627, 1995

Gonzalez, R. C. and Wintz P., *Digital Image Processing*, 2nd edn, Addison Wesley, Reading, MA, 1987

Hartley, R. L. V., A More Symmetrical Fourier Analysis Applied to Transmission Problems, *Proc. IRE*, **144**, pp. 144–150, 1942

Ifeachor, E. C. and Jervis, B. W., *Digital Signal Processing*, 2nd edn, Prentice Hall, Hemel Hempstead, 2002

Jain, A. K. *Fundamentals of Computer Vision*, Prentice Hall International (UK), Hemel Hempstead, 1989

Karhunen, K., Über Lineare Methoden in der Wahrscheinlich-Keitsrechnung, *Ann. Acad. Sci. Fennicae*, Ser A.I.37, 1947 (Translation in I. Selin, On Linear Methods in Probability Theory, Doc. T-131, The RAND Corporation, Santa Monica, CA, 1960)

Lades, M., Vorbruggen, J. C., Buhmann, J., Lange, J., Madsburg, C. V. D., Wurtz, R. P. and Konen, W., Distortion Invariant Object Recognition in the Dynamic Link Architecture, *IEEE Trans. Comput.*, **42**, pp. 300–311, 1993

Laine, A. and Fan, J., Texture Classification by Wavelet Packet Signatures, *IEEE Trans. PAMI*, **15**, pp. 1186–1191, 1993

Lex, T. C. O. L. T. (!!), *Who is Fourier, A Mathematical Adventure*, Language Research Foundation, Boston, MA, US, 1995

Loéve, M., Fonctions Alétoires de Seconde Ordre, In: P. Levy (Ed.), *Processus Stochastiques et Mouvement Brownien*, Hermann, Paris, 1948

Mallat, S., *A Wavelet Tour of Signal Processing*, 2nd edn, Academic Press, New York, 1999

Meade, M. L. and Dillon, C. R., *Signals and Systems, Models and Behaviour*, Van Nostrand Reinhold (UK), Wokingham, 1986

Oppenheim, A. V., Schafer, R. W. and Buck, J. R. *Digital Signal Processing*, 2nd edn, Prentice Hall International (UK), Hemel Hempstead, 1999

Pratt, W. K., *Digital Image Processing: PIKS Scientific Inside*, 4th edn, Wiley, New York, 2007

Press, W. H., Teukolsky, S. A., Vetterling, W. T. and Flannery, B. P., *Numerical Recipes in C++: The Art of Scientific Computing*, 2nd edn, Cambridge University Press, Cambridge, 2002

Rabiner, L. R. and Gold, B., *Theory and Application of Digital Signal Processing*, Prentice Hall, Englewood Cliffs, NJ, 1975

da Silva, E. A. B. and Ghanbari, M., On the Performance of Linear Phase Wavelet Transforms in Low Bit-Rate Image Coding, *IEEE Trans. Image Process.*, **5**(5), pp. 689–704, 1996

Walsh, J. L., A Closed Set of Normal Orthogonal Functions, *Am. J. Math.*, **45**(1), pp. 5–24, 1923

Woods, J. W., *Multidimensional Signal, Image, and Video Processing and Coding*, Academic Press, New York, 2006

Wornell, G. W., *Signal Processing with Fractals, A Wavelet-Based Approach*, Prentice Hall, Upper Saddle River, NJ, 1996

3

Basic image processing operations

3.1 Overview

We shall now start to process digital images. First, we shall describe the brightness variation in an image using its histogram. We shall then look at operations that manipulate the image so as to change the histogram, processes that shift and scale the result (making the image brighter or dimmer, in different ways). We shall also consider thresholding techniques that turn an image from grey level to binary. These are called single point operations. After that, we shall move to group operations where the group is those points found inside a template. Some of the most common operations on the groups of points are statistical, providing images where each point is the result of, say, averaging the neighbourhood of each point in the original image. We shall

Table 3.1 Overview of Chapter 3

Main topic	Sub topics	Main points
Image description	Portray *variation* in image brightness content as a graph/*histogram*.	Histograms, image contrast.
Point operations	Calculate *new* image points as a *function* of the point at the same place in the original image. The functions can be *mathematical*, or can be computed from the image itself and will change the image's histogram. Finally, *thresholding* turns an image from *grey* level to a *binary* (black and white) representation.	Histogram manipulation; intensity mapping: addition, inversion, scaling, logarithm, exponent. Intensity normalization; histogram equalization. Thresholding and optimal thresholding.
Group operations	Calculate new image points as a function of *neighbourhood* of the point at the same place in the original image. The functions can be *statistical*, including: mean (average); median and mode. *Advanced* filtering techniques, including feature preservation. *Morphological* operators process an image according to *shape*, starting with binary and moving to grey-level operations.	Template convolution (including frequency domain implementation). Statistical operators: direct averaging, median filter and mode filter. Anisotropic diffusion for image smoothing. Other operators: force field transform. Mathematical morphology: hit or miss transform, erosion, dilation (including grey-level operators) and Minkowski operators.

see how the statistical operations can reduce noise in the image, which is of benefit to the feature extraction techniques to be considered later. As such, these basic operations are usually for preprocessing for later feature extraction or to improve display quality.

3.2 Histograms

The intensity *histogram* shows how individual brightness levels are occupied in an image; the *image contrast* is measured by the range of brightness levels. The histogram plots the number of pixels with a particular brightness level against the brightness level. For 8 bit pixels, the brightness ranges from zero (black) to 255 (white). Figure 3.1 shows an image of an eye and its histogram. The histogram (Figure 3.1b) shows that not all the grey levels are used and the lowest and highest intensity levels are close together, reflecting moderate *contrast*. The histogram has a region between 100 and 120 brightness values which contains the dark portions of the image, such as the hair (including the eyebrow) and the eye's iris. The brighter points relate mainly to the skin. If the image was darker, overall, the histogram would be concentrated towards black. If the image was brighter, but with lower contrast, then the histogram would be thinner and concentrated near the whiter brightness levels.

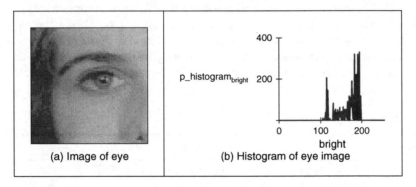

Figure 3.1 An image and its histogram

This histogram shows us that we have not used all available grey levels. Accordingly, we could stretch the image to use them all, and the image would become clearer. This is essentially cosmetic attention to make the image's appearance better. Making the appearance better, especially in view of later processing, is the focus of many basic image processing operations, as will be covered in this chapter. The histogram can also reveal whether there is much noise in the image, if the ideal histogram is known. We might want to remove this noise, not only to improve the appearance of the image, but also to ease the task of (and to present the target better for) later feature extraction techniques. This chapter concerns these basic operations which can improve the appearance and quality of images.

The histogram can be evaluated by the operator histogram, in Code 3.1. The operator first initializes the histogram to zero. Then, the operator works by counting up the number of image points that have an intensity at a particular value. These counts for the different values form the overall histogram. The counts are then returned as the two-dimensional (2D) histogram (a vector of the count values), which can be plotted as a graph (Figure 3.1b).

```
histogram(pic):= | for bright∈0..255
                      pixels_at_level_bright←0
                  for x∈0..cols(pic)-1
                    for y∈0..rows(pic)-1
                      | level←pic_y,x
                      | pixels_at_level_level←pixels_at_level_level+1
                  pixels_at_level
```

Code 3.1 Evaluating the histogram

3.3 Point operators

3.3.1 Basic point operations

The most basic operations in image processing are point operations where each pixel value is replaced with a new value obtained from the old one. If we want to increase the brightness to stretch the contrast we can simply multiply all pixel values by a scalar, say by 2 to double the range. Conversely, to reduce the contrast (although this is not usual) we can divide all point values by a scalar. If the overall brightness is controlled by a *level, l,* (e.g. the brightness of global light) and the range is controlled by a *gain, k,* the brightness of the points in a new picture, **N**, can be related to the brightness in old picture, **O**, by:

$$\mathbf{N}_{x,y} = k \times \mathbf{O}_{x,y} + l \qquad \forall x, y \in 1, N \tag{3.1}$$

This is a point operator that replaces the brightness at points in the picture according to a linear brightness relation. The level controls overall brightness and is the minimum value of the output picture. The gain controls the contrast, or range, and if the gain is greater than unity, the output range will be increased. This process is illustrated in Figure 3.2. So the image of the eye, processed by $k = 1.2$ and $l = 10$, will become brighter (Figure 3.2a) and with better contrast, although in this case the brighter points are mostly set near to white (255). These factors can be seen in its histogram (Figure 3.2b).

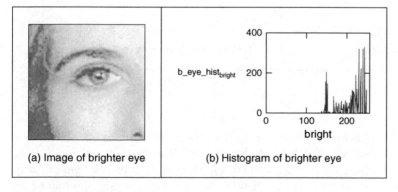

(a) Image of brighter eye   (b) Histogram of brighter eye

Figure 3.2 Brightening an image

The basis of the implementation of point operators was given earlier, for addition in Code 1.3. The stretching process can be displayed as a mapping between the input and output ranges, according to the specified relationship, as in Figure 3.3. Figure 3.3(a) is a mapping where the output is a direct copy of the input (this relationship is the dotted line in Figure 3.3c and d); Figure 3.3(b) is the mapping for brightness *inversion* where dark parts in an image become bright and vice versa. Figure 3.3(c) is the mapping for *addition* and Figure 3.3(d) is the mapping for *multiplication* (or *division*, if the slope was less than that of the input). In these mappings, if the mapping produces values that are smaller than the expected minimum (say negative when zero represents black), or larger than a specified maximum, then a *clipping* process can be used to set the output values to a chosen level. For example, if the relationship between input and output aims to produce output points with intensity value greater than 255, as used for white, the output value can be set to white for these points, as it is in Figure 3.3(c).

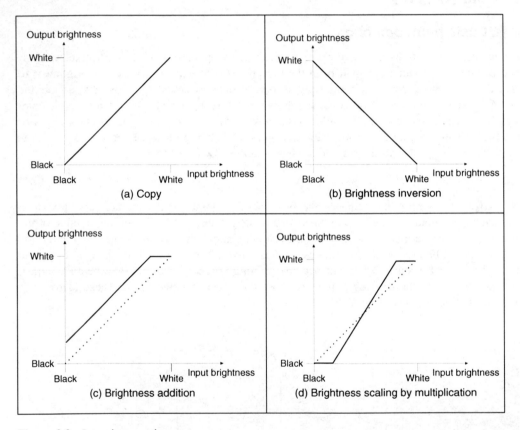

Figure 3.3 Intensity mappings

The *sawtooth* operator is an alternative form of the linear operator and uses a repeated form of the linear operator for chosen intervals in the brightness range. The sawtooth operator is used to emphasize local contrast change (as in images where regions of interest can be light or dark). This is illustrated in Figure 3.4, where the range of brightness levels is mapped into four linear regions by the sawtooth operator (Figure 3.4b). This remaps the intensity in the eye image to

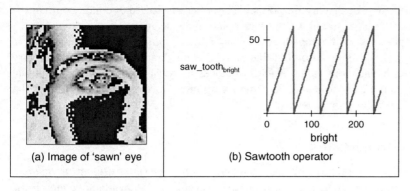

(a) Image of 'sawn' eye

saw_tooth_bright

(b) Sawtooth operator

Figure 3.4 Applying the sawtooth operator

highlight local intensity variation, as opposed to global variation, in Figure 3.4(a). The image is now presented in regions, where the region selection is controlled by the intensity of its pixels.

Finally, rather than simple multiplication we can use arithmetic functions such as logarithm to reduce the range or exponent to increase it. This can be used, say, to equalize the response of a camera, or to compress the range of displayed brightness levels. If the camera has a known exponential performance, and outputs a value for brightness which is proportional to the exponential of the brightness of the corresponding point in the scene of view, the application of a *logarithmic point operator* will restore the original range of brightness levels. The effect of replacing brightness by a scaled version of its natural logarithm (implemented as $N_{x,y} = 20\ln(100O_{x,y})$) is shown in Figure 3.5(a); the effect of a scaled version of the exponent (implemented as $N_{x,y} = 20\exp(O_{x,y}/100)$) is shown in Figure 3.5(b). The scaling factors were chosen to ensure that the resulting image can be displayed since the logarithm or exponent greatly reduces or magnifies pixel values, respectively. This can be seen in the results: Figure 3.5(a) is dark with a small range of brightness levels, whereas Figure 3.5(b) is much brighter, with greater contrast. Naturally, application of the logarithmic point operator will change any *multiplicative* changes in brightness to become *additive*. As such, the logarithmic operator can find application in reducing the effects of multiplicative intensity change. The logarithm operator is often used to

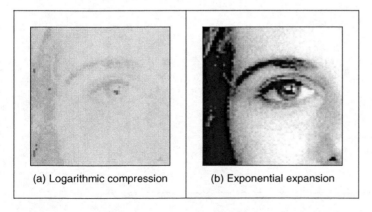

(a) Logarithmic compression

(b) Exponential expansion

Figure 3.5 Applying exponential and logarithmic point operators

compress Fourier transforms, for display purposes. This is because the d.c. component can be very large with contrast, too large to allow the other points to be seen.

In hardware, point operators can be implemented using *look-up tables* (LUTs), which exist in some framegrabber units. LUTs give an output that is programmed, and stored, in a table entry that corresponds to a particular input value. If the brightness response of the camera is known, it is possible to preprogram a LUT to make the camera response equivalent to a uniform or flat response across the range of brightness levels.

3.3.2 Histogram normalization

Popular techniques to stretch the range of intensities include *histogram (intensity) normalization*. Here, the original histogram is stretched, and shifted, to cover all the 256 available levels. If the original histogram of old picture $\mathbf{O}$ starts at $\mathbf{O}_{min}$ and extends up to $\mathbf{O}_{max}$ brightness levels, then we can scale up the image so that the pixels in the new picture $\mathbf{N}$ lie between a minimum output level $\mathbf{N}_{min}$ and a maximum level $\mathbf{N}_{max}$, simply by scaling up the input intensity levels according to:

$$\mathbf{N}_{x,y} = \frac{\mathbf{N}_{max} - \mathbf{N}_{min}}{\mathbf{O}_{max} - \mathbf{O}_{min}} \times (\mathbf{O}_{x,y} - \mathbf{O}_{min}) + \mathbf{N}_{min} \qquad \forall x, y \in 1, N \qquad (3.2)$$

A Matlab implementation of intensity normalization, appearing to mimic Matlab's `imagesc` function, the `normalize` function in Code 3.2, uses an output ranging from $\mathbf{N}_{min} = 0$ to $\mathbf{N}_{max} = 255$. This is scaled by the input range that is determined by applying the `max` and the `min` operators to the input picture. Note that in Matlab, a 2D array needs double application of the `max` and `min` operators, whereas in Mathcad `max(image)` delivers the maximum. Each point

```
function normalized=normalize(image)
%Histogram normalization to stretch from black to white

%Usage: [new image]=normalize(image)
%Parameters: image-array of integers
%Author: Mark S. Nixon

%get dimensions
[rows,cols]=size(image);

%set minimum
minim=min(min(image));

%work out range of input levels
range=max(max(image))-minim;

%normalize the image
for x=1:cols %address all columns
  for y=1:rows %address all rows
    normalized(y,x)=floor((image(y,x)-minim)*255/range);
  end
end
```

Code 3.2 Intensity normalization

in the picture is then scaled as in Equation 3.2 and the floor function is used to ensure an integer output.

The process is illustrated in Figure 3.6, and can be compared with the original image and histogram in Figure 3.1. An intensity normalized version of the eye image is shown in Figure 3.6(a), which now has better contrast and appears better to the human eye. Its histogram (Figure 3.6b) shows that the intensity now ranges across all available levels (there is actually one black pixel!).

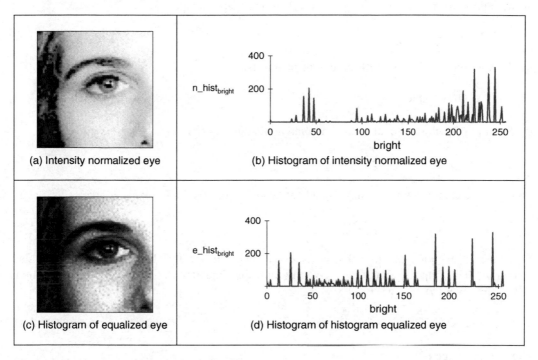

(a) Intensity normalized eye

(b) Histogram of intensity normalized eye

(c) Histogram of equalized eye

(d) Histogram of histogram equalized eye

Figure 3.6 Illustrating intensity normalization and histogram equalization

3.3.3 Histogram equalization

Histogram equalization is a *non-linear* process aimed to highlight image brightness in a way particularly suited to human visual analysis. Histogram equalization aims to change a picture in such a way as to produce a picture with a *flatter* histogram, where all levels are equiprobable. In order to develop the operator, we can first inspect the histograms. For a range of M levels then the histogram plots the points per level against level. For the input (old) and the output (new) image, the number of points per level is denoted as $\mathbf{O}(l)$ and $\mathbf{N}(l)$ (for $0 \leq l \leq M$), respectively. For square images, there are N^2 points in the input and the output image, so the sum of points per level in each should be equal:

$$\sum_{l=0}^{M} \mathbf{O}(l) = \sum_{l=0}^{M} \mathbf{N}(l) \tag{3.3}$$

Also, this should be the same for an arbitrarily chosen level p, since we are aiming for an output picture with a uniformly flat histogram. So the cumulative histogram up to level p should be transformed to cover up to the level q in the new histogram:

$$\sum_{l=0}^{p} \mathbf{O}(l) = \sum_{l=0}^{q} \mathbf{N}(l) \tag{3.4}$$

Since the output histogram is uniformly flat, the cumulative histogram up to level p should be a fraction of the overall sum. So the number of points per level in the output picture is the ratio of the number of points to the range of levels in the output image:

$$\mathbf{N}(l) = \frac{N^2}{\mathbf{N}_{max} - \mathbf{N}_{min}} \tag{3.5}$$

So the cumulative histogram of the output picture is:

$$\sum_{l=0}^{q} \mathbf{N}(l) = q \times \frac{N^2}{\mathbf{N}_{max} - \mathbf{N}_{min}} \tag{3.6}$$

By Equation 3.4 this is equal to the cumulative histogram of the input image, so:

$$q \times \frac{N^2}{\mathbf{N}_{max} - \mathbf{N}_{min}} = \sum_{l=0}^{p} \mathbf{O}(l) \tag{3.7}$$

This gives a mapping for the output pixels at level q, from the input pixels at level p as:

$$q = \frac{\mathbf{N}_{max} - \mathbf{N}_{min}}{N^2} \times \sum_{l=0}^{p} \mathbf{O}(l) \tag{3.8}$$

This gives a mapping function that provides an output image that has an approximately flat histogram. The mapping function is given by phrasing Equation 3.8 as an equalizing function (E) of the level (q) and the image ($\mathbf{O}$) as

$$E(q, \mathbf{O}) = \frac{\mathbf{N}_{max} - \mathbf{N}_{min}}{N^2} \times \sum_{l=0}^{p} \mathbf{O}(l) \tag{3.9}$$

The output image is then

$$\mathbf{N}_{x,y} = E(\mathbf{O}_{x,y}, \mathbf{O}) \tag{3.10}$$

The result of equalizing the eye image is shown in Figure 3.6. The intensity equalized image, Figure 3.6(c) has much better defined features (especially around the eyes) than in the original version (Figure 3.1). The histogram (Figure 3.6d) reveals the non-linear mapping process whereby white and black are not assigned equal weight, as they were in intensity normalization. Accordingly, more pixels are mapped into the darker region and the brighter intensities become better spread, consistent with the aims of histogram equalization.

Its performance can be very convincing since it is well mapped to the properties of human vision. If a linear brightness transformation is applied to the original image then the equalized histogram will be the same. If we replace pixel values with ones computed according to Equation 3.1, the result of histogram equalization will not change. An alternative interpretation is that if we equalize images (before further processing) then we need not worry about any brightness transformation in the original image. This is to be expected, since the linear operation of the brightness change in Equation 3.2 does not change the overall shape of the histogram, only its size and position. However, noise in the image acquisition process will affect the shape of the original histogram, and hence the equalized version. So the equalized histogram of a picture will not be the same as the equalized histogram of a picture with some noise added to it. You cannot avoid noise in electrical systems, however well you design a system to reduce its effect. Accordingly, histogram equalization finds little use in generic image processing systems, although it can be potent in *specialized* applications. For these reasons, intensity normalization is often preferred when a picture's histogram requires manipulation.

In implementation, the function `equalize` in Code 3.3, we shall use an output range where $N_{min} = 0$ and $N_{max} = 255$. The implementation first determines the cumulative histogram for each level of the brightness histogram. This is then used as a LUT for the new output brightness at that level. The LUT is used to speed implementation of Equation 3.9, since it can be precomputed from the image to be equalized.

```
equalize(pic) :=   range ← 255
                   number ← rows(pic).cols(pic)
                   for bright ∈ 0..255
                      pixels_at_level_bright ← 0
                   for x ∈ 0..cols(pic)-1
                      for y ∈ 0..rows(pic)-1
                         pixels_at_level_pic_y,x ← pixels_at_level_pic_y,x +1


                   sum ← 0
                   for level ∈ 0..255
                      sum ← sum+pixels_at_level_level

                      hist_level ← floor[ (range/number)·sum+0.00001 ]


                   for x ∈ 0..cols(pic)-1
                      for y ∈ 0..rows(pic)-1
                         newpic_y,x ← hist_pic_y,x


                   newpic
```

Code 3.3   Histogram equalization

An alternative argument against use of histogram equalization is that it is a non-linear process and is irreversible. We cannot return to the original picture after equalization, and we cannot separate the histogram of an unwanted picture. In contrast, intensity normalization is a linear process and we can return to the original image, should we need to, or separate pictures, if required.

3.3.4 Thresholding

The last point operator of major interest is called *thresholding*. This operator selects pixels that have a particular value, or are within a specified range. It can be used to find objects within a picture if their brightness level (or range) is known. This implies that the object's brightness must be known as well. There are two main forms: uniform and adaptive thresholding. In *uniform thresholding*, pixels above a specified level are set to white, those below the specified level are set to black. Given the original eye image, Figure 3.7 shows a thresholded image where all pixels *above* 160 brightness levels are set to white, and those below 160 brightness levels are set to black. By this process, the parts pertaining to the facial skin are separated from the background; the cheeks, forehead and other bright areas are separated from the hair and eyes. This can therefore provide a way of isolating points of interest.

Figure 3.7 Thresholding the eye image

Uniform thresholding clearly requires knowledge of the grey level, or the target features might not be selected in the thresholding process. If the level is not known, histogram equalization or intensity normalization can be used, but with the restrictions on performance stated earlier. This is, of course, a problem of image interpretation. These problems can only be solved by simple approaches, such as thresholding, for very special cases. In general, it is often prudent to investigate the more sophisticated techniques of feature selection and extraction, to be covered later. Before that, we shall investigate group operators, which are a natural counterpart to point operators.

There are more advanced techniques, known as *optimal thresholding*. These usually seek to select a value for the threshold that separates an object from its background. This suggests that the object has a different range of intensities to the background, in order that an appropriate threshold can be chosen, as illustrated in Figure 3.8. Otsu's method (Otsu, 1979) is one of the most popular techniques of optimal thresholding; there have been surveys (Sahoo et al., 1988; Lee et al., 1990; Glasbey, 1993) which compare the performance different methods can achieve. Essentially, Otsu's technique maximizes the likelihood that the threshold is chosen so as to split the image between an object and its background. This is achieved by selecting a threshold that gives the best separation of classes, for all pixels in an image. The theory is beyond the scope of this section and we shall merely survey its results and give their implementation. The basis is use of the normalized histogram where the number of points at each level is divided by the total number of points in the image. As such, this represents a probability distribution for the intensity levels as

$$p(l) = \frac{\mathbf{N}(l)}{N^2} \tag{3.11}$$

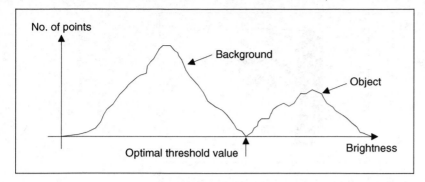

Figure 3.8 Optimal thresholding

This can be used to compute the zero- and first-order cumulative moments of the normalized histogram up to the kth level as

$$\omega(k) = \sum_{l=1}^{k} p(l) \tag{3.12}$$

and

$$\mu(k) = \sum_{l=1}^{k} l \cdot p(l) \tag{3.13}$$

The total mean level of the image is given by

$$\mu T - \sum_{l=1}^{N_{max}} l \cdot p(l) \tag{3.14}$$

The variance of the class separability is then the ratio

$$\sigma_B^2(k) = \frac{(\mu T \cdot \omega(k) - \mu(k))^2}{\omega(k)(1 - \omega(k))} \qquad \forall k \in 1, N_{max} \tag{3.15}$$

The optimal threshold is the level for which the variance of class separability is at its maximum, namely the optimal threshold T_{opt} is that for which the variance

$$\sigma_B^2(T_{opt}) = \max_{1 \le k < N_{max}} \left(\sigma_B^2(k) \right) \tag{3.16}$$

A comparison of uniform thresholding with optimal thresholding is given in Figure 3.9 for the eye image. The threshold selected by Otsu's operator is actually slightly lower than the value selected manually, and so the thresholded image does omit some detail around the eye, especially in the eyelids. However, the selection by Otsu is *automatic*, as opposed to *manual*, and this can be to application advantage in automated vision. Consider for example the need to isolate the human figure in Figure 3.10(a). This can be performed automatically by Otsu as shown in Figure 3.10(b). Note, however, that there are some extra points, due to illumination, which have appeared in the resulting image together with the human subject. It is easy to remove the isolated points, as we will see later, but more difficult to remove the connected ones. In this instance, the size of the human shape could be used as information to remove the extra points, although you might like to suggest other factors that could lead to their removal.

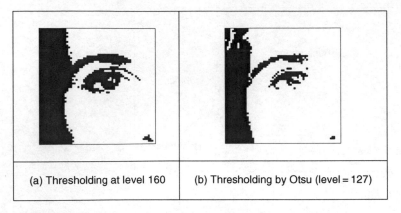

| (a) Thresholding at level 160 | (b) Thresholding by Otsu (level = 127) |

Figure 3.9 Thresholding the eye image: manual and automatic

The code implementing Otsu's technique is given in Code 3.4, which follows Equations 3.11–3.16 to provide the results in Figures 3.9 and 3.10. Here, the histogram function of Code 3.1 is used to give the normalized histogram. The remaining code refers directly to the earlier description of Otsu's technique.

```
ω(k,histogram):= Σ from l=1 to k  histogram_{l-1}

μ(k,histogram):= Σ from l=1 to k  l·histogram_{l-1}

μT(histogram):= Σ from l=1 to 256  l·histogram_{l-1}

Otsu(image):= | image_hist ← histogram(image) / (rows(image)·cols(image))
              | for k∈1..255
              |     values_k ← ((μT(image_hist)·ω(k,image_hist) − μ(k,image_hist))²) / (ω(k,image_hist)·(1 − ω(k,image_hist)))
              | find_value(max(values),values)
```

Code 3.4 Optimal thresholding by Otsu's technique

So far, we have considered *global* techniques, methods that operate on the entire image. There are also *locally adaptive* techniques that are often used to binarize document images before character recognition. As mentioned before, surveys of thresholding are available, and one (more recent) approach (Rosin, 2001) targets thresholding of images whose histogram is unimodal (has a single peak). One survey (Trier and Jain, 1995) compares global and local techniques with reference to document image analysis. These techniques are often used in statistical pattern recognition: the thresholded object is classified according to its statistical properties. However, these techniques find less use in image interpretation, where a common paradigm is that there is more than one object in the scene, such as Figure 3.7 where the thresholding operator has

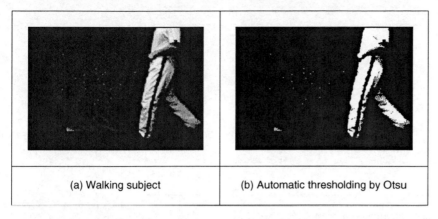

| (a) Walking subject | (b) Automatic thresholding by Otsu |

Figure 3.10 Thresholding an image of a walking subject

selected many objects of potential interest. As such, only uniform thresholding is used in many vision applications, since objects are often occluded (hidden), and many objects have similar ranges of pixel intensity. Accordingly, more sophisticated metrics are required to separate them, by using the uniformly thresholded image, as discussed in later chapters. Further, the operation to process the thresholded image, say to fill in the holes in the silhouette or to remove the noise on its boundary or outside, is *morphology*, which is covered in Section 3.6.

3.4 Group operations

3.4.1 Template convolution

Group operations calculate new pixel values from a pixel's neighbourhood by using a 'grouping' process. The group operation is usually expressed in terms of *template convolution*, where the template is a set of weighting coefficients. The template is usually square, and its size is usually odd to ensure that it can be positioned appropriately. The size is usually used to describe the template; a 3×3 template is 3 pixels wide by 3 pixels long. New pixel values are calculated by placing the template at the point of interest. Pixel values are multiplied by the corresponding weighting coefficient and added to an overall sum. The sum (usually) evaluates a new value for the centre pixel (where the template is centred) and this becomes the pixel in a new output image. If the template's position has not yet reached the end of a line, the template is then moved horizontally by one pixel and the process repeats.

This is illustrated in Figure 3.11, where a new image is calculated from an original one, by template convolution. The calculation obtained by template convolution for the centre pixel of the template in the original image becomes the point in the output image. Since the template cannot extend beyond the image, the new image is smaller than the original image because a new value cannot be computed for points in the border of the new image. When the template reaches the end of a line, it is repositioned at the start of the next line. For a 3×3 neighbourhood, nine weighting coefficients w_t are applied to points in the original image to calculate a point in the new image. The position of the new point (at the centre) is shaded in the template.

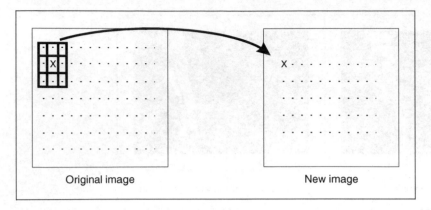

Figure 3.11 Template convolution process

To calculate the value in new image, $\mathbf{N}$, at point with coordinates x,y, the template in Figure 3.12 operates on an original image $\mathbf{O}$ according to:

$$\mathbf{N}_{x,y} = \begin{array}{llll} w_0 \times \mathbf{O}_{x-1,y-1} & + & w_1 \times \mathbf{O}_{x,y-1} & + & w_2 \times \mathbf{O}_{x+1,y-1} & + \\ w_3 \times \mathbf{O}_{x-1,y} & + & w_4 \times \mathbf{O}_{x,y} & + & w_5 \times \mathbf{O}_{x+1,y} & + \\ w_6 \times \mathbf{O}_{x-1,y+1} & + & w_7 \times \mathbf{O}_{x,y+1} & + & w_8 \times \mathbf{O}_{x+1,y+1} \end{array} \quad \forall x, y \in 2, N-1$$

$$(3.17)$$

w_0	w_1	w_2
w_3	w_4	w_5
w_6	w_7	w_8

Figure 3.12 3×3 Template and weighting coefficients

Note that we cannot ascribe values to the picture's borders. This is because when we place the template at the border, parts of the template fall outside the image and have no information from which to calculate the new pixel value. The width of the border equals half the size of the template. To calculate values for the border pixels, we now have three choices:

- set the border to black (or deliver a smaller picture)
- assume (as in Fourier) that the image replicates to infinity along both dimensions and calculate new values by cyclic shift from the far border
- calculate the pixel value from a smaller area.

None of these approaches is optimal. The results here use the first option and set border pixels to black. Note that in many applications the object of interest is imaged centrally or, at least, imaged within the picture. As such, the border information is of little consequence to the remainder of the process. Here, the border points are set to black, by starting functions with a zero function which sets all the points in the picture initially to black (0).

An alternative representation for this process is given by using the convolution notation as

$$\mathbf{N} = \mathbf{W} * \mathbf{O} \qquad\qquad (3.18)$$

where $\mathbf{N}$ is the new image which results from convolving the template $\mathbf{W}$ (of weighting coefficients) with the image $\mathbf{O}$.

The Matlab implementation of a general template convolution operator convolve is given in Code 3.5. This function accepts, as arguments, the picture image and the template to be convolved with it, template. The result of template convolution is a picture convolved. The operator first initializes the temporary image temp to black (zero brightness levels). Then

```
function convolved=convolve(image,template)
%New image point brightness convolution of template with image
%Usage:[new image]=convolve(image,template of point values)
%Parameters:image-array of points
% template-array of weighting coefficients
%Author: Mark S. Nixon

%get image dimensions
[irows,icols]=size(image);

%get template dimensions
[trows,tcols]=size(template);

%set a temporary image to black
temp(1:irows,1:icols)=0;

%half of template rows is
trhalf=floor(trows/2);
%half of template cols is
tchalf=floor(tcols/2);

%then convolve the template
for x=trhalf+1:icols-trhalf %address all columns except border
  for y=tchalf+1:irows-tchalf %address all rows except border
    sum=0;
    for iwin=1:trows %address template columns
      for jwin=1:tcols %address template rows
        sum=sum+image(y+jwin-tchalf-1,x+iwin-trhalf-1)*
                      template(jwin,iwin);
      end
    end
    temp(y,x)=sum;
  end
end

%finally, normalize the image
convolved=normalize(temp);
```

Code 3.5 Template convolution operator

the size of the template is evaluated. These give the range of picture points to be processed in the outer `for` loops that give the coordinates of all points resulting from template convolution. The template is convolved at each picture point by generating a running summation of the pixel values within the template's window multiplied by the respective template weighting coefficient. Finally, the resulting image is normalized to ensure that the brightness levels are occupied appropriately.

Template convolution is usually implemented in software. It can also be implemented in hardware and requires a two-line store, together with some further latches, for the (input) video data. The output is the result of template convolution, summing the result of multiplying weighting coefficients by pixel values. This is called pipelining, since the pixels essentially move along a pipeline of information. Note that two line stores can be used if the video fields only are processed. To process a full frame, one of the fields must be stored since it is presented in interlaced format.

Processing can be analogue, using operational amplifier circuits and charge coupled device (CCD) for storage along bucket brigade delay lines. Finally, an alternative implementation is to use a parallel architecture: for multiple instruction multiple data (MIMD) architectures, the picture can be split into blocks (spatial partitioning); single instruction multiple data (SIMD) architectures can implement template convolution as a combination of shift and add instructions.

3.4.2   Averaging operator

For an *averaging operator*, the template weighting functions are unity (or 1/9 to ensure that the result of averaging nine white pixels is white, not more than white!). The template for a 3×3 averaging operator, implementing Equation 3.17, is given by the template in Figure 3.13, where

1/9	1/9	1/9
1/9	1/9	1/9
1/9	1/9	1/9

Figure 3.13 3×3 averaging operator template coefficients

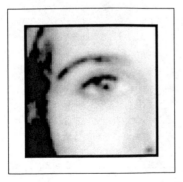

Figure 3.14   Applying direct averaging

the location of the point of interest is again shaded. The result of averaging the eye image with a 3×3 operator is shown in Figure 3.14. This shows that much of the detail has now disappeared, revealing the broad image structure. The eyes and eyebrows are now much clearer from the background, but the fine detail in their structure has been removed.

For a general implementation, Code 3.6, we can define the width of the operator as `winsize`, the template size is `winsize × winsize`. We then form the average of all points within the area covered by the template. This is normalized (divided) by the number of points in the template's window. This is a direct implementation of a general averaging operator (i.e. without using the template convolution operator in Code 3.5).

```
ave(pic,winsize):= | new ← zero(pic)
                   |
                   | half ← floor ( winsize / 2 )
                   |
                   | for x∈half..cols(pic)-half-1
                   |   for y∈half..rows(pic)-half-1
                   |
                   |                        ⎡  winsize-1  winsize-1                            ⎤
                   |                        ⎢   Σ          Σ     pic_(y+iwin-half,x+jwin-half) ⎥
                   |                        ⎢  iwin=0     jwin=0                               ⎥
                   |     new_(y,x) ← floor  ⎢ ─────────────────────────────────────────────── ⎥
                   |                        ⎣              (winsize·winsize)                   ⎦
                   | new
```

Code 3.6 Direct averaging

To implement averaging by using the template convolution operator, we need to define a template. This is illustrated for direct averaging in Code 3.7, even though the simplicity of the direct averaging template usually precludes such implementation. The application of this template is also shown in Code 3.7. (Note that there are averaging operators in Mathcad and Matlab that can also be used for this purpose.)

```
averaging_template(winsize):= | sum ← winsize.winsize
                              | for y∈0..winsize-1
                              |   for x∈0..winsize-1
                              |     template_(y,x) ← 1
                              | template
                              | ─────────
                              | sum
smoothed:=tm_conv(p,averaging_template(3))
```

Code 3.7 Direct averaging by template convolution

The effect of averaging is to reduce noise; this is its advantage. An associated disadvantage is that averaging causes blurring which reduces detail in an image. It is also a *low-pass* filter since its effect is to allow low spatial frequencies to be retained, and to suppress high-frequency components. A larger template, say 3×3 or 5×5, will remove more noise (high frequencies) but reduce the level of detail. The size of an averaging operator is then equivalent to the reciprocal of the bandwidth of a low-pass filter that it implements

Smoothing was earlier achieved by low-pass filtering via the Fourier transform (Section 2.8). The Fourier transform gives an alternative method to implement template convolution and to speed it up, for larger templates. In Fourier transforms, the process that is dual to *convolution* is *multiplication* (as in Section 2.3). So template convolution (denoted*) can be implemented by multiplying the Fourier transform of the template $\Im(\mathbf{T})$ by the Fourier transform of the picture, $\Im(\mathbf{P})$, to which the template is to be applied. The result needs to be inverse transformed to return to the picture domain.

$$\mathbf{P} * \mathbf{T} = \Im^{-1}\left(\Im\left(\mathbf{P}\right) \times \Im\left(\mathbf{T}\right)\right) \tag{3.19}$$

The transform of the template and the picture need to be the same size before we can perform the point-by-point multiplication. Accordingly, the image containing the template is *zero padded* before its transform, which simply means that zeros are added to the template in positions which lead to a template of the same size as the image. The process is illustrated in Code 3.8 and starts by calculation of the transform of the zero-padded template. The convolution routine then multiplies the transform of the template by the transform of the picture point by point (using the vectorize operator, symbolized by the arrow above the operation). When the routine is invoked, it is supplied with a transformed picture. The resulting transform is reordered before inverse transformation, to ensure that the image is presented correctly. (Theoretical study of this process is presented in Section 5.3.2, where we show how the same process can be used to find shapes in images.)

```
conv(pic,temp):= pic_spectrum←Fourier(pic)
                 temp_spectrum←Fourier(temp)
                 ─────────────────────────────────────────→
                 convolved_spectrum←(pic_spectrum.temp_spectrum)
                 result←inv_Fourier(rearrange(convolved_spectrum))
                 result

new_smooth:=conv(p,square)
```

Code 3.8 Template convolution by the Fourier transform

Code 3.8 is simply a different implementation of direct averaging. It achieves the same result, but by transform domain calculus. It can be faster to use the transform rather than the direct implementation. The computational cost of a 2D fast Fourier transform (FFT) is of the order of $N^2 \log(N)$. If the transform of the template is precomputed, there are two transforms required and there is one multiplication for each of the N^2 transformed points. The total cost of the Fourier implementation of template convolution is then of the order of

$$C_{\text{FFT}} = 4N^2 \log(N) + N^2 \tag{3.20}$$

The cost of the direct implementation for an $m \times m$ template is then m^2 multiplications for each image point, so the cost of the direct implementation is of the order of

$$C_{\text{dir}} = N^2 m^2 \tag{3.21}$$

For $C_{dir} < C_{FFT}$, we require:

$$N^2 m^2 < 4N^2 \log(N) + N^2 \tag{3.22}$$

If the direct implementation of template matching is faster than its Fourier implementation, we need to choose m so that

$$m^2 < 4\log(N) + 1 \tag{3.23}$$

This implies that, for a 256×256 image, a direct implementation is fastest for 3×3 and 5×5 templates, whereas a transform calculation is faster for larger ones. An alternative analysis (Campbell, 1969) has suggested that (Gonzalez and Wintz, 1987) 'if the number of non-zero terms in (the template) is less than 132 then a direct implementation...is more efficient than using the FFT approach'. This implies a considerably larger template than our analysis suggests. This is in part due to higher considerations of complexity than our analysis has included. There are further considerations in the use of transform calculus, the most important being the use of windowing (such as Hamming or Hanning) operators to reduce variance in high-order spectral estimates. This implies that template convolution by transform calculus should perhaps be used when large templates are involved, and then only when speed is critical. If speed is indeed critical, it might be better to implement the operator in dedicated hardware, as described earlier.

3.4.3 On different template size

Templates can be larger than 3×3. Since they are usually centred on a point of interest, to produce a new output value at that point, they are usually of odd dimension. For reasons of speed, the most common sizes are 3×3, 5×5 and 7×7. Beyond this, say 9×9, many template points are used to calculate a single value for a new point, and this imposes high computational cost, especially for large images. (For example, a 9×9 operator covers nine times more points than a 3×3 operator.) Square templates have the same properties along both image axes. Some implementations use vector templates (a line), either because their properties are desirable in a particular application, or for reasons of speed.

The effect of larger averaging operators is to smooth the image more, to remove more detail while giving greater emphasis to the large structures. This is illustrated in Figure 3.15. A 5×5 operator (Figure 3.15a) retains more detail than a 7×7 operator (Figure 3.15b), and much more than a 9×9 operator (Figure 3.15c). Conversely, the 9×9 operator retains only the largest structures such as the eye region (and virtually removing the iris), whereas this is retained more

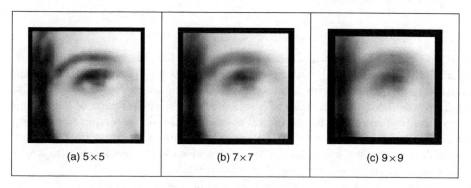

(a) 5×5   (b) 7×7   (c) 9×9

Figure 3.15   Illustrating the effect of window size

by the operators of smaller size. Note that the larger operators leave a larger border (since new values cannot be computed in that region) and this can be seen in the increase in border size for the larger operators, in Figure 3.15(b) and (c).

3.4.4 Gaussian averaging operator

The *Gaussian averaging operator* has been considered to be optimal for image smoothing. The template for the Gaussian operator has values set by the Gaussian relationship. The Gaussian *function g* at coordinates x, y is controlled by the *variance* σ^2 according to:

$$g(x, y, \sigma) = \frac{1}{2\pi\sigma^2} e^{-\left(\frac{x^2+y^2}{2\sigma^2}\right)} \tag{3.24}$$

Equation 3.24 gives a way to calculate coefficients for a Gaussian template which is then convolved with an image. The effects of selection of Gaussian templates of differing size are shown in Figure 3.16. The Gaussian function essentially removes the influence of points greater than 3σ in (radial) distance from the centre of the template. The 3×3 operator (Figure 3.16a) retains many more of the features than those retained by direct averaging (Figure 3.14). The effect of larger size is to remove more detail (and noise) at the expense of losing features. This is reflected in the loss of internal eye component by the 5×5 and the 7×7 operators in Figure 3.16(b) and (c), respectively.

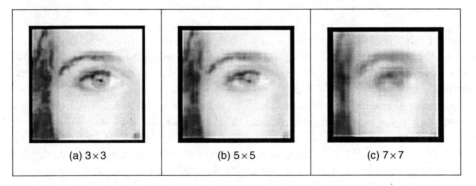

| (a) 3×3 | (b) 5×5 | (c) 7×7 |

Figure 3.16 Applying Gaussian averaging

A surface plot of the 2D Gaussian function of Equation 3.24 has the famous bell shape, as shown in Figure 3.17. The values of the function at discrete points are the values of a Gaussian template. Convolving this template with an image gives Gaussian averaging: the point in the averaged picture is calculated from the sum of a region where the central parts of the picture are weighted to contribute more than the peripheral points. The size of the template essentially dictates appropriate choice of the variance. The variance is chosen to ensure that template coefficients drop to near zero at the template's edge. A common choice for the template size is 5×5 with variance unity, giving the template shown in Figure 3.18.

This template is then convolved with the image to give the Gaussian blurring function. It is possible to give the Gaussian blurring function antisymmetric properties by scaling the x and y

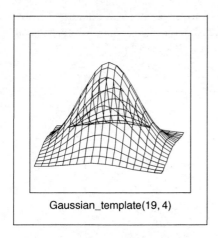

Gaussian_template(19, 4)

Figure 3.17 Gaussian function

0.002	0.013	0.220	0.013	0.002
0.013	0.060	0.098	0.060	0.013
0.220	0.098	0.162	0.098	0.220
0.013	0.060	0.098	0.060	0.013
0.002	0.013	0.220	0.013	0.002

Figure 3.18 Template for the 5×5 Gaussian averaging operator ($\sigma = 1.0$)

coordinates. This can find application when an object's shape, and orientation, is known before image analysis.

By reference to Figure 3.16 it is clear that the Gaussian filter can offer improved performance compared with direct averaging: more features are retained while the noise is removed. This can be understood by Fourier transform theory. In Section 2.4.2 (Chapter 2) we found that the Fourier transform of a *square* is a 2D *sinc* function. This has a frequency response where the magnitude of the transform does not reduce in a smooth manner and has regions where the transform becomes negative, called sidelobes. These can have undesirable effects since there are high frequencies that contribute *more* than some lower ones, which is a bit paradoxical in low-pass filtering to remove noise. In contrast, the Fourier transform of a Gaussian function is another Gaussian function, which decreases smoothly without these sidelobes. This can lead to better performance since the contributions of the frequency components reduce in a controlled manner.

In a software implementation of the Gaussian operator, we need a function implementing Equation 3.24, the `Gaussian_template` function in Code 3.9. This is used to calculate the coefficients of a template to be centred on an image point. The two arguments are `winsize`, the (square) operator's size, and the standard deviation σ that controls its width, as discussed earlier. The operator coefficients are normalized by the sum of template values, as before. This summation is stored in sum, which is initialized to zero. The centre of the square template is then evaluated as half the size of the operator. Then, all template coefficients are calculated by a version of Equation 3.24 which specifies a weight relative to the centre coordinates. Finally, the normalized template coefficients are returned as the Gaussian template. The operator is used in template convolution, via `convolve`, as in direct averaging (Code 3.5).

```
function template=gaussian_template(winsize,sigma)
%Template for Gaussian averaging

%Usage:[template]=gaussian_template(number, number)

%Parameters: winsize-size of template (odd, integer)
% sigma-variance of Gaussian function
%Author: Mark S. Nixon

%centre is half of window size
centre=floor(winsize/2)+1;

%we'll normalize by the total sum
sum=0;

%so work out the coefficients and the running total
for i=1:winsize
  for j=1:winsize
    template(j,i)=exp(-(((j-centre)*(j-centre))+((i-centre)
                     *(i-centre)))/(2*sigma*sigma))
    sum=sum+template(j,i);
  end
end

%and then normalize
template=template/sum;
```

Code 3.9   Gaussian template specification

3.5   Other statistical operators

3.5.1   More on averaging

The averaging process is actually a statistical operator since it aims to estimate the mean of a local neighbourhood. The error in the process is high; for a population of N samples, the statistical error is of the order of:

$$\text{error} = \frac{\text{mean}}{\sqrt{N}} \tag{3.25}$$

Increasing the averaging operator's size improves the error in the estimate of the mean, but at the expense of fine detail in the image. The average is an estimate optimal for a signal corrupted by additive *Gaussian noise* (see Appendix 3, Section 11.1). The estimate of the mean maximizes the probability that the noise has its mean value, namely zero. According to the *central limit theorem*, the result of adding many noise sources together is a Gaussian distributed noise source. In images, noise arises in sampling, in quantization, in transmission and in processing. By the central limit theorem, the result of these (independent) noise sources is that image noise can be assumed to be Gaussian. In fact, image noise is not necessarily Gaussian distributed, giving rise to more statistical operators. One of these is the *median* operator, which has demonstrated capability to reduce noise while retaining feature boundaries (in contrast to smoothing, which blurs both noise and the boundaries), and the *mode* operator, which can be viewed as optimal for a number of noise sources, including *Rayleigh noise*, but is very difficult to determine for small, discrete, populations.

3.5.2 Median filter

The *median* is another frequently used statistic; the median is the centre of a rank-ordered distribution. The median is usually taken from a template centred on the point of interest. Given the arrangement of pixels in Figure 3.19(a), the pixel values are arranged into a vector format (Figure 3.19b). The vector is then sorted into ascending order (Figure 3.19c). The median is the central component of the sorted vector; this is the fifth component since we have nine values.

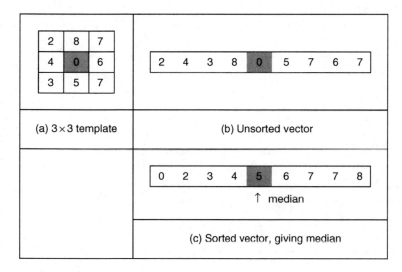

Figure 3.19   Finding the median from a 3×3 template

The median operator is usually implemented using a template. Here, we shall consider a 3×3 template. Accordingly, we need to process the nine pixels in a template centred on a point with coordinates (x, y). In a Mathcad implementation, these nine points can be extracted into vector format using the operator `unsorted` in Code 3.10. This requires a integer pointer to nine values, `x1`. The modulus operator is then used to ensure that the correct nine values are extracted.

```
x1:=0..8
unsorted_x1:=p_{x+mod(x1,3)-1,x+floor(x1/3)-1}
```

Code 3.10   Reformatting a neighbourhood into a vector

We then arrange the nine pixels, within the template, in ascending order using the Mathcad sort function (Code 3.11). This gives the rank ordered list and the median is the central component of the sorted vector, in this case the fifth component (Code 3.12). These functions can then be grouped to give the full median operator as in Code 3.13.

```
sorted:=sort(unsorted)
```

Code 3.11 Using the Mathcad sort function

```
our_median:=sorted₄
```

Code 3.12 Determining the median

```
med(pic):= newpic←zero(pic)
            for x∈1..cols(pic)-2
              for y∈1..rows(pic)-2
                for x1∈ 0..8

                  unsorted_{x1} ← pic_{y+mod(x1,3)-1,x+floor(x1/3)-1}

                  sorted← sort(unsorted)
                  newpic_{y,x}← sorted₄

            newpic
```

Code 3.13 Determining the median

The median can be taken from *larger* template sizes. It is available as the median operator in Mathcad, but only for square matrices. The development here has aimed not only to demonstrate how the median operator works, but also to provide a basis for further development. The rank ordering process is computationally demanding (*slow*) and motivates study into the deployment of fast algorithms, such as Quicksort (e.g. Huang et al., 1979, is an early approach). The computational demand has also motivated use of template shapes, other than a square. A selection of alternative shapes is shown in Figure 3.20. Common alternative shapes include a cross or a line (horizontal or vertical), centred on the point of interest; these shapes can afford much faster operation since they cover fewer pixels. The basis of the arrangement presented here could be used for these alternative shapes, if required.

The median has a well-known ability to remove *salt and pepper noise*. This form of noise, arising from say decoding errors in picture transmission systems, can cause isolated white and black points to appear within an image. It can also arise when rotating an image, when points

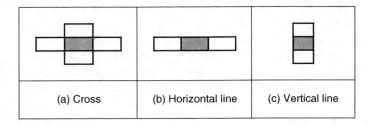

(a) Cross	(b) Horizontal line	(c) Vertical line

Figure 3.20 Alternative template shapes for median operator

remain unspecified by a standard rotation operator (Appendix 2), as in a texture image, rotated by 10° in Figure 3.21(a). When a median operator is applied, the salt and pepper noise points will appear at either end of the rank-ordered list and are removed by the median process, as shown in Figure 3.21(b). The median operator has practical advantage, owing to its ability to retain edges (the boundaries of shapes in images) while suppressing the noise contamination. As such, like direct averaging, it remains a worthwhile member of the stock of standard image processing tools. For further details concerning properties and implementation, see Hodgson et al. (1985). (Note that practical implementation of image rotation is a computer graphics issue, and is usually done by *texture mapping*; further details can be found in Hearn and Baker, 1997.)

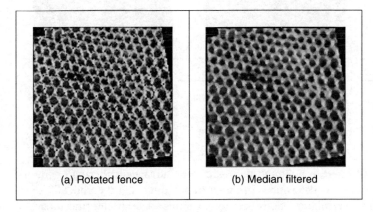

(a) Rotated fence	(b) Median filtered

Figure 3.21 Illustrating median filtering

Finding the *background* to an image is an example application of statistical operators. Say we have a sequence of images of a walking subject, and we want to be able to find the background (so that we can separate the walking subject from it), such as the sequence of images shown in Figure 3.22(a)–(f) where a subject is walking from left to right. We can average the images so as to find the background. If we form a *temporal average*, an image where each point is the average of the points in the same position in each of the six images, we achieve a result where the walking subject appears to be in the background, but very faintly, as in Figure 3.22(g). The shadow occurs since the walking subject's influence on image brightness is reduced by one-sixth, but it is still there. We could use more images, the ones in between the ones we already have, and then the shadow will become much fainter. We can also include spatial averaging as in Section 3.3.2, to reduce further the effect of the walking subject, as shown in Figure 3.22(h). This gives *spatiotemporal averaging*. For this, we have not required any more images, but the penalty paid for the better improvement in the estimate of the background is lack of detail. We cannot see the numbers in the clock, because of the nature of spatial averaging. However, if we form the background image by taking the median of the six images, a *temporal median*, we get a much better estimate of the background, as shown in Figure 3.22(i). A lot of the image detail is retained, while the walking subject disappears. In this case, for a *sequence* of images where the target walks in front of a *static* background, the median is the most appropriate operator. If we did not have a sequence, we could just average the single image with a large operator and that could provide some estimate of the background.

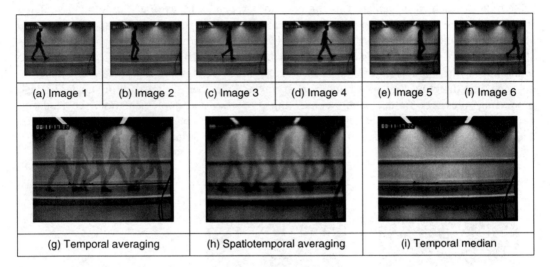

(a) Image 1	(b) Image 2	(c) Image 3	(d) Image 4	(e) Image 5	(f) Image 6

(g) Temporal averaging	(h) Spatiotemporal averaging	(i) Temporal median

Figure 3.22 Background estimation by mean and median filtering

3.5.3 Mode filter

The *mode* is the final statistic of interest, although there are more advanced filtering operators to come. The mode is very difficult to determine for small populations and theoretically does not even exist for a continuous distribution. Consider, for example, determining the mode of the pixels within a square 5×5 template. It is possible for all 25 pixels to be different, so each could be considered to be the mode. As such, we are forced to estimate the mode: the truncated median filter, as introduced by Davies (1988), aims to achieve this. The *truncated median filter* is based on the premise that for many non-Gaussian distributions, the order of the mean, the median and the mode is the same for many images, as illustrated in Figure 3.23. Accordingly, if we truncate the distribution (i.e. remove part of it, where the part selected to be removed in Figure 3.23 is from the region beyond the mean) then the median of the truncated distribution will approach the mode of the original distribution.

The implementation of the truncated median, `trun_med`, operator is given in Code 3.14. The operator first finds the mean and the median of the current window. The distribution of intensity of points within the current window is truncated on the side of the mean so that the median now bisects the distribution of the remaining points (as such, not affecting symmetrical

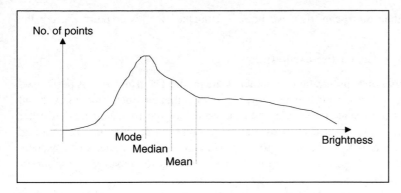

Figure 3.23 Arrangement of mode, median and mean

```
trun_med(p,wsze):= | newpic←zero(p)
                   |          ⎛ wsze ⎞
                   | ha←floor ⎜ ──── ⎟
                   |          ⎝  2   ⎠
                   | for x∈ha..cols(p)−ha−1
                   |   for y∈ha..rows(p)−ha−1
                   |     | win←submatrix(p,y−ha,y+ha,x−ha,x+ha)
                   |     | med←median(win)
                   |     | ave←mean(win)
                   |     | upper←2·med−min(win)
                   |     | lower←2·med−max(win)
                   |     | cc←0
                   |     | for i∈0..wsze−1
                   |     |   for j∈0..wsze−1
                   |     |     | if (win_{j,i}<upper)·(med<ave)
                   |     |     |   | trun_{cc}←win_{j,i}
                   |     |     |   | cc←cc+1
                   |     |     | if (win_{j,i}>lower)·(med>ave)
                   |     |     |   | trun_{cc}←win_{j,i}
                   |     |     |   | cc←cc+1
                   |     | newpic_{y,x}←median(trun) if cc>0
                   |     | newpic_{y,x}←med otherwise
                   |
                   | newpic
```

Code 3.14 The truncated median operator

distributions). So that the median bisects the remaining distribution, if the median is less than the mean, the point at which the distribution is truncated, *upper*, is

$$upper = median + (median - \min(distribution))$$
$$= 2 \cdot median - \min(distribution) \tag{3.26}$$

If the median is greater than the mean, then we need to truncate at a lower point (before the mean), *lower*, given by

$$lower = 2 \cdot median - \max(distribution) \tag{3.27}$$

The median of the remaining distribution then approaches the mode. The truncation is performed by storing pixels values in a vector `trun`. A pointer, `cc`, is incremented each time a new point is stored. The median of the truncated vector is then the output of the truncated median filter at that point. The window is placed at each possible image point, as in template convolution. However, there can be several iterations at each position to ensure that the mode is approached. In practice, only a few iterations are usually required for the median to converge to the mode. The window size is usually large, say 7×7 or 9×9 or even more.

The action of the operator is illustrated in Figure 3.24 when applied to a 128×128 part of the ultrasound image (Figure 1.1c), from the centre of the image and containing a cross-sectional view of an artery. Ultrasound results in particularly noisy images, in part because the scanner is usually external to the body. The noise is multiplicative Rayleigh noise, for which the mode is the optimal estimate. This noise obscures the artery which appears in cross-section in Figure 3.24(a); the artery is basically elliptical in shape. The action of the 9×9 truncated median operator (Figure 3.24b) is to remove noise while retaining feature boundaries, whereas a larger operator shows better effect (Figure 3.24c).

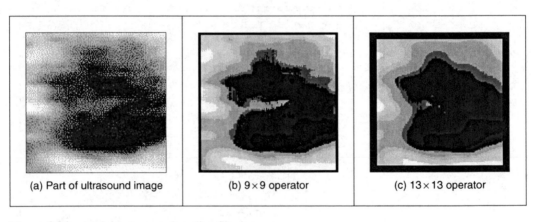

| (a) Part of ultrasound image | (b) 9×9 operator | (c) 13×13 operator |

Figure 3.24   Applying truncated median filtering

Close examination of the result of the truncated median filter shows that a selection of boundaries are preserved which are not readily apparent in the original ultrasound image. This is one of the known properties of median filtering: an ability to reduce noise while retaining feature boundaries. There have been many other approaches to speckle filtering; the most popular include direct averaging (Shankar, 1986), median filtering, adaptive (weighted) median filtering (Loupas and McDicken, 1987) and unsharp masking (Bamber and Daft, 1986).

3.5.4 Anisotropic diffusion

The most advanced form of smoothing is achieved by preserving the *boundaries* of the image features in the smoothing process (Perona and Malik, 1990). This is one of the advantages of the median operator and a disadvantage of the Gaussian smoothing operator. The process is called *anisotropic diffusion*, by virtue of its basis. Its result is illustrated in Figure 3.25(b), where the feature boundaries (such as those of the eyebrows or the eyes) in the smoothed image are crisp and the skin is more matt in appearance. This implies that we are filtering within the features and not at their edges. By way of contrast, the Gaussian operator result in Figure 3.25(c) smooths not just the skin but also the boundaries (the eyebrows in particular seem quite blurred), giving a less pleasing and less useful result. Since we shall later use the boundary information to interpret the image, its preservation is of much interest.

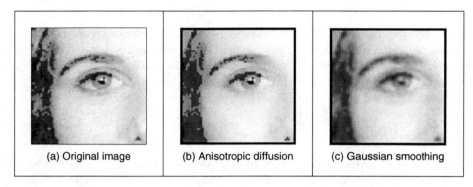

| (a) Original image | (b) Anisotropic diffusion | (c) Gaussian smoothing |

Figure 3.25 Filtering by anisotropic diffusion and the Gaussian operator

As ever, there are some parameters to select to control the operation, so we shall consider the technique's basis so as to guide their selection. Further, it is computationally more complex than Gaussian filtering. The basis of anisotropic diffusion is, however, rather complex, especially here, and invokes concepts of low-level feature extraction which are covered in the next chapter. One strategy you might use is to mark this page, then go ahead and read Sections 4.1 and 4.2, and then return here. Alternatively, you could just plough on, since that is exactly what we shall do. The complexity is due to the process not only invoking low-level feature extraction (to preserve feature boundaries) but also, as its basis, invoking concepts of *heat flow*, as well as introducing the concept of *scale space*. So it will certainly be a hard read for many, but comparison of Figure 3.25(b) with Figure 3.25(c) shows that it is well worth the effort.

The essential idea of scale space is that there is a *multiscale* representation of images, from low resolution (a coarsely sampled image) to high resolution (a finely sampled image). This is inherent in the sampling process, where the coarse image is the structure and the higher resolution increases the level of detail. As such, we can derive a multiscale set of images by convolving an original image with a Gaussian function, by Equation 3.24

$$I_{x,y}(\sigma) = I_{x,y}(0) * g(x, y, \sigma) \tag{3.28}$$

where $I_{x,y}(0)$ is the original image, $g(x,\ y,\ \sigma)$ is the Gaussian template derived from Equation 3.24, and $I_{x,y}(\sigma)$ is the image at level σ. The coarser level corresponds to larger values of the standard deviation σ; conversely, the finer detail is given by smaller values. We have already seen that the larger values of σ reduce the detail and are then equivalent to an image at a coarser scale, so this is a different view of the same process. The difficult part is that the family of images derived this way can equivalently be viewed as the solution of the heat equation

$$\partial I / \partial t = \Delta I_{x,y}(t) \tag{3.29}$$

where Δ denotes del, the (directional) gradient operator from vector algebra, and with the initial condition that $I_0 = I_{x,y}(0)$. The heat equation itself describes the temperature T changing with time t as a function of the thermal diffusivity (related to conduction) κ as

$$\partial T / \partial t = \kappa \nabla^2 T \qquad (3.30)$$

and in one-dimensional (1D) form this is

$$\partial T / \partial t = \kappa \frac{\partial^2 T}{\partial x^2} \qquad (3.31)$$

so the temperature measured along a line is a function of time, distance, the initial and boundary conditions, and the properties of a material. The relation of this with image processing is clearly an enormous ouch! There are clear similarities between Equations 3.31 and 3.29. They have the same functional form and this allows for insight, analysis and parameter selection. The heat equation (Equation 3.29) is the anisotropic diffusion equation

$$\partial I / \partial t = \nabla \cdot \left(c_{x,y}(t) \nabla I_{x,y}(t) \right) \qquad (3.32)$$

where $\nabla \cdot$ is the divergence operator (which essentially measures how the density within a region changes), with diffusion coefficient $c_{x,y}$. The diffusion coefficient applies to the local change in the image $\nabla I_{x,y}(t)$ in different directions. If we have a lot of local change, we seek to retain it since the amount of change is the amount of boundary information. The diffusion coefficient indicates how much importance we give to local change: how much of it is retained. (The equation reduces to isotropic diffusion – Gaussian filtering – if the diffusivity is constant, since $\nabla c = 0$.) There is no explicit solution to this equation. By approximating differentiation by differencing (this is explored more in Section 4.2), the rate of change of the image between time step t and time step $t+1$, we have

$$\partial I / \partial t = I(t+1) - I(t) \qquad (3.33)$$

This implies that we have an iterative solution, and for later consistency we shall denote the image I at time step $t+1$ as $I^{<t+1>} = I(t+1)$, so we then have

$$I^{<t+1>} - I^{<t>} = \nabla \cdot \left(c_{x,y}(t) \nabla I_{x,y}^{<t>} \right) \qquad (3.34)$$

and again by approximation, using differences evaluated this time over the four compass directions north, south, east and west, we have

$$\nabla_N(I_{x,y}) = I_{x,y-1} - I_{x,y} \qquad (3.35)$$

$$\nabla_S(I_{x,y}) = I_{x,y+1} - I_{x,y} \qquad (3.36)$$

$$\nabla_E(I_{x,y}) = I_{x-1,y} - I_{x,y} \qquad (3.37)$$

$$\nabla_W(I_{x,y}) = I_{x+1,y} - I_{x,y} \qquad (3.38)$$

The template and weighting coefficients for these are shown in Figure 3.26.

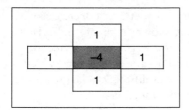

Figure 3.26 Approximations by spatial difference in anisotropic diffusion

When we use these as an approximation to the right-hand side in Equation 3.34, we then have $\nabla \cdot \left(c_{x,y}(t)\nabla I_{x,y}^{<I>}\right) = \lambda\left(cN_{x,y}\nabla_N(I) + cS_{x,y}\nabla_S(I) + cE_{x,y}\nabla_E(I) + cW_{x,y}\nabla_W(I)\right)$, which gives

$$I^{<I+1>} - I^{<I>} = \lambda\left(cN_{x,y}\nabla_N(I) + cS_{x,y}\nabla_S(I) + cE_{x,y}\nabla_E(I) + cW_{x,y}\nabla_W(I)\right)\big|I = I_{x,y}^{<I>} \quad (3.39)$$

where $0 \leq \lambda \leq 1/4$ and where $cN_{x,y}$, $cS_{x,y}$, $cE_{x,y}$ and $cW_{x,y}$ denote the conduction coefficients in the four compass directions. By rearrangement of this we obtain the equation that we shall use for the anisotropic diffusion operator

$$I^{<I+1>} = I^{<I>} + \lambda\left(cN_{x,y}\nabla_N(I) + cS_{x,y}\nabla_S(I) + cE_{x,y}\nabla_E(I) + cW_{x,y}\nabla_W(I)\right)\big|I = I_{x,y}^{<I>} \quad (3.40)$$

This shows that the solution is *iterative*: images at *one* time step (denoted by $^{<I+1>}$) are computed from images at the *previous* time step (denoted $^{<I>}$), given the initial condition that the first image is the original (noisy) image. Change (in time and in space) has been approximated as the difference between two adjacent points, which gives the iterative equation and shows that the new image is formed by adding a controlled amount of the local change consistent with the main idea: that the smoothing process retains some of the boundary information.

We are not finished yet, though, since we need to find values for $cN_{x,y}$, $cS_{x,y}$, $cE_{x,y}$ and $cW_{x,y}$. These are chosen to be a function of the difference along the compass directions, so that the boundary (edge) information is preserved. In this way we seek a function that tends to zero with increase in the difference (an edge or boundary with greater contrast) so that diffusion does not take place across the boundaries, keeping the edge information. As such, we seek

$$cN_{x,y} = g\left(\|\nabla_N(I)\|\right)$$

$$cS_{x,y} = g\left(\|\nabla_S(I)\|\right)$$

$$cE_{x,y} = g\left(\|\nabla_E(I)\|\right) \quad (3.41)$$

$$cW_{x,y} = g\left(\|\nabla_W(I)\|\right)$$

and one function that can achieve this is

$$g(x, k) = e^{-x^2/k^2} \quad (3.42)$$

[There is potential confusion with using the same symbol as for the Gaussian function (Equation 3.24), but we have followed the original authors' presentation.] This function clearly has the desired properties since when the values of the differences ∇ are large the function g is very small; conversely, when ∇ is small then g tends to unity. k is another parameter whose value we have to choose: it controls the rate at which the conduction coefficient decreases with increasing difference magnitude. The effect of this parameter is shown in Figure 3.27. Here, the solid line is for the smaller value of k and the dotted one is for a larger value. Evidently, a larger value of k means that the contribution of the difference reduces less than for a smaller value of k. In both cases, the resulting function is near unity for small differences and near zero for large differences, as required. An alternative to this is to use the function

$$g2(x, k) = \frac{1}{1 + \frac{x^2}{k^2}}$$

(3.43)

which has similar properties to the function in Equation 3.42.

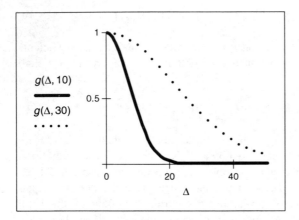

Figure 3.27 Controlling the conduction coefficient in anisotropic diffusion

This all looks rather complicated, so let's recap. First, we want to filter an image by retaining boundary points. These are retained according to the value of k chosen in Equation 3.42. This function is operated in the four compass directions, to weight the brightness difference in each direction (Equation 3. 41). These contribute to an iterative equation which calculates a new value for an image point by considering the contribution from its four neighbouring points (Equation 3.40). This needs the choice of one parameter, λ. Further, we need to choose the number of iterations for which calculation proceeds. For information, Figure 3.25(b) was calculated over 20 iterations and we need to use sufficient iterations to ensure that convergence has been achieved. We also need to choose values for k and λ. By analogy, k is the conduction coefficient and low values preserve edges and high values allow diffusion (conduction) to occur; and how much smoothing can take place. The two parameters are interrelated, although λ largely controls the amount of smoothing. Given that low values of either parameter means that no filtering effect is observed, we can investigate their effect by setting one parameter to a high value and varying the other. In Figure 3.28(a)–(c) we use a high value of k, which means that edges are not preserved, and we can observe that different values of λ control the amount of smoothing. (A discussion of how this Gaussian filtering process is achieved can be inferred from Section 4.2.4.) Conversely, we can see how different values for k control the level of edge preservation in Figure 3.28(d)–(f), where some structures around the eye are not preserved for larger values of k.

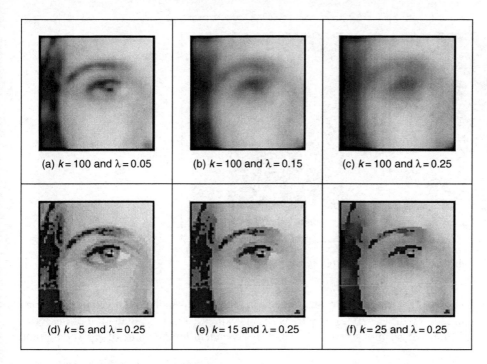

(a) $k = 100$ and $\lambda = 0.05$	(b) $k = 100$ and $\lambda = 0.15$	(c) $k = 100$ and $\lambda = 0.25$
(d) $k = 5$ and $\lambda = 0.25$	(e) $k = 15$ and $\lambda = 0.25$	(f) $k = 25$ and $\lambda = 0.25$

Figure 3.28   Applying anisotropic diffusion

The original presentation of anisotropic diffusion (Perona and Malik, 1990) is extremely lucid and well worth a read if you consider selecting this technique. It has greater detail on formulation and on analysis of results than space here allows for (and is suitable at this stage). Among other papers on this topic, one (Black et al., 1998) studied the choice of conduction coefficient leading to a function which preserves sharper edges and improves automatic termination. As ever, with techniques that require much computation there have been approaches that speed implementation or achieve similar performance more rapidly (e.g. Fischl and Schwartz, 1999).

3.5.5   Force field transform

There are many more image filtering operators; we have so far covered those that are among the most popular. Others offer alternative insight, sometimes developed in the context of a specific application. By way of example, Hurley developed a transform called the force field transform (Hurley et al., 2002, 2005) which uses an analogy to gravitational force. The transform pretends that each pixel exerts a force on its neighbours which is inversely proportional to the square of the distance between them. This generates a force field where the net force at each point is the aggregate of the forces exerted by all the other pixels on a 'unit test pixel' at that point. This very large-scale summation affords very powerful averaging which reduces the effect of noise. The approach was developed in the context of ear biometrics, recognizing people by their ears, which has unique advantage as a biometric in that the shape of people's ears does not change with age, and of course, unlike a face, ears do not smile! The force field transform of an ear (Figure 3.29a) is shown in Figure 3.29(b). Here, the averaging process is reflected in the reduction of the effects of hair. The transform itself has highlighted ear structures, especially the top of the ear and the lower 'keyhole' (the notch).

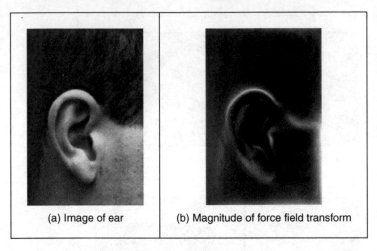

| (a) Image of ear | (b) Magnitude of force field transform |

Figure 3.29 Illustrating the force field transform

The image shown is the magnitude of the force field. The transform itself is a vector operation, and includes direction (Hurley, 2002). The transform is expressed as the calculation of the force $\mathbf{F}$ between two points at positions $\mathbf{r}_i$ and $\mathbf{r}_j$, which is dependent on the value of a pixel at point $\mathbf{r}_i$ as

$$\mathbf{F}_i\left(\mathbf{r}_j\right) = \mathbf{P}\left(\mathbf{r}_i\right) \frac{\mathbf{r}_i - \mathbf{r}_j}{\left|\mathbf{r}_i - \mathbf{r}_j\right|^3} \tag{3.44}$$

which assumes that the point $\mathbf{r}_j$ is of unit 'mass'. This is a directional force (which is why the inverse square law is expressed as the ratio of the difference to its magnitude cubed) and the magnitude and directional information has been exploited to determine an ear 'signature' by which people can be recognized. In application, Equation 3.44 can be used to define the coefficients of a template that is convolved with an image (implemented by the FFT to improve speed); as with many of the techniques that have been covered in this chapter; a Mathcad implementation is also given (Hurley et al., 2002). Note that this transform exposes low-level features (the boundaries of the ears), which is the focus of the next chapter. How we can determine shapes is a higher level process, and the processes by which we infer or recognize identity from the low- and the high-level features will be covered in Chapter 8.

3.5.6 Comparison of statistical operators

The different image filtering operators are shown by way of comparison in Figure 3.30. All operators are 5×5 and are applied to the earlier ultrasound image (Figure 3.24a). Figure 3.30(a)–(d) are the result of the mean (direct averaging), Gaussian averaging, median and truncated median, respectively. We have just shown the advantages of anisotropic diffusion compared with Gaussian smoothing, so we will not repeat them here. Each operator shows a different performance: the mean operator removes much noise, but blurs feature boundaries; Gaussian averaging retains more features, but shows little advantage over direct averaging (it is not Gaussian-distributed noise anyway); the median operator retains some noise, but with clear feature boundaries; and the truncated median removes more noise, but along with picture detail. Clearly, the increased size of the truncated median template, by the results in Figure 3.24(b)

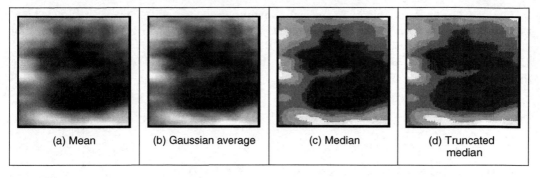

| (a) Mean | (b) Gaussian average | (c) Median | (d) Truncated median |

Figure 3.30 Comparison of filtering operators

and (c), can offer improved performance. This is to be expected since, by increasing the size of the truncated median template, we are essentially increasing the size of the distribution from which the mode is found.

As yet, however, we have not yet studied any quantitative means to evaluate this comparison. We can only perform subjective appraisal of the images in Figure 3.30. This appraisal has been phrased in terms of the contrast boundaries perceived in the image, and on the basic shape that the image presents. Accordingly, better appraisal is based on the use of feature extraction. Boundaries are the low-level features studied in the next chapter; shape is a high-level feature studied in Chapter 5.

3.6 Mathematical morphology

Mathematical morphology analyses images by using operators developed using set theory (Serra, 1986; Serra and Soile, 1994). It was originally developed for binary images and was extended to include grey-level data. The word morphology concerns shapes: in mathematical morphology we process images according to shape, by treating both as sets of points. In this way, morphological operators define *local transformations* that change pixel values that are represented as *sets*. The ways in which pixel values are changed is formalized by the definition of the *hit or miss transformation*.

In the hit and miss transformation, an object represented by a set X is examined through a structural element represented by a set B. Different structuring elements are used to change the operations on the set X. The hit or miss transformation is defined as the point operator

$$X \otimes B = \left\{ x \,\middle|\, B_x^1 \subset X \cap B_x^2 \subset X^c \right\} \tag{3.45}$$

In this equation, x represents one element of X, that is a pixel in an image. The symbol X^c denotes the complement of X (the set of image pixels that is not in the set X) and the *structuring element B* is represented by two parts, B^1 and B^2, that are applied to the set X or to its complement X^c. The structuring element is a shape and this is how mathematical morphology operations process images according to shape properties. The operation of B^1 on X is a *hit*; the operation of B^2 on X^c is a *miss*. The subindex x in the structural element indicates that it is moved to the position of the element x. That is, in a manner similar to other group operators, B defines a window that is moved through the image.

Figure 3.31 illustrates a binary image and a structuring element. Image pixels are divided into those belonging to X and those belonging to its complement X^c. The figure shows a structural element and it decomposition into the two sets B^1 and B^2. Each subset is used to analyse the set X and its complement. Here, we use black for the elements of B^1 and white for B^2 to indicate that they are applied to X and X^c, respectively.

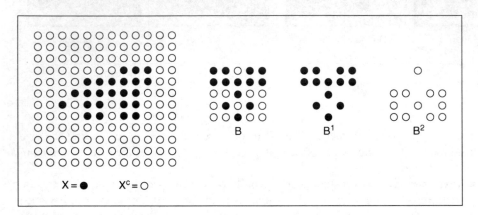

Figure 3.31 Image and structural element

Equation 3.45 defines a process that moves the structural element B to be placed at each pixel in the image and it performs a pixel-by-pixel comparison against the template B. If the value of the image is the same as that of the structuring element, then the image's pixel forms part of the resulting set $X \otimes B$. An important feature of this process is that is not invertible. That is, information is removed to suppress or enhance geometrical features in an image.

3.6.1 Morphological operators

The simplest forms of morphological operators are defined when either B^1 or B^2 are empty. When B^1 is empty Equation 3.45 defines an *erosion* (reduction) and when B^2 is empty it defines a *dilation* (increase). That is, an erosion operation is given by

$$X \ominus B = \left\{ x \,\middle|\, B_x^1 \subset X \right\} \tag{3.46}$$

and a dilation is given by

$$X \oplus B = \left\{ x \,\middle|\, B_x^2 \subset X^c \right\} \tag{3.47}$$

In the erosion operator, the hit or miss transformation establishes that a pixel x belongs to the eroded set if each point of the element B^1 translated to x is on X. Since all the points in B^1 need to be in X, this operator removes the pixels at the borders of objects in the set X. Thus, it erodes or shrinks the set. One of the most common applications of this is to remove noise in thresholded images. This is illustrated in Figure 3.32, where in (a) we have a noisy binary image, the image is eroded in (b), removing noise but making the letters smaller, and this is corrected by *dilation* in (c). We shall show how we can use shape to improve this filtering process: put the morph into morphology.

Figure 3.33 illustrates the operation of the erosion operator. Figure 3.33(a) contains a 3×3 template that defines the structural element B^1. The centre pixel is the origin of the set.

| (a) Original image | (b) Erosion | (c) Dilation |

Figure 3.32 Filtering by morphology

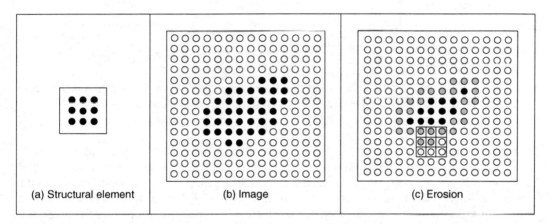

| (a) Structural element | (b) Image | (c) Erosion |

Figure 3.33 Example of the erosion operator

Figure 3.33(b) shows an image containing a region of black pixels that defines the set X. Figure 3.33(c) shows the result of the erosion. The eroded set is formed from black pixels only and we use grey to highlight the pixels that were removed from X by the erosion operator. For example, when the structural element is moved to the position shown as a grid in Figure 3.33(c), the central pixel is removed since only five pixels of the structural element are in X.

The dilation operator defined in Equation 3.47 establishes that a point belongs to the dilated set when all the points in B^2 are in the complement. This operator erodes or shrinks the complement and when the complement is eroded, the set X is dilated.

Figure 3.34 illustrates a dilation process. The structural element shown in Figure 3.34(a) defines the set B^2. We indicate its elements in white since it should be applied to the complement of X. Figure 3.34(b) shows an image example and Figure 3.34(c) the result of the dilation. The black and grey pixels belong to the dilation of X. We use grey to highlight the pixels that are added to the set. During the dilation, we place the structural element on each pixel in the complement. These are the white pixels in Figure 3.34(b). When the structural element is not fully contained, it is removed form the complement, so it becomes part of X. For example, when the structural element is moved to the position shown as a grid in Figure 3.34(c), the central pixel is removed from the complement since one of the pixels in the template is in X.

There is an alternative formulation for the dilation operator that defines the transformation over the set X instead to its complement. This definition is obtained by observing that when all

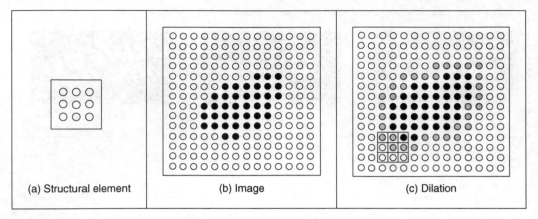

| (a) Structural element | (b) Image | (c) Dilation |

Figure 3.34 Example of the dilation operator

elements of B^2 are in X^c, this is equivalent to none of the elements in the negation of B^2 being in X. That is, dilation can also be written as the intersection of translated sets, as

$$X \oplus B = \left\{ x \,\middle|\, x \in \neg B_x^2 \right\}$$ (3.48)

Here, the symbol $\neg$ denotes negation and it changes the structural element from being applied to the complement to the set. For example, the negation of the structural element in Figure 3.34(a) is the set in Figure 3.33(a). Thus, Equation 3.48 defines a process where a point is added to the dilated set when at least one element of $\neg B^2$ is in X. For example, when the structural element is at the position shown in Figure 3.34(c), one element in X is in the template, thus the central point is added to the dilation.

Neither dilation nor erosion specifies a required shape for the structuring element. In general, it is defined to be square or circular, but other shapes such as a cross or a triangle can be used. Changes in the shape will produce subtle changes in the results, but the main feature of the structural element is given by its size, since this determines the strength of the transformation. In general, applications prefer to use small structural elements (for speed) and perform a succession of transformations until a desirable result is obtained. Other operators can be defined by sequences of erosions and dilations. For example, the *opening operator* is defined by an erosion followed by a dilation. That is,

$$X \circ B = (X \ominus B) \oplus B$$ (3.49)

Similarly, a *closing operator* is defined by a dilation followed of an erosion. That is,

$$X \bullet B = (X \oplus B) \ominus B$$ (3.50)

Closing and opening operators are generally used as filters that remove dots characteristic of pepper noise and to smooth the surface of shapes in images. These operators are generally applied in succession and the number of times they are applied depends on the structural element size and image structure.

In addition to filtering, morphological operators can be used to develop other image processing techniques. For example, edges can be detected by subtracting the original image and the one obtained by an erosion or dilation. Other example is the computation of skeletons that are thin representations of a shape. A skeleton can be computed as the union of subtracting images obtained by applying erosions and openings with structural elements of increasing sizes.

3.6.2 Grey-level morphology

In the definition in Equation 3.45 pixels belong to either the set X or its complement. Thus, it only applies to binary images. *Greyscale* or *grey-level morphology* extends Equation 3.45 to represent functions as sets, thus morphology operators can be applied to grey-level images. There are two alternative representations of functions as sets: the cross-section (Serra, 1986; Serra and Soile, 1994) and the umbra (Sternberg, 1986). The cross-section representation uses multiple thresholds to obtain a pile of binary images. Thus, the definition of Equation 3.45 can be applied to grey-level images by considering a collection of binary images as a stack of binary images formed at each threshold level. The formulation and implementation of this approach is cumbersome since it requires multiple structural elements and operators over the stack. The *umbra approach* is more intuitive and it defines sets as the points contained below functions. The umbra of a function $f(x)$ consists of all points that satisfy $f(x)$. That is,

$$U(X) = \{(x, z) \,|\, z < f(x)\} \tag{3.51}$$

Here, x represents a pixel and $f(x)$ its grey level. Thus, the space (x, z) is formed by the combination of all pixels and grey levels. For images, x is defined in two dimensions, thus all the points of the form (x, z) define a cube in 3D space. An *umbra* is a collection of points in this 3D space. Notice that morphological definitions are for discrete sets, thus the function is defined at discrete points and for discrete grey levels.

Figure 3.35 illustrates the concept of an umbra. For simplicity we show $f(x)$ as a 1D function. In Figure 3.35(a), the umbra is drawn as a collection of points below the curve. The complement of the umbra is denoted as $U^c(X)$ and it is given by the points on and above the curve. The union of $U(X)$ and $U^c(X)$ defines all the image points and grey-level values (x, z). In grey-level morphology, images and structural elements are represented by umbrae. Figure 3.35(b) illustrates the definition of two structural elements. The first example defines a structural element for the umbra, that is B^1. Similar to an image function, the umbra of the structural elements is defined by the points under the curve. The second example in Figure 3.35(b) defines a structural element for the complement, that is B^2. Similar to the complement of the umbra, this operator defines the points on and over the curve.

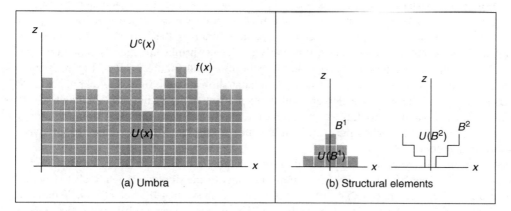

(a) Umbra   (b) Structural elements

Figure 3.35 Grey-level morphology

The hit or miss transformation in Equation 3.45 is extended to grey-level functions by considering the *inclusion operator* in the umbrae. That is,

$$U(X \otimes B) = \left\{ (x, z) \,\middle|\, U\left(B^1_{x,z}\right) \subset U(X) \cap U\left(B^2_{x,z}\right) \subset U^c(X) \right\} \tag{3.52}$$

Similar to the binary case, this equation defines a process that evaluates the inclusion of the translated structural element B. At difference of the binary definition, the structural element is translated along the pixels and grey-level values; that is, to the points (x, z). Thus, a point (x, z) belongs to the umbra of the hit or miss transformation, if the umbrae of the elements B^1 and B^2 translated to (x, z) are included in the umbra and its complement, respectively. The inclusion operator is defined for the umbra and its complement in different ways. An umbra is contained in other umbra if corresponding values of its function are equal or lower. For the complement, an umbra is contained if corresponding values of its function are equal or greater.

We can visualize the process in Equation 3.52 by translating the structural element in the example in Figure 3.35. To determine whether a point (x, z) is in the transformed set, we move the structural element B^1 to the point and see whether its umbra fully intersects $U(X)$. If that is the case, the umbra of the structural element is contained in the umbra of the function and $U(B^1_{x,t}) \subset U(X)$ is true. Similarly, to test for $U(B^2_{x,t}) \subset U^c(X)$, we move the structural element B^2 and see whether it is contained in the upper region of the curve. If both conditions are true, then the point where the operator is translated belongs to the umbra of the hit or miss transformation.

3.6.3 Grey-level erosion and dilation

Based on the generalization in Equation 3.52, it is possible to reformulate operators developed for binary morphology so they can be applied to grey-level data. The *erosion* and *dilation* defined in Equations 3.46 and 3.47 are generalized to grey-level morphology as

$$U(X \ominus B) = \left\{ (x, z) \,\middle|\, U\left(B^1_{x,z}\right) \subset U(X) \right\} \tag{3.53}$$

and

$$U(X \oplus B) = \left\{ (x, z) \,\middle|\, U\left(B^2_{x,z}\right) \subset U^c(X) \right\} \tag{3.54}$$

The erosion operator establishes that the point (x, z) belongs to the umbra of the eroded set if each point of the umbra of the element B^1 translated to the point (x, z) is under the umbra of X. A common way to visualize this process is to think that we move the structural element upwards in the grey-level axis. The erosion border is the highest point we can reach without going out of the umbra. Similar to the binary case, this operator removes the borders of the set X by increasing the separation in holes. Thus, it actually erodes or shrinks the structures in an image. Figure 3.36(a) illustrates the erosion operator for the image in Figure 3.35(a). Figure 3.36(a) shows the result of the erosion for the structural element shown on the right. For clarity we have marked the origin of the structure element with a black spot. In the result, only the black pixels form the eroded set, and we use grey to highlight the pixels that were removed from the umbra of X. It is easy to see that when the structural element is translated to a point that is removed, its umbra intersects $U^c(X)$.

Analogous to binary morphology, the dilation operator can be seen as an erosion of the complement of the umbra of X. That is, a point belongs to the dilated set when all the points in the umbra of B^2 are in $U^c(X)$. This operator erodes or shrinks the set $U^c(X)$. When the complement is eroded, the umbra of X is dilated. The dilation operator fills holes decreasing the separation between prominent structures. This process is illustrated in Figure 3.36(b) for the

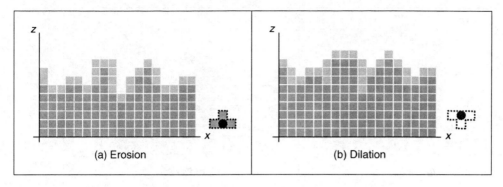

Figure 3.36 Grey-level operators

example in Figure 3.36(a). The structural element used is shown to the right in Figure 3.36(b). In the results, the black and grey pixels belong to the dilation. We use grey to highlight points that are added to the set. Points are removed from the complement and added to $U(X)$ by translating the structural element looking for points where the structural element is not fully included in $U^c(X)$. It is easy to see that when the structural element is translated to a point that is added to the dilation, its umbra intersects $U(X)$.

Similar to Equation 3.48, dilation can be written as intersection of translated sets, thus it can be defined as an operator on the umbra of an image. That is,

$$U(X \oplus B) = \{(x, z) \mid (x, z) \in U(\neg B^2_{x,z})\} \tag{3.55}$$

The negation changes the structural element from being applied to the complement of the umbra to the umbra. That is, it changes the sing of the umbra to be defined below the curve. For the example in Figure 3.36(b), it easy to see that if the structural element $\neg B^2$ is translated to any point added during the dilation, it intersects the umbra at least in one point.

3.6.4 Minkowski operators

Equations 3.53, 3.54 and 3.55 require the computation of intersections of the pixels of a structural element that is translated to all the points in the image and for each grey-level value. Thus, its computation involves significant processing. However, some simplifications can be made. For the erosion process in Equation 3.53, the value of a pixel can be simply computed by comparing the grey-level values of the structural element and corresponding image pixels. The highest position that we can translate the structural element without intersecting the complement is given by the minimum value of the difference between the grey level of the image pixel and the corresponding pixel in the structural element. That is,

$$\ominus(x) = \min_i \{f(x - i) - B(i)\} \tag{3.56}$$

Here, $B(i)$ denotes the value of the ith pixel of the structural element. Figure 3.37(a) illustrates a numerical example for this equation. The structural element has three pixels with values 0, 1 and 0, respectively. The subtractions for the position shown in Figure 3.37(a) are $4 - 0 = 4$, $6 - 1 = 5$ and $7 - 0 = 7$. Thus, the minimum value is 4. As shown in Figure 3.37(a), this corresponds to the highest grey-level value that we can move up to the structural element and it is still fully contained in the umbra of the function.

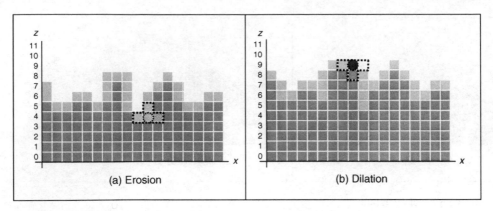

Figure 3.37 Example of Minkowski difference and addition

Similar to Equation 3.56, the dilation can be obtained by comparing the grey-level values of the image and the structural element. For the dilation we have

$$\oplus (x) = \max_i \{ f(x-i) + B(i) \} \tag{3.57}$$

Figure 3.37(b) illustrates a numerical example of this equation. For the position of the structural element in Figure 3.37(b), the summation gives the values $8 + 0 = 8$, $8 + 1 = 9$ and $4 + 0 = 4$. As shown in the figure, the maximum value of 9 corresponds to the point where the structural element still intersects the umbra; therefore, this point should be added to the dilation.

Equations 3.56 and 3.57 are known as the *Minkowski operators* and they formalize set operations as summations and differences. Thus, they provide definitions very useful for computer implementations. Code 3.15 shows the implement of the erosion operator based on Equation 3.56. Similar to Code 3.5, the value pixels in the output image are obtained by translating the operator

```
function eroded = Erosion(image,template)
%Implementation of erosion operator
%Parameters: Template and image array of points

%get the image and template dimensions
[irows,icols]=size(image);
[trows,tcols]=size(template);

%create result image
eroded(1:irows,1:icols)=uint8(0);

%half of template
trhalf=floor(trows/2);
tchalf=floor(tcols/2);

%Erosion
for x=trhalf+1:icols-trhalf %columns in the image except border
    for y=tchalf+1:irows-tchalf %rows in the image except border
        min=256;
        for iwin=1:tcols      %template columns
            for jwin=1:trows      %template rows
                xi=x-trhalf-1+iwin;
                yi=y-tchalf-1+jwin;
```

```
            sub=double(image(xi,yi))-double(template(iwin,jwin));
            if sub<min & sub>0
                min=sub;
            end
        end
    end
        eroded(x,y)=uint8(min);
    end
end
```

Code 3.15 Erosion implementation

along the image pixels. The code subtracts the value of corresponding image and template pixels
and it sets the value of the pixel in the output image to the minima.

Code 3.16 shows the implement of the dilation operator based on Equation 3.57. This code
is similar to Code 3.15, but corresponding values of the image and the structural element are
added, and the maximum value is set as the result of the dilation.

```
function dilated = Dilation(image,template)
%Implementation of dilation operator
%Parameters: Template and image array of points

%get the image and template dimensions
[irows,icols]=size(image);
[trows,tcols]=size(template);

%create result image
dilated(1:irows,1:icols)=uint8(0);

%half of template
trhalf=floor(trows/2);
tchalf=floor(tcols/2);

%Dilation
for x=trhalf+1:icols-trhalf %columns in the image except border
    for y=tchalf+1:irows-tchalf %rows in the image except border
        max=0;
        for iwin=1:tcols     %template columns
            for jwin=1:trows     %template rows
                xi=x-trhalf-1+iwin;
                yi=y-tchalf-1+jwin;
                sub=double(image(xi,yi))+double(template(iwin,jwin));
                if sub>max & sub>0
                    max=sub;
                end
            end
        end
        dilated(x,y)=uint8(max);
    end
end
```

Code 3.16 Dilation implementation

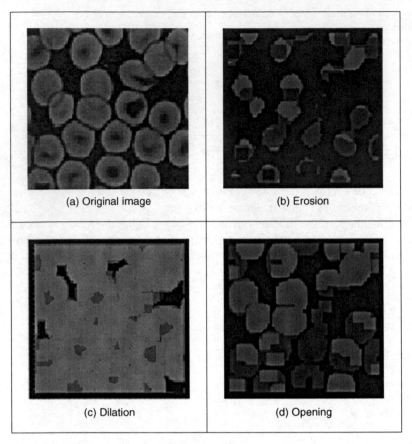

(a) Original image   (b) Erosion

(c) Dilation   (d) Opening

Figure 3.38   Examples of morphology operators

Figure 3.38 shows an example of the results obtained from the erosion and dilation using Codes 3.15 and 3.16. The original image shown in Figure 3.38(a) has 128×128 pixels and we used a flat structural element defined by an image with 9×9 pixels set to zero. For its simplicity, flat structural elements are very common in applications and they are generally set to zero to avoid creating offsets in the grey levels. In Figure 3.38, we can see that the erosion operation reduces the objects in the image while dilation expands white regions. We also used the erosion and dilation in succession to perform the opening show in Figure 3.38(d). The opening operation has a tendency to form regular regions of similar size to the original image while removing peaks and small regions. The strength of the operators is defined by the size of the structural elements. In these examples we use a fixed size and we can see that it strongly modifies regions during dilation and erosion. Elaborate techniques have combined multiresolution structures and morphological operators to analyse an image with operators of different sizes (Montiel et al., 1995).

3.7   Further reading

Many texts cover basic point and group operators in much detail; in particular, some texts give many more examples, such as Russ (1995) and Seul (2000). Books with a C implementation often concentrate on more basic techniques, including low-level image processing (Lindley, 1991; Parker, 1994). Some of the more advanced texts include more coverage of low-level operators (Rosenfeld and Kak, 1982; Castleman,1996). Parker (1994) includes C code for nearly all the low-level operations in this chapter and Seul (2000) has code too, and there is Matlab code in Gonzalez (2003). For study of the effect of the median operator on image data, see Bovik et al. (1987). Some of the newer techniques receive little treatment in the established literature, except for Chan and Shen (2005; with extensive coverage of noise filtering too). The truncated median filter is covered again in Davies (2005). For further study of the effects of different statistical operators on ultrasound images, see Evans and Nixon (1995, 1996). The concept of scale-space allows for considerably more refined analysis than is given here and while we shall revisit it later, it is rather unsuited to an introductory text. It was originally introduced by Witkin (1983) and further developed by others, including Koenderink (1984) who also considers the heat equation. There is even a series of conferences devoted to scale-space and *morphology*.

3.8 References

Bamber, J. C. and Daft, C., Adaptive Filtering for Reduction of Speckle in Ultrasonic Pulse-Echo Images, *Ultrasonics*, **24**(3), pp. 41–44, 1986

Black, M. J., Sapiro, G., Marimont, D. H. and Meeger, D., Robust Anisotropic Diffusion, *IEEE Trans. Image Process.*, **7**(3), pp. 421–432, 1998

Bovik, A. C., Huang, T. S. and Munson, D. C., The Effect of Median Filtering on Edge Estimation and Detection, *IEEE Trans. PAMI*, **9**(2), pp. 181–194, 1987

Campbell, J. D., *Edge Structure and the Representation of Pictures*, PhD Thesis, University of Missouri, Columbia, MO, 1969

Castleman, K. R., *Digital Image Processing*, Prentice Hall, Englewood Cliffs, NJ, 1996

Chan, T. and Shen, J., *Image Processing and Analysis: Variational, PDE, Wavelet, and Stochastic Methods*, Society for Industrial and Applied Mathematics, Philadelphia, PA, USA, 2005

Davies, E. R., On the Noise Suppression Characteristics of the Median, Truncated Median and Mode Filters, *Pattern Recog. Lett.*, **7**(2), pp. 87–97, 1988

Davies, E. R., *Machine Vision: Theory, Algorithms and Practicalities*, 3rd edn, Morgan Kaufmann (Elsevier), Amsterdam, Netherlands, 2005

Evans, A. N. and Nixon M. S., Mode Filtering to Reduce Ultrasound Speckle for Feature Extraction, *Proc. IEE – Vision, Image and Signal Processing*, **142**(2), pp. 87–94, 1995

Evans, A. N. and Nixon M. S., Biased Motion-Adaptive Temporal Filtering for Speckle Reduction in Echocardiography, *IEEE Trans. Med. Imaging*, **15**(1), pp. 39–50, 1996

Fischl, B. and Schwartz, E. L., Adaptive Nonlocal Filtering: A Fast Alternative to Anisotropic Diffusion for Image Enhancement, *IEEE Trans. PAMI*, **21**(1), pp. 42–48, 1999

Glasbey, C. A., An Analysis of Histogram-Based Thresholding Algorithms, *CVGIP–Graphical Models Image Process.*, **55**(6), pp. 532–537, 1993

Gonzalez, R. C. and Wintz P., *Digital Image Processing*, 2nd edn, Addison Wesley, Reading, MA, 1987

Gonzalez, R. C., Woods, R. E. and Eddins, S., *Digital Image Processing using MATLAB*, 1st edn, Prentice Hall, 2003

Hearn, D. and Baker, M. P., *Computer Graphics C Version*, 2nd edn, Prentice Hall, Upper Saddle River, NJ, 1997

Hodgson, R. M., Bailey, D. G., Naylor, M. J., Ng, A. and McNeill, S. J., Properties, Implementations and Applications of Rank Filters, *Image Vision Comput.*, **3**(1), pp. 3–14, 1985

Huang, T., Yang, G. and Tang, G., A Fast Two-Dimensional Median Filtering Algorithm, *IEEE Trans. ASSP*, **27**(1), pp. 13–18, 1979

Hurley, D. J., Nixon, M. S. and Carter, J. N., Force Field Energy Functionals for Image Feature Extraction, *Image Vision Comput.*, **20**, pp. 311–317, 2002

Hurley, D. J., Nixon, M. S. and Carter, J. N., Force Field Feature Extraction for Ear Biometrics, *Comput. Vision Image Understand.*, **98**(3), pp. 491–512, 2005

Koenderink, J., The Structure of Images, *Biol. Cybern.*, **50**, pp. 363–370, 1984

Lee, S. A., Chung, S. Y. and Park, R. H., A Comparative Performance Study of Several Global Thresholding Techniques for Segmentation, *CVGIP*, **52**, pp. 171–190, 1990

Lindley, C. A., *Practical Image Processing in C*, Wiley & Sons, New York, 1991

Loupas, T. and McDicken, W. N., Noise Reduction in Ultrasound Images by Digital Filtering, *Br. J. Radiol.*, **60**, pp. 389–392, 1987

Montiel, M. E., Aguado, A. S., Garza, M. and Alarcón, J., Image Manipulation using M-filters in a Pyramidal Computer Model, *IEEE Trans. PAMI*, **17**(11), pp. 1110–1115, 1995

Otsu, N., A Threshold Selection Method from Gray-Level Histograms, *IEEE Trans. SMC*, **9**(1), pp. 62–66, 1979

Parker, J. R., *Practical Computer Vision using C*, Wiley & Sons, New York, 1994

Perona, P. and Malik, J., Scale-Space and Edge Detection using Anisotropic Diffusion, *IEEE Trans. PAMI*, **17**(7), pp. 629–639, 1990

Rosenfeld, A. and Kak, A. C., *Digital Picture Processing*, 2nd edn, Vols 1 and 2, Academic Press, Orlando, FL, 1982

Rosin, P. L., Unimodal Thresholding, *Pattern Recog.*, **34**(11), pp. 2083–2096, 2001

Russ, J. C., *The Image Processing Handbook*, 4th edn, CRC Press (IEEE Press), Boca Raton, FL, 1995

Sahoo, P. K., Soltani, S., Wong, A. K. C. and Chen, Y. C., Survey of Thresholding Techniques, *CVGIP*, **41**(2), pp. 233–260, 1988

Serra J., Introduction to Mathematical Morphology, *CVGIP*, **35**, pp. 283–305, 1986

Serra, J. P. and Soille, P. (eds), *Mathematical Morphology and its Applications to Image Processing*, Kluwer Academic, Springer, NY, USA, 1994

Seul, M., O'Gorman, L. and Sammon, M. J., *Practical Algorithms for Image Analysis: Descriptions, Examples, and Code*, Cambridge University Press, Cambridge, 2000

Shankar, P. M., Speckle Reduction in Ultrasound B Scans using Weighted Averaging in Spatial Compounding, *IEEE Trans. Ultrasonics Ferroelectrics Frequency Control*, **33**(6), pp. 754–758, 1986

Sternberg, S. R., Gray Scale Morphology, *CVGIP*, **35**, pp. 333–355, 1986

Trier, O. D. and Jain, A. K., Goal-Directed Evaluation of Image Binarization Methods, *IEEE Trans. PAMI*, **17**(12), pp. 1191–1201, 1995

Witkin, A., Scale-Space Filtering: A New Approach to Multi-Scale Description, *Proc. Int. Joint Conf. Artificial Intelligence*, pp. 1019–1021, 1983

4

Low-level feature extraction (including edge detection)

4.1 Overview

We shall define *low-level features* to be those basic features that can be extracted automatically from an image without any shape information (information about *spatial* relationships). As such, thresholding is a form of low-level feature extraction performed as a point operation. All of these approaches can be used in high-level feature extraction, where we find shapes in images. It is well known that we can recognize people from caricaturists' portraits. That is the first low-level feature we shall encounter. It is called *edge detection* and it aims to produce a *line drawing*, like one of a face in Figure 4.1(a) and (d), something akin to a caricaturist's sketch, although without the exaggeration a caricaturist would imbue. There are very basic techniques and more advanced ones and we shall look at some of the most popular approaches. The first order detectors are equivalent to first order differentiation, and the second order edge detection operators are equivalent to a one-higher level of differentiation. An alternative form of edge detection is called phase congruency and we shall again see the frequency domain used to aid analysis, this time for low-level feature extraction.

We shall also consider corner detection, which can be thought of as detecting those points where lines bend very sharply with high curvature, as for the aeroplane in Figure 4.1(b) and (e). These are another low-level feature that again can be extracted automatically from the image. These are largely techniques for *localized feature extraction*, in this case the curvature, and the more modern approaches extend to the detection of localized regions or *patches* of interest. Finally, we shall investigate a technique that describes *motion*, called optical flow. This is illustrated in Figure 4.1(c) and (f) with the optical flow from images of a walking man: the bits that are moving fastest are the brightest points, like the hands and the feet. All of these can provide a set of points, albeit points with different properties, but all are suitable for grouping for shape extraction. Consider a square box moving through a sequence of images. The edges are the perimeter of the box; the corners are the apices; the flow is how the box moves. All these can be collected together to find the moving box. We shall start with the edge detection techniques, with the first order operators, which accords with the chronology of development. The first order techniques date back more than 30 years.

115

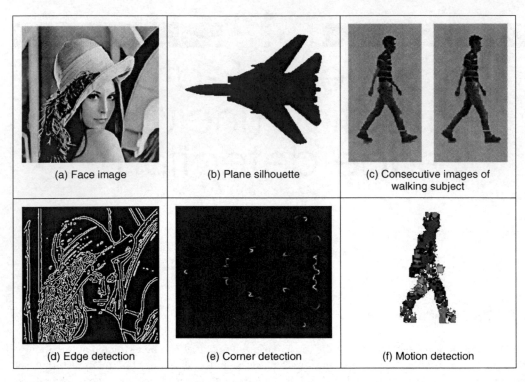

(a) Face image (b) Plane silhouette (c) Consecutive images of walking subject

(d) Edge detection (e) Corner detection (f) Motion detection

Figure 4.1 Low-level feature detection

Table 4.1 Overview of Chapter 4

Main topic	Sub topics	Main points
First order edge detection	What is an edge and how we detect it. The equivalence of operators to first order differentiation and the insight this brings. The need for filtering and more sophisticated first order operators.	Difference operation; Roberts Cross, smoothing, Prewitt, Sobel, Canny. Basis of the operators and frequency domain analysis.
Second order edge detection	Relationship between first and second order differencing operations. The basis of a second order operator. The need to include filtering and better operations.	Second order differencing; Laplacian, zero-crossing detection; Marr–Hildreth, Laplacian of Gaussian, difference of Gaussian. Scale space.
Other edge operators	Alternative approaches and performance aspects. Comparing different operators.	Other noise models; Spacek. Other edge models; Petrou.
Phase congruency	Inverse Fourier transform; phase for feature extraction. Alternative form of edge and feature detection	Frequency domain analysis; detecting a range of features; photometric invariance, wavelets.

(Continued)

Table 4.1 (Continued)

Main topic	Sub topics	Main points
Localized feature extraction	Finding localized low-level features; extension from curvature to patches. Nature of curvature and computation from: edge information; by change in intensity; and by correlation. Motivation of patch detection and principles of modern approaches.	Planar curvature; corners. Curvature estimation by: change in edge direction; intensity change; Harris corner detector. Modern feature detectors: SIFT operator and saliency.
Optical flow Estimation	Movement and the nature of optical flow. Estimating the optical flow by differential approach. Need for other approaches (including matching regions).	Detection by differencing. Optical flow; aperture problem; smoothness constraint. Differential approach; Horn and Schunk method; correlation.

4.2 First order edge detection operators

4.2.1 Basic operators

Many approaches to image interpretation are based on edges, since analysis based on edge detection is insensitive to change in the overall illumination level. Edge detection highlights image *contrast*. Detecting contrast, which is difference in intensity, can emphasize the boundaries of features within an image, since this is where image contrast occurs. This is how human vision can perceive the perimeter of an object, since the object is of different intensity to its surroundings. Essentially, the boundary of an object is a step change in the intensity levels. The edge is at the position of the step change. To detect the edge position we can use *first order* differentiation, since this emphasizes change; first order differentiation gives no response when applied to signals that do not change. The first edge detection operators to be studied here are group operators which aim to deliver an output that approximates the result of first order differentiation.

A change in intensity can be revealed by differencing adjacent points. Differencing horizontally adjacent points will detect *vertical* changes in intensity and is often called a *horizontal edge-detector* by virtue of its action. A horizontal operator will not show up *horizontal* changes in intensity since the difference is zero. (This is the form of edge detection used within the anisotropic diffusion smoothing operator in the previous chapter.) When applied to an image **P** the action of the horizontal edge-detector forms the difference between two horizontally adjacent points, as such detecting the vertical edges, **Ex**, as:

$$\mathbf{Ex}_{x,y} = \left| \mathbf{P}_{x,y} - \mathbf{P}_{x+1,y} \right| \qquad \forall x \in 1, N-1; \ y \in 1, N \tag{4.1}$$

To detect horizontal edges we need a *vertical edge-detector* which differences vertically adjacent points. This will determine *horizontal* intensity changes, but not *vertical* ones, so the vertical edge-detector detects the *horizontal* edges, **Ey**, according to:

$$\mathbf{Ey}_{x,y} = \left| \mathbf{P}_{x,y} - \mathbf{P}_{x,y+1} \right| \qquad \forall x \in 1, N; \ y \in 1, N-1 \tag{4.2}$$

Figure 4.2(b) and (c) show the application of the vertical and horizontal operators to the synthesized image of the square in Figure 4.2(a). The left-hand vertical edge in Figure 4.2(b)

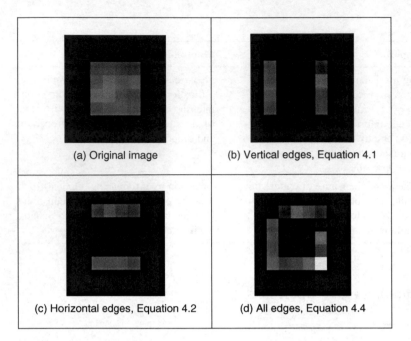

(a) Original image	(b) Vertical edges, Equation 4.1
(c) Horizontal edges, Equation 4.2	(d) All edges, Equation 4.4

Figure 4.2 First order edge detection

appears to be beside the square by virtue of the forward differencing process. Likewise, the upper edge in Figure 4.2(c) appears above the original square.

Combining the two gives an operator **E** that can detect vertical and horizontal edges *together*. That is,

$$\mathbf{E}_{x,y} = \left| \mathbf{P}_{x,y} - \mathbf{P}_{x+1,y} + \mathbf{P}_{x,y} - \mathbf{P}_{x,y+1} \right| \qquad \forall x, y \in 1, N-1 \tag{4.3}$$

which gives:

$$\mathbf{E}_{x,y} = \left| 2 \times \mathbf{P}_{x,y} - \mathbf{P}_{x+1,y} - \mathbf{P}_{x,y+1} \right| \qquad \forall x, y \in 1, N-1 \tag{4.4}$$

Equation 4.4 gives the coefficients of a differencing template which can be convolved with an image to detect all the edge points, such as those shown in Figure 4.2(d). As in the previous chapter, the current point of operation (the position of the point we are computing a new value for) is shaded. The template shows only the weighting coefficients and not the modulus operation. Note that the bright point in the lower right corner of the edges of the square in Figure 4.2(d) is much brighter than the other points. This is because it is the only point to be detected as an edge by both the vertical and the horizontal operators and is therefore much brighter than the other edge points. In contrast, the top left-hand corner point is detected by neither operator and so does not appear in the final image.

The template in Figure 4.3 is convolved with the image to detect edges. The direct implementation of this operator, i.e. using Equation 4.4 rather than template convolution, is given in Code 4.1. Template convolution could be used, but it is unnecessarily complex in this case.

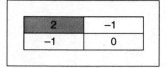

Figure 4.3 Template for first order difference

```
edge(pic):= newpic←zero(pic)
             for x∈0.. cols(pic)-2
               for y∈0.. rows(pic)-2
                 newpic_y,x← |2·pic_y,x-pic_y,x+1-pic_y+1,x|
             newpic
```

Code 4.1 First order edge detection

Uniform thresholding (Section 3.3.4) is often used to select the brightest points, following application of an edge detection operator. The threshold level controls the number of selected points; too high a level can select too few points, whereas too low a level can select too much noise. Often, the threshold level is chosen by experience or by experiment, but it can be determined automatically by considering edge data (Venkatesh and Rosin, 1995), or empirically (Haddon, 1988). For the moment, let us concentrate on the development of edge detection operators, rather than on their application.

4.2.2 Analysis of the basic operators

Taylor series analysis reveals that differencing adjacent points provides an estimate of the first order derivative at a point. If the difference is taken between points separated by Δx then by Taylor expansion for $f(x + \Delta x)$ we obtain:

$$f(x+\Delta x) = f(x) + \Delta x \times f'(x) + \frac{\Delta x^2}{2!} \times f''(x) + O(\Delta x^3) \qquad (4.5)$$

By rearrangement, the first order derivative $f'(x)$ is:

$$f'(x) = \frac{f(x+\Delta x) - f(x)}{\Delta x} - O(\Delta x) \qquad (4.6)$$

This shows that the difference between adjacent points is an estimate of the first order derivative, with error $O(\Delta x)$. This error depends on the size of the interval Δx and on the complexity of the curve. When Δx is large this error can be significant. The error is also large when the high-order derivatives take large values. In practice, the short sampling of image pixels and the reduced high-frequency content make this approximation adequate. However, the error can be reduced by spacing the differenced points by one pixel. This is equivalent to computing the first order difference delivered by Equation 4.1 at two adjacent points, as a new horizontal difference **Exx**, where

$$\mathbf{Exx}_{x,y} = \mathbf{Ex}_{x+1,y} + \mathbf{Ex}_{x,y} = \mathbf{P}_{x+1,y} - \mathbf{P}_{x,y} + \mathbf{P}_{x,y} - \mathbf{P}_{x-1,y} = \mathbf{P}_{x+1,y} - \mathbf{P}_{x-1,y} \qquad (4.7)$$

This is equivalent to incorporating spacing to detect the edges **Exx** by:

$$\mathbf{Exx}_{x,y} = \left| \mathbf{P}_{x+1,y} - \mathbf{P}_{x-1,y} \right| \qquad \forall x \in 2, N-1; y \in 1, N \tag{4.8}$$

To analyse this, again by Taylor series, we expand $f(x - \Delta x)$ as:

$$f(x - \Delta x) = f(x) - \Delta x \times f'(x) + \frac{\Delta x^2}{2!} \times f''(x) - O(\Delta x^3) \tag{4.9}$$

By differencing Equation 4.9 from Equation 4.5, we obtain the first order derivative as:

$$f'(x) = \frac{f(x + \Delta x) - f(x - \Delta x)}{2\Delta x} - O(\Delta x^2) \tag{4.10}$$

Equation 4.10 suggests that the estimate of the first order difference is now the difference between points separated by one pixel, with error $O(\Delta x^2)$. If $\Delta x < 1$, this error is clearly smaller than the error associated with differencing adjacent pixels, in Equation 4.6. Again, averaging has reduced noise, or error. The template for a horizontal edge detection operator is given in Figure 4.4(a). This template gives the vertical edges detected at its centre pixel. A transposed version of the template gives a vertical edge detection operator (Figure 4.4b).

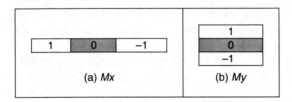

Figure 4.4 Templates for improved first order difference

The *Roberts cross operator* (Roberts, 1965) was one of the earliest edge detection operators. It implements a version of basic first order edge detection and uses two templates which difference pixel values in a diagonal manner, as opposed to along the axes' directions. The two templates are called M^+ and M^- and are given in Figure 4.5.

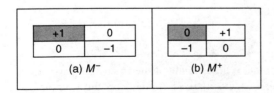

Figure 4.5 Templates for Roberts cross operator

In implementation, the maximum value delivered by application of these templates is stored as the value of the edge at that point. The edge point $\mathbf{E}_{x,y}$ is then the maximum of the two values derived by convolving the two templates at an image point $\mathbf{P}_{x,y}$:

$$\mathbf{E}_{x,y} = \max \left\{ \left| M^+ * \mathbf{P}_{x,y} \right|, \left| M^- * \mathbf{P}_{x,y} \right| \right\} \qquad \forall x, y \in 1, N-1 \tag{4.11}$$

The application of the Roberts cross operator to the image of the square is shown in Figure 4.6. The two templates provide the results in Figure 4.6(a) and (b) and the result delivered by the Roberts operator is shown in Figure 4.6(c). Note that the corners of the square now appear in the edge image, by virtue of the diagonal differencing action, whereas they were less apparent in Figure 4.2(d) (where the top left corner did not appear).

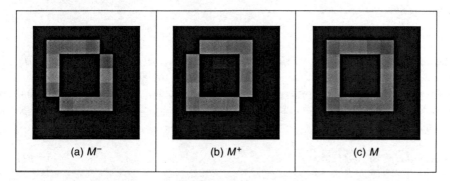

(a) M^- (b) M^+ (c) M

Figure 4.6   Applying the Roberts cross operator

An alternative to taking the maximum is simply to *add* the results of the two templates together to combine horizontal and vertical edges. There are of course more varieties of edges and it is often better to consider the two templates as providing components of an *edge vector*: the strength of the edge along the horizontal and vertical axes. These give components of a vector and can be added in a vectorial manner (which is perhaps more usual for the Roberts operator). The *edge magnitude* is the *length* of the vector, and the *edge direction* is the vector's *orientation*, as shown in Figure 4.7.

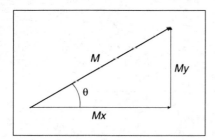

Figure 4.7   Edge detection in vectorial format

4.2.3   Prewitt edge detection operator

Edge detection is akin to differentiation. Since it detects change it is bound to respond to *noise*, as well as to step-like changes in image intensity (its frequency domain analogue is high-pass filtering, as illustrated in Figure 2.26c). It is therefore prudent to incorporate *averaging* within the edge detection process. We can then extend the vertical template, Mx, along three rows,

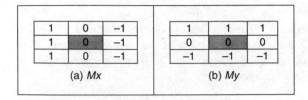

Figure 4.8 Templates for Prewitt operator

and the horizontal template, *My*, along three columns. These give the *Prewitt edge detection operator* (Prewitt and Mendelsohn, 1966), which consists of two templates (Figure 4.8).

This gives two results: the rate of change of brightness along each axis. As such, this is the vector illustrated in Figure 4.7: the edge magnitude, *M*, is the length of the vector and the edge direction, θ, is the angle of the vector:

$$M(x, y) = \sqrt{Mx(x, y)^2 + My(x, y)^2} \tag{4.12}$$

$$\theta(x, y) = \tan^{-1}\left(\frac{My(x, y)}{Mx(x, y)}\right) \tag{4.13}$$

Again, the signs of *Mx* and *My* can be used to determine the appropriate quadrant for the edge direction. A Mathcad implementation of the two templates of Figure 4.8 is given in Code 4.2. In this code, both templates operate on a 3×3 subpicture (which can be supplied, in Mathcad, using the submatrix function). Again, template convolution could be used to implement this operator, but (as with direct averaging and basic first order edge detection) it is less suited to simple templates. Also, the provision of edge magnitude and direction would require extension of the template convolution operator given earlier (Code 3.5).

$\text{Prewitt33\_x(pic)} := \sum_{y=0}^{2} pic_{y,0} - \sum_{y=0}^{2} pic_{y,2}$	$\text{Prewitt33\_y(pic)} := \sum_{x=0}^{2} pic_{0,x} - \sum_{x=0}^{2} pic_{2,x}$
(a) *Mx*	(b) *My*

Code 4.2 Implementing the Prewitt operator

When applied to the image of the square (Figure 4.9a), we obtain the edge magnitude and direction (Figure 4.9b and d, respectively, where part d does not include the border points, only the edge direction at processed points). The edge direction in Figure 4.9(d) is shown measured in degrees, where 0° and 360° are horizontal, to the right, and 90° is vertical, upwards. Although the regions of edge points are wider owing to the operator's averaging properties, the edge data is clearer than the earlier first order operator, highlighting the regions where intensity changed in a more reliable fashion (compare, for example, the upper left corner of the square which was not revealed earlier). The direction is less clear in an image format and is better exposed by Mathcad's *vector* format in Figure 4.9(c). In vector format, the edge direction data is clearly

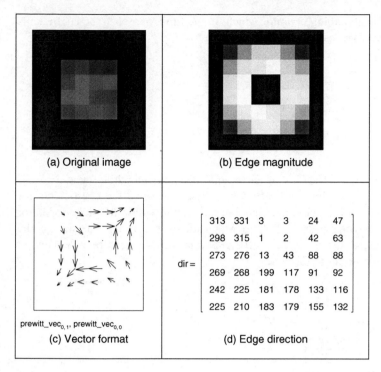

| | (a) Original image | | | | (b) Edge magnitude | | |

$$dir = \begin{bmatrix} 313 & 331 & 3 & 3 & 24 & 47 \\ 298 & 315 & 1 & 2 & 42 & 63 \\ 273 & 276 & 13 & 43 & 88 & 88 \\ 269 & 268 & 199 & 117 & 91 & 92 \\ 242 & 225 & 181 & 178 & 133 & 116 \\ 225 & 210 & 183 & 179 & 155 & 132 \end{bmatrix}$$

prewitt_vec$_{0,1}$, prewitt_vec$_{0,0}$

(c) Vector format
(d) Edge direction

Figure 4.9 Applying the Prewitt operator

less well defined at the corners of the square (as expected, since the first order derivative is discontinuous at these points).

4.2.4 Sobel edge detection operator

When the weight at the central pixels, for both Prewitt templates, is doubled, this gives the famous *Sobel edge detection operator* which, again, consists of two masks to determine the edge in vector form. The Sobel operator was the most popular edge detection operator until the development of edge detection techniques with a theoretical basis. It proved popular because it gave, overall, a better performance than other contemporaneous edge detection operators, such as the Prewitt operator. The templates for the Sobel operator can be found in Figure 4.10.

The Mathcad implementation of these masks is very similar to the implementation of the Prewitt operator (Code 4.2), again operating on a 3×3 subpicture. This is the standard formulation of the Sobel templates, but how do we form larger templates, say for 5×5 or 7×7? Few textbooks state its original derivation, but it has been attributed (Heath et al., 1997) as originating from a PhD thesis (Sobel, 1970). Unfortunately, a theoretical basis that can be used to calculate the coefficients of larger templates is rarely given. One approach to a theoretical basis is to consider the optimal forms of averaging and of differencing. Gaussian averaging has already been stated to give optimal averaging. The binomial expansion gives the integer coefficients of a series that, in the limit, approximates the normal distribution. Pascal's triangle gives sets of coefficients for a smoothing operator which, in the limit, approach the coefficients of a Gaussian smoothing operator. Pascal's triangle is then:

Window size

2				1		1	
3			1		2		1
4		1		3		3		1
5	1		4		6		4		1

This gives the (unnormalized) coefficients of an optimal discrete smoothing operator (it is essentially a Gaussian operator with integer coefficients). The rows give the coefficients for increasing template, or window, size. The coefficients of smoothing within the Sobel operator (Figure 4.10) are those for a window size of 3. In Mathcad, by specifying the size of the smoothing window as winsize, the template coefficients $smooth_{x\_win}$ can be calculated at each window point x_win according to Code 4.3.

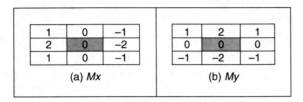

1	0	−1
2	0	−2
1	0	−1

(a) *Mx*

1	2	1
0	0	0
−1	−2	−1

(b) *My*

Figure 4.10 Templates for Sobel operator

$$smooth_{x\_win} := \frac{(winsize-1)!}{(winsize-1-x\_win)! \cdot x\_win!}$$

Code 4.3 Smoothing function

The differencing coefficients are given by Pascal's triangle for subtraction:

Window size

2				1		−1	
3			1		0		−1
4		1		1		−1		−1
5	1		2		0		−2		−1

This can be implemented by subtracting the templates derived from two adjacent expansions for a smaller window size. Accordingly, we require an operator that can provide the coefficients of Pascal's triangle for arguments which are a window size n and a position k. The operator is the Pascal(k,n) operator in Code 4.4.

$$\text{Pascal}(k,n) := \left| \begin{array}{l} \dfrac{n!}{(n-k)! \cdot k!} \quad \text{if} \, (k \geq 0) \cdot (k \leq n) \\[2ex] 0 \quad \text{otherwise} \end{array} \right.$$

Code 4.4 Pascal's triangle

The differencing template, diff_{x_win}, is then given by the difference between two Pascal expansions, as given in Code 4.5.

$$\text{diff}_{x\_win} := \text{Pascal}(x\_win, winsize-2) - \text{Pascal}(x\_win-1, winsize-2)$$

Code 4.5 Differencing function

These give the coefficients of optimal differencing and optimal smoothing. This *general* form of the Sobel operator combines optimal smoothing along one axis, with optimal differencing along the other. This general form of the Sobel operator is then given in Code 4.6, which combines the differencing function along one axis, with smoothing along the other.

$$\text{Sobel\_x}(pic) := \sum_{x\_win=0}^{winsize-1} \sum_{y\_win=0}^{winsize-1} \text{smooth}_{y\_win} \cdot \text{diff}_{x\_win} \cdot \text{pic}_{y\_win, x\_win}$$

(a) Mx

$$\text{Sobel\_y}(pic) := \sum_{x\_win=0}^{winsize-1} \sum_{y\_win=0}^{winsize-1} \text{smooth}_{x\_win} \cdot \text{diff}_{y\_win} \cdot \text{pic}_{y\_win, x\_win}$$

(b) My

Code 4.6 Generalized Sobel templates

This generates a template for the Mx template for a Sobel operator, given for 5×5 in Code 4.7.

$$\text{Sobel\_template\_x} = \begin{bmatrix} 1 & 2 & 0 & -2 & -1 \\ 4 & 8 & 0 & -8 & -4 \\ 6 & 12 & 0 & -12 & -6 \\ 4 & 8 & 0 & -8 & -4 \\ 1 & 2 & 0 & -2 & -1 \end{bmatrix}$$

Code 4.7 5×5 Sobel template Mx

All template-based techniques can be larger than 5×5 so, as with any group operator, there is a 7×7 Sobel, and so on. The virtue of a larger edge detection template is that it involves more smoothing to reduce noise, but edge blurring becomes a great problem. The estimate of edge direction can be improved with more smoothing since it is particularly sensitive to noise. There are circular edge operators designed specifically to provide accurate edge direction data.

The Sobel templates can be invoked by operating on a matrix of dimension equal to the window size, from which edge magnitude and gradient are calculated. The Sobel function (Code 4.8) convolves the generalized Sobel template (of size chosen to be winsize) with the picture supplied as argument, to give outputs which are the images of edge magnitude and direction, in vector form.

```
Sobel(pic,winsize):=
        │ w2←floor (winsize/2)
        │
        │ edge_mag←zero(pic)
        │ edge_dir←zero(pic)
        │ for x∈w2.. cols(pic)-1-w2
        │    for y∈w2.. rows(pic)-1-w2
        │       │ x_mag←Sobel_x(submatrix(pic,y-w2,y+w2,x-w2,x+w2))
        │       │ y_mag←Sobel_y(submatrix(pic,y-w2,y+w2,x-w2,x+w2))
        │       │ edge_mag_{y,x}←floor (magnitude(x_mag,y_mag)/mag_normalise)
        │       │ edge_dir_{y,x}←direction(x_mag,y_mag)
        │ (edge_mag edge_dir)
```

Code 4.8   Generalized Sobel operator

The results of applying the 3×3 Sobel operator can be seen in Figure 4.11. The original face image (Figure 4.11a) has many edges in the hair and in the region of the eyes. This is shown in the edge magnitude image (Figure 4.11b). When this is thresholded at a suitable value, many edge points are found (Figure 4.11c). Note that in areas of the image where the brightness

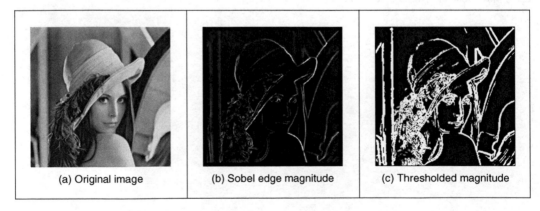

(a) Original image   (b) Sobel edge magnitude   (c) Thresholded magnitude

Figure 4.11   Applying the Sobel operator

remains fairly constant, such as the cheek and shoulder, there is little change, which is reflected by low edge magnitude and few points in the thresholded data.

The Sobel edge direction data can be arranged to point in different ways, as can the direction provided by the Prewitt operator. If the templates are inverted to be of the form shown in Figure 4.12, the edge direction will be inverted around both axes. If only one of the templates is inverted, the measured edge direction will be inverted around the chosen axis.

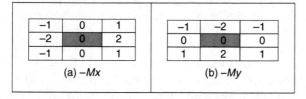

Figure 4.12 Inverted templates for Sobel operator

This gives *four* possible directions for measurement of the edge direction provided by the Sobel operator, two of which (for the templates of Figures 4.10 and 4.12) are illustrated in Figure 4.13(a) and (b), respectively, where inverting the *Mx* template does not highlight

Figure 4.13 Alternative arrangements of edge direction

discontinuity at the corners. (The edge magnitude of the Sobel applied to the square is not shown, but is similar to that derived by application of the Prewitt operator; Figure 4.9b). By swapping the Sobel templates, the measured edge direction can be arranged to be normal to the edge itself (as opposed to tangential data along the edge). This is illustrated in Figure 4.13(c) and (d) for swapped versions of the templates given in Figures 4.10 and 4.12, respectively. The rearrangement can lead to simplicity in algorithm construction when finding shapes, as to be shown later. Any algorithm that uses edge direction for finding shapes must know precisely which arrangement has been used, since the edge direction can be used to speed algorithm performance, but it must map precisely to the expected image data if used in that way.

Detecting edges by *template convolution* again has a frequency domain interpretation. The magnitude of the Fourier transform of a 5×5 Sobel template of Code 4.7 is given in Figure 4.14. The Fourier transform is given in relief in Figure 4.14(a) and as a contour plot in Figure 4.14(b). The template is for horizontal differencing action, My, which highlights vertical change. Accordingly, its transform reveals that it selects vertical spatial frequencies, while smoothing the horizontal ones. The horizontal frequencies are selected from a region near the origin (*low-pass* filtering), whereas the vertical frequencies are selected away from the origin (*high-pass*). This highlights the action of the Sobel operator; combining smoothing of the spatial frequencies along one axis with differencing of the other. In Figure 4.14, the smoothing is of horizontal spatial frequencies, while the differencing is of vertical spatial frequencies.

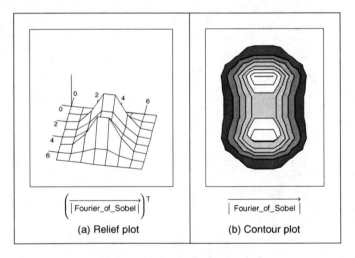

$$\left(\left|\overrightarrow{Fourier\_of\_Sobel}\right|\right)^{T}$$

(a) Relief plot

$$\left|\overrightarrow{Fourier\_of\_Sobel}\right|$$

(b) Contour plot

Figure 4.14   Fourier transform of the Sobel operator

An alternative frequency domain analysis of the Sobel can be derived via the z-transform operator. This is more the domain of signal processing courses in electronic and electrical engineering, and is included here for completeness and for linkage with signal processing. Essentially, z^{-1} is a unit time-step delay operator, so z can be thought of a unit (time-step) advance, so $f(t - \tau) = z^{-1} f(t)$ and $f(t + \tau) = z f(t)$, where τ is the sampling interval. Given that we have two spatial axes x and y, we can express the Sobel operator of Figure 4.12(a) using delay and advance via the z-transform notation along the two axes as

$$S(x, y) = \begin{matrix} -z_x^{-1} z_y^{-1} & +0+ & z_x z_y^{-1} \\ -2z_x^{-1} & +0+ & 2z_x \\ -z_x^{-1} z_y & +0+ & z_x z_y \end{matrix} \tag{4.14}$$

including zeros for the null template elements. Given that there is a standard substitution (by conformal mapping, evaluated along the frequency axis) $z^{-1} = e^{-j\omega t}$ to transform from the time (z) domain to the frequency domain (ω), we have

$$
\begin{aligned}
Sobel\left(\omega_x, \omega_y\right) &= -e^{-j\omega_x t}e^{-j\omega_y t} + e^{j\omega_x t}e^{-j\omega_y t} - 2e^{-j\omega_x t} + 2e^{j\omega_x t} - e^{-j\omega_x t}e^{j\omega_y t} + e^{j\omega_x t}e^{j\omega_y t} \\
&= \left(e^{-j\omega_y t} + 2 + e^{j\omega_y t}\right)\left(-e^{-j\omega_x t} + e^{j\omega_x t}\right) \\
&= \left(e^{-\frac{j\omega_y t}{2}} + e^{\frac{j\omega_y t}{2}}\right)^2\left(-e^{-j\omega_x t} + e^{j\omega_x t}\right) \qquad (4.15) \\
&= 8j\cos^2\left(\tfrac{\omega_y t}{2}\right)\sin\left(\omega_x t\right)
\end{aligned}
$$

where the transform *Sobel* is a function of spatial frequency, ω_x, ω_y, along the x and the y axes. This conforms rather well to the separation between smoothing along one axis (the first part of Equation 4.15) and differencing along the other; here by differencing (high-pass) along the x-axis and averaging (low-pass) along the y-axis. This provides an analytic form of the function shown in Figure 4.14; the relationship between the discrete Fourier transform (DFT), and this approach is evident by applying the DFT relationship (Equation 2.15) to the components of the Sobel operator.

4.2.5 Canny edge detection operator

The *Canny edge detection operator* (Canny, 1986) is perhaps the most popular edge detection technique at present. It was formulated with three main objectives:

- *optimal* detection with no spurious responses
- *good* localization with minimal distance between detected and true edge position
- *single* response to eliminate multiple responses to a single edge.

The first requirement aims to *reduce* the response to noise. This can be effected by optimal smoothing; Canny was the first to demonstrate that Gaussian filtering is optimal for edge detection (within his criteria). The second criterion aims for accuracy: edges are to be detected, in the right place. This can be achieved by a process of *non-maximum suppression* (which is equivalent to peak detection). Non-maximum suppression retains only those points at the top of a ridge of edge data, while suppressing all others. This results in thinning: the output of non-maximum suppression is thin lines of edge points, in the right place. The third constraint concerns location of a single edge point in response to a change in brightness. This is because more than one edge can be denoted to be present, consistent with the output obtained by earlier edge operators.

Canny showed that the Gaussian operator was optimal for image smoothing. Recalling that the Gaussian operator $g(x, y, \sigma)$ is given by:

$$
g\left(x, y, \sigma\right) = e^{\frac{-\left(x^2 + y^2\right)}{2\sigma^2}} \qquad (4.16)
$$

by differentiation, for unit vectors $U_x = [1, 0]$ and $U_y = [0, 1]$ along the coordinate axes, we obtain:

$$\nabla g(x, y) = \frac{\partial g(x, y, \sigma)}{\partial x} U_x + \frac{\partial g(x, y, \sigma)}{\partial y} U_y$$

$$= -\frac{x}{\sigma^2} e^{\frac{-(x^2 + y^2)}{2\sigma^2}} U_x - \frac{y}{\sigma^2} e^{\frac{-(x^2 + y^2)}{2\sigma^2}} U_y$$

(4.17)

Equation 4.17 gives a way to calculate the coefficients of a *derivative of Gaussian* template that combines first order differentiation with Gaussian smoothing. This is a smoothed image, and so the edge will be a ridge of data. To mark an edge at the correct point (and to reduce multiple response), we can convolve an image with an operator which gives the first derivative in a direction normal to the edge. The maximum of this function should be the peak of the edge data, where the gradient in the original image is sharpest, and hence the location of the edge. Accordingly, we seek an operator, G_n, which is a first derivative of a Gaussian function g in the direction of the normal, $\mathbf{n}_\perp$:

$$G_n = \frac{\partial g}{\partial \mathbf{n}_\perp}$$

(4.18)

where $\mathbf{n}_\perp$ can be estimated from the first order derivative of the Gaussian function g convolved with the image $\mathbf{P}$, and scaled appropriately as:

$$\mathbf{n}_\perp = \frac{\nabla(\mathbf{P} * g)}{|\nabla(\mathbf{P} * g)|}$$

(4.19)

The location of the true edge point is then at the maximum point of G_n convolved with the image. This maximum is when the differential (along $\mathbf{n}_\perp$) is zero:

$$\frac{\partial(G_n * \mathbf{P})}{\partial \mathbf{n}_\perp} = 0$$

(4.20)

By substitution of Equation 4.18 in Equation 4.20,

$$\frac{\partial^2(G * \mathbf{P})}{\partial \mathbf{n}_\perp^2} = 0$$

(4.21)

Equation 4.21 provides the basis for an operator which meets one of Canny's criteria, namely that edges should be detected in the correct place. This is non-maximum suppression, which is equivalent to retaining peaks (and thus equivalent to differentiation perpendicular to the edge), which thins the response of the edge detection operator to give edge points that are in the right place, without multiple response and with minimal response to noise. However, it is virtually impossible to achieve an exact implementation of Canny given the requirement to estimate the normal direction.

A common approximation is, as illustrated in Figure 4.15:

1. Use Gaussian smoothing (as in Section 3.4.4) (Figure 4.15a).
2. Use the Sobel operator (Figure 4.15b).
3. Use non-maximal suppression (Figure 4.15c).
4. Threshold with hysteresis to connect edge points (Figure 4.15d).

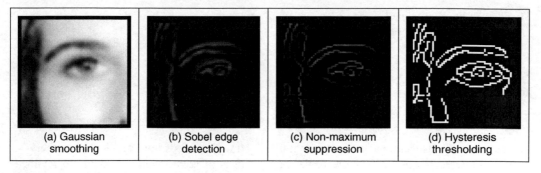

| (a) Gaussian smoothing | (b) Sobel edge detection | (c) Non-maximum suppression | (d) Hysteresis thresholding |

Figure 4.15 Stages in Canny edge detection

Note that the first two stages can be combined using a version of Equation 4.17, but are separated here so that all stages in the edge detection process can be shown clearly. An alternative implementation of Canny's approach (Deriche, 1987) used Canny's criteria to develop two-dimensional (2D) recursive filters, claiming performance and implementation advantage over the approximation here.

Non-maximum suppression essentially locates the highest points in the edge magnitude data. This is performed by using edge direction information, to check that points are at the peak of a ridge. Given a 3×3 region, a point is at a maximum if the gradient at either side of it is less than the gradient at the point. This implies that we need values of gradient along a line that is normal to the edge at a point. This is illustrated in Figure 4.16, which shows the neighbouring points to the point of interest, $\mathbf{P}_{x,y}$, the edge direction at $\mathbf{P}_{x,y}$ and the normal to the edge direction at $\mathbf{P}_{x,y}$. The point $\mathbf{P}_{x,y}$ is to be marked as maximum if its gradient, $M(x, y)$, exceeds the gradient at points 1 and 2, M_1 and M_2, respectively. Since we have a discrete neighbourhood, M_1 and

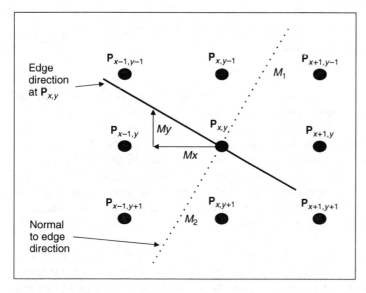

Figure 4.16 Interpolation in non-maximum suppression

Low-level feature extraction (including edge detection) 131

M_2 need to be interpolated, First order interpolation using Mx and My at $\mathbf{P}_{x,y}$, and the values of Mx and My for the neighbours gives:

$$M_1 = \frac{My}{Mx}M(x+1, \ y-1) + \frac{Mx - My}{Mx}M(x, \ y-1) \tag{4.22}$$

and

$$M_2 = \frac{My}{Mx}M(x-1, \ y+1) + \frac{Mx - My}{Mx}M(x, \ y+1) \tag{4.23}$$

The point $\mathbf{P}_{x,y}$ is then marked as a maximum if $M(x, y)$ exceeds both M_1 and M_2, otherwise it is set to zero. In this manner the peaks of the ridges of edge magnitude data are retained, while those not at the peak are set to zero. The implementation of non-maximum suppression first requires a function that generates the coordinates of the points between which the edge magnitude is interpolated. This is the function get_coords in Code 4.9, which requires the angle of the normal to the edge direction, returning the coordinates of the points beyond and behind the normal.

```
get_coords(angle):=  δ←0.000000000000001

                     x1←ceil[(cos(angle+ π/8) · √2)-0.5-δ]

                     y1←ceil[(-sin(angle- π/8) · √2)-0.5-δ]

                     x2←ceil[(cos(angle- π/8) · √2)-0.5-δ]

                     y2←ceil[(-sin(angle- π/8) · √2)-0.5-δ]

                     (x1 y1 x2 y2)
```

Code 4.9 Generating coordinates for interpolation

The non-maximum suppression operator, non_max in Code 4.10, then interpolates the edge magnitude at the two points either side of the normal to the edge direction. If the edge magnitude at the point of interest exceeds these two then it is retained, otherwise it is discarded. Note that the potential singularity in Equations 4.22 and 4.23 can be avoided by use of multiplication in the magnitude comparison, as opposed to division in interpolation, as it is in Code 4.10. In practice, however, this implementation, Codes 4.9 and 4.10, can suffer from numerical imprecision and ill-conditioning. Accordingly, it is better to implement a hand-crafted interpretation of Equations 4.22 and 4.23 applied separately to the four quadrants. This is too lengthy to be included here, but a version is included with the Worksheets for Chapter 4.

The transfer function associated with *hysteresis thresholding* is shown in Figure 4.17. Points are set to white once the upper threshold is exceeded and set to black when the lower threshold is reached. The arrows reflect possible movement: there is only one way to change from black to white and vice versa.

```
non_max(edges):=  for i∈1..cols(edges₀,₀)-2
                    for j∈1..rows(edges₀,₀)-2
```

$$Mx \leftarrow (edges_{0,0})_{j,i}$$

$$My \leftarrow (edges_{0,1})_{j,i}$$

$$o \leftarrow atan\left(\frac{Mx}{My}\right) \text{ if } My \neq 0$$

$$\left(o \leftarrow \frac{\pi}{2}\right) \text{ if } (My=0) \cdot (Mx>0)$$

$$o \leftarrow \frac{-\pi}{2} \text{ otherwise}$$

$$adds \leftarrow get\_coords(o)$$

$$M1 \leftarrow \left[My \cdot (edges_{0,2})_{j+adds_{0,1}, i+adds_{0,0}} \cdots \atop + (Mx-My) \cdot (edges_{0,2})_{j+adds_{0,3}, i+adds_{0,2}} \right]$$

$$adds \leftarrow get\_coords(o+\pi)$$

$$M2 \leftarrow \left[My \cdot (edges_{0,2})_{j+adds_{0,1}, i+adds_{0,0}} \cdots \atop + (Mx-My) \cdot (edges_{0,2})_{j+adds_{0,3}, i+adds_{0,2}} \right]$$

$$isbigger \leftarrow \left[\left[Mx \cdot (edges_{0,2})_{j,i} > M1 \right] \cdot \left[Mx \cdot (edges_{0,2})_{j,i} \geq M2 \right]\right] \cdots$$
$$+ \left[\left[Mx \cdot (edges_{0,2})_{j,i} < M1 \right] \cdot \left[Mx \cdot (edges_{0,2})_{j,i} \leq M2 \right]\right]$$

$$new\_edge_{j,i} \leftarrow (edges_{0,2})_{j,i} \text{ if } isbigger$$

$$new\_edge_{j,i} \leftarrow 0 \text{ otherwise}$$

```
new_edge
```

Code 4.10   Non-maximum suppression

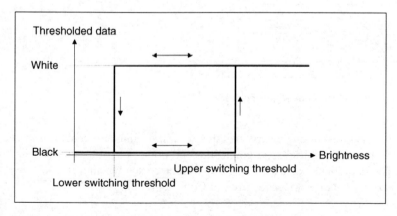

Figure 4.17   Hysteresis thresholding transfer function

The application of non-maximum suppression and hysteresis thresholding is illustrated in Figure 4.18. This contains a ridge of edge data, the edge magnitude. The action of non-maximum suppression is to select the points along the top of the ridge. Given that the top of the ridge initially exceeds the upper threshold, the thresholded output is set to white until the peak of

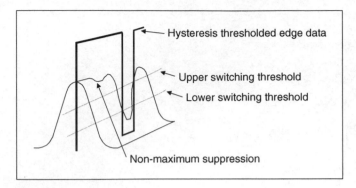

Figure 4.18 Action of non-maximum suppression and hysteresis thresholding

the ridge falls beneath the lower threshold. The thresholded output is then set to black until the peak of the ridge exceeds the upper switching threshold.

Hysteresis thresholding requires two thresholds, an *upper* and a *lower* threshold. The process starts when an edge point from non-maximum suppression is found to exceed the upper threshold. This is labelled as an edge point (usually white, with a value of 255) and forms the first point of a line of edge points. The neighbours of the point are then searched to determine whether or not they exceed the lower threshold, as in Figure 4.19. Any neighbour that exceeds the lower threshold is labelled as an edge point and its neighbours are then searched to determine whether or not they exceed the lower threshold. In this manner, the first edge point found (the one that exceeded the upper threshold) becomes a *seed* point for a search. Its neighbours, in turn, become seed points if they exceed the lower threshold, and so the search extends, along branches arising from neighbours that exceeded the lower threshold. For each branch, the search terminates at points that have no neighbours above the lower threshold.

≥ lower	≥ lower	≥ lower
≥ lower	seed ≥ upper	≥ lower
≥ lower	≥ lower	≥ lower

Figure 4.19 Neighbourhood search for hysteresis thresholding

In implementation, hysteresis thresholding clearly requires *recursion*, since the length of any branch is unknown. Having found the initial *seed* point, the seed point is set to white and its neighbours are searched. The coordinates of each point are checked to see whether it is within the picture size, according to the operator check, given in Code 4.11.

$$\text{check}(xc,yc,pic) := \begin{vmatrix} 1 & \text{if } (xc \geq 1) \cdot (xc \leq \text{cols}(pic) - 2) \cdot (yc \geq 1) \cdot (yc \leq \text{rows}(pic) - 2) \\ 0 & \text{otherwise} \end{vmatrix}$$

Code 4.11 Checking points are within an image

The neighbourhood (as in Figure 4.19) is then searched by a function connect (Code 4.12) which is fed with the non-maximum suppressed edge image, the coordinates of the seed point whose connectivity is under analysis and the lower switching threshold. Each of the neighbours is searched if its value exceeds the lower threshold, and the point has not already been labelled as white (otherwise the function would become an infinite loop). If both conditions are satisfied (and the point is within the picture) then the point is set to white and becomes a seed point for further analysis. This implementation tries to check the seed point as well, even though it has already been set to white. The operator could be arranged not to check the current seed point, by direct calculation without the for loops, and this would be marginally faster. Including an extra Boolean constraint to inhibit check of the seed point would only slow the operation. The connect routine is recursive: it is called again by the new seed point.

```
connect(x,y,nedg,low):= | for x1∈x-1.. x+1
                        |     for y1∈y-1.. y+1
                        |         if(nedg_{y1,x1}≥low)·(nedg_{y1,x1}≠255)·check
                        |         (x1,y1,nedg)
                        |             | nedg_{y1,x1}←255
                        |             | nedg←connect(x1,y1,nedg,low)
                        | nedg
```

Code 4.12 Connectivity analysis after seed point location

The process starts with the point that exceeds the upper threshold. When such a point is found, it is set to white and it becomes a seed point where connectivity analysis starts. The calling operator for the connectivity analysis, hyst_thr, which starts the whole process, is given in Code 4.13. When hyst_thr is invoked, its arguments are the coordinates of the point of current interest, the non-maximum suppressed edge image, n_edg (which is eventually delivered as the hysteresis thresholded image), and the upper and lower switching thresholds, upp and low, respectively. For *display* purposes, this operator requires a later operation to remove points which have not been set to white (to remove those points which are below the upper threshold and which are not connected to points above the lower threshold). This is rarely used in application since the points set to white are the only ones of interest in later processing.

```
hyst_thr(n_edg,upp,low):= | for x∈1.. cols(n_edg)-2
                          |     for y∈1.. rows(n_edg)-2
                          |         if[(n_edg_{y,x}≥upp)·(n_edg_{y,x}≠255)]
                          |             | n_edg_{y,x}←255
                          |             | n_edg←connect(x,y,n_edg,low)
                          | n_edg
```

Code 4.13 Hysteresis thresholding operator

A comparison with the results of *uniform* thresholding is shown in Figure 4.20. Figure 4.20(a) shows the result of hysteresis thresholding of a Sobel edge detected image of the eye with an upper threshold set to 40 pixels, and a lower threshold of 10 pixels. Figure 4.20(b) and (c) show the result of uniform thresholding applied to the image with thresholds of 40 pixels and 10 pixels, respectively. Uniform thresholding can select too few points if the threshold is too high, and too many if it is too low. Hysteresis thresholding selects *all* the points in Figure 4.20(b), and *some* of those in Figure 4.20(c), those connected to the points in (b). In particular, part of the nose is partly present in Figure 4.20(a), whereas it is absent in Figure 4.20(b) and masked by too many edge points in Figure 4.20(c). Also, the eyebrow is more complete in (a), whereas it is only partial in (b) and complete (but obscured) in (c). Hysteresis thresholding therefore has an ability to detect major features of interest in the edge image, in an improved manner to uniform thresholding.

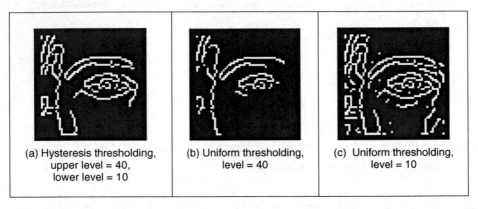

(a) Hysteresis thresholding, upper level = 40, lower level = 10

(b) Uniform thresholding, level = 40

(c) Uniform thresholding, level = 10

Figure 4.20   Comparing hysteresis thresholding with uniform thresholding

The action of the Canny operator on a larger image is shown in Figure 4.21, in comparison with the result of the Sobel operator. Figure 4.21(a) is the original image of a face, Figure 4.21(b) is the result of the Canny operator (using a 5×5 Gaussian operator with $\sigma = 1.0$ and with upper and lower thresholds set appropriately) and Figure 4.21(c) is the result of a 3×3 Sobel operator with uniform thresholding. The retention of major detail by the Canny operator is very clear; the face is virtually recognizable in Figure 4.21(b), whereas it is less clear in Figure 4.21(c).

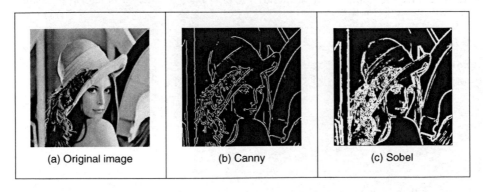

(a) Original image

(b) Canny

(c) Sobel

Figure 4.21   Comparing Canny with Sobel

4.3 Second order edge detection operators

4.3.1 Motivation

First order edge detection is based on the premise that differentiation highlights change; image intensity changes in the region of a feature boundary. The process is illustrated in Figure 4.22, where Figure 4.22(a) is a cross-section through image data. The result of *first order* edge detection, $f'(x) = df/dx$ in Figure 4.22(b), is a *peak* where the rate of change of the original signal, $f(x)$ in Figure 4.22(a), is greatest. There are higher order derivatives; applied to the same cross-section of data, the *second order* derivative, $f''(x) = d^2 f/dx^2$ in Figure 4.22(c), is greatest where the rate of change of the signal is greatest and zero when the rate of change is constant. The rate of change is constant at the peak of the first order derivative. This is where there is a *zero-crossing* in the second order derivative, where it changes sign. Accordingly, an alternative to first order differentiation is to apply second order differentiation and then find zero-crossings in the second order information.

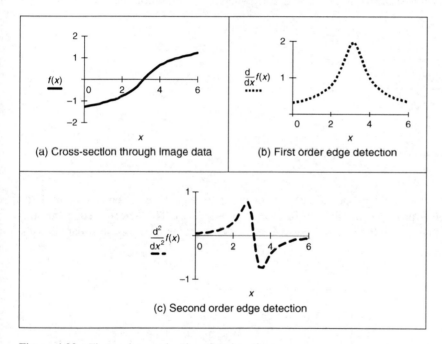

Figure 4.22 First and second order edge detection

4.3.2 Basic operators: the Laplacian

The *Laplacian operator* is a template which implements second order differencing. The second order differential can be approximated by the difference between two adjacent first order differences:

$$f''(x) \cong f'(x) - f'(x+1) \tag{4.24}$$

which, by Equation 4.6, gives

$$f''(x) \cong -f(x) + 2f(x+1) - f(x+2) \tag{4.25}$$

This gives a horizontal second order template, as given in Figure 4.23.

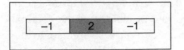

Figure 4.23 Horizontal second order template

When the horizontal second order operator is combined with a vertical second order difference we obtain the full Laplacian template, given in Figure 4.24. Essentially, this computes the difference between a point and the average of its four direct neighbours. This was the operator used earlier in anisotropic diffusion (Section 3.5.4), where it is an approximate solution to the heat equation.

0	−1	0
−1	**4**	−1
0	−1	0

Figure 4.24 Laplacian edge detection operator

Application of the Laplacian operator to the image of the square is given in Figure 4.25. The original image is provided in numeric form in Figure 4.25(a). The detected edges are the *zero-crossings* in Figure 4.25(b) and can be seen to lie between the edge of the square and its background.

$$p = \begin{bmatrix} 1 & 2 & 3 & 4 & 1 & 1 & 2 & 1 \\ 2 & 2 & 3 & 0 & 1 & 2 & 2 & 1 \\ 3 & 0 & 38 & 39 & 37 & 36 & 3 & 0 \\ 4 & 1 & 40 & 44 & 41 & 42 & 2 & 1 \\ 1 & 2 & 43 & 44 & 40 & 39 & 1 & 3 \\ 2 & 0 & 39 & 41 & 42 & 40 & 2 & 0 \\ 1 & 2 & 0 & 2 & 2 & 3 & 1 & 1 \\ 0 & 2 & 1 & 3 & 1 & 0 & 4 & 2 \end{bmatrix}$$

(a) Image data

$$L = \begin{bmatrix} 0 & 0 & 0 & 0 & 0 & 0 & 0 & 0 \\ 0 & 1 & -31 & -47 & -36 & -32 & 0 & 0 \\ 0 & -44 & 70 & 37 & 31 & 60 & -28 & 0 \\ 0 & -42 & 34 & 12 & 1 & 50 & -39 & 0 \\ 0 & -37 & 47 & 8 & -6 & 33 & -42 & 0 \\ 0 & -45 & 72 & 37 & 45 & 74 & -34 & 0 \\ 0 & 5 & -44 & -38 & -40 & -31 & -6 & 0 \\ 0 & 0 & 0 & 0 & 0 & 0 & 0 & 0 \end{bmatrix}$$

(b) After Laplacian operator

Figure 4.25 Edge detection via the Laplacian operator

138 *Feature Extraction and Image Processing*

An alternative structure to the template in Figure 4.24 is one where the central weighting is 8 and the neighbours are all weighted as -1. This includes a different form of image information, so the effects are slightly different. (Essentially, this now computes the difference between a pixel and the average of its neighbouring points, including the corners.) In both structures, the central weighting can be negative and that of the four or the eight neighbours can be positive, without loss of generality. It is important to ensure that the sum of template coefficients is zero, so that edges are not detected in areas of uniform brightness. One advantage of the Laplacian operator is that it is *isotropic* (like the Gaussian operator): it has the same properties in each direction. However, as yet it contains *no* smoothing and will again respond to noise, more so than a first order operator since it is differentiation of a higher order. As such, the Laplacian operator is rarely used in its basic form. Smoothing can use the averaging operator described earlier, but a more optimal form is Gaussian smoothing. When this is incorporated with the Laplacian we obtain a Laplacian of Gaussian (LoG) operator, which is the basis of the Marr–Hildreth approach, to be considered next. A clear disadvantage with the Laplacian operator is that edge direction is not available. It does, however, impose low computational cost, which is its main advantage. Although interest in the Laplacian operator abated with rising interest in the Marr–Hildreth approach, a non-linear Laplacian operator was developed (Vliet and Young, 1989) and shown to have good performance, especially in low-noise situations.

4.3.3 Marr–Hildreth operator

The *Marr–Hildreth* approach (Marr and Hildreth, 1980) again uses Gaussian filtering. In principle, we require an image which is the second differential ∇^2 of a Gaussian operator $g(x, y)$ convolved with an image **P**. This convolution process can be separated as:

$$\nabla^2 (g(x, y) * \mathbf{P}) = \nabla^2 (g(x, y)) * \mathbf{P} \tag{4.26}$$

Accordingly, we need to compute a template for $\nabla^2 (g(x, y))$ and convolve this with the image. By further differentiation of Equation 4.17, we achieve a LoG operator:

$$\nabla^2 g(x, y) - \frac{\partial^2 g(x, y, \sigma)}{\partial x^2} U_x + \frac{\partial^2 g(x, y, \sigma)}{\partial y^2} U_y$$

$$= \frac{\partial \nabla g(x, y, \sigma)}{\partial x} U_x + \frac{\partial \nabla g(x, y, \sigma)}{\partial y} U_y$$

$$= \left(\frac{x^2}{\sigma^2} - 1 \right) \frac{e^{\frac{-(x^2+y^2)}{2\sigma^2}}}{\sigma^2} + \left(\frac{y^2}{\sigma^2} - 1 \right) \frac{e^{\frac{-(x^2+y^2)}{2\sigma^2}}}{\sigma^2} \tag{4.27}$$

$$= \frac{1}{\sigma^2} \left(\frac{(x^2 + y^2)}{\sigma^2} - 2 \right) e^{\frac{-(x^2+y^2)}{2\sigma^2}}$$

This is the basis of the Marr–Hildreth operator. Equation 4.27 can be used to calculate the coefficients of a template which, when convolved with an image, combines Gaussian smoothing with second order differentiation. The operator is sometimes called a 'Mexican hat' operator, since its surface plot is the shape of a sombrero, as illustrated in Figure 4.26.

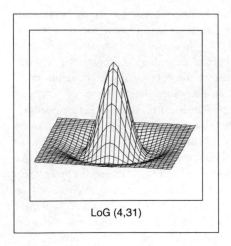

LoG (4,31)

Figure 4.26   Shape of Laplacian of Gaussian operator

The calculation of the Laplacian of Gaussian can be approximated by the *difference of Gaussian*, where the difference is formed from the result of convolving two Gaussian filters with differing variance (Marr, 1982; Lindeberg, 1994).

$$\sigma \nabla^2 g\,(x, y, \sigma) = \frac{\partial g}{\partial \sigma} \approx \frac{g\,(x, y, k\sigma) - g\,(x, y, \sigma)}{k\sigma - \sigma} \tag{4.28}$$

where $g(x, y, \sigma)$ is the Gaussian function and k is a constant. Although similarly named, the *derivative of Gaussian* (Equation 4.17) is a *first* order operator including Gaussian smoothing, $\nabla g(x, y)$. It does seem counter-intuitive that the difference of two smoothing operators should lead to second order edge detection. The approximation is illustrated in Figure 4.27, where in one dimension, two Gaussian distributions of different variance are subtracted to form a one-dimensional (1D) operator whose cross-section is equivalent to the shape of the LoG operator (a cross-section of Figure 4.26).

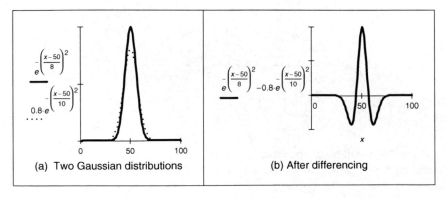

(a) Two Gaussian distributions   (b) After differencing

Figure 4.27   Approximating the Laplacian of Gaussian by difference of Gaussian

140   *Feature Extraction and Image Processing*

The implementation of Equation 4.27 to calculate template coefficients for the LoG operator is given in Code 4.14. The function includes a normalization function which ensures that the sum of the template coefficients is unity, so that edges are not detected in area of uniform brightness. This is in contrast with the earlier Laplacian operator (where the template coefficients summed to zero), since the LoG operator includes smoothing within the differencing action, whereas the Laplacian is pure differencing. The template generated by this function can then be used within template convolution. The Gaussian operator again suppresses the influence of points away from the centre of the template, basing differentiation on those points nearer the centre; the standard deviation, σ, is chosen to ensure this action. Again, it is isotropic consistent with Gaussian smoothing.

$$
\text{LoG}(\sigma, \text{size}) := \left| \begin{array}{l} cx \leftarrow \dfrac{\text{size}-1}{2} \\[2mm] cy \leftarrow \dfrac{\text{size}-1}{2} \\[2mm] \text{for } x \in 0 .. \text{ size}-1 \\ \quad \text{for } y \in 0 .. \text{ size}-1 \\ \qquad \left| \begin{array}{l} nx \leftarrow x-cx \\ ny \leftarrow y-cy \\[2mm] \text{template}_{y,x} \leftarrow \dfrac{1}{\sigma^2} \cdot \left(\dfrac{nx^2+ny^2}{\sigma^2} - 2 \right) \cdot e^{-\left(\frac{nx^2+ny^2}{2 \cdot \sigma^2}\right)} \end{array} \right. \\[4mm] \text{template} \leftarrow \text{normalize(template)} \\ \text{template} \end{array} \right.
$$

Code 4.14 Implementation of the Laplacian of Gaussian operator

Determining the zero-crossing points is a major difficulty with this approach. A variety of techniques can be used, including manual determination of zero-crossing or a least squares fit of a plane to local image data, which is followed by determination of the point at which the plane crosses zero, if it does. The former is too simplistic, whereas the latter is quite complex (see Section 11.2, Appendix 3).

The approach here is much simpler: given a local 3×3 area of an image, this is split into quadrants. These are shown in Figure 4.28, where each quadrant contains the centre pixel. The first quadrant contains the four points in the upper left corner and the third quadrant contains the four points in the upper right. If the average of the points in any quadrant differs in sign from the average in any other quadrant, there must be a zero-crossing at the centre point. In `zerox`, Code 4.15, the average intensity in each quadrant is then evaluated, giving four values and `int0`, `int1`, `int2` and `int3`. If the maximum value of these points is positive, and the minimum value is negative, there must be a zero-crossing within the neighbourhood. If one exists, the output image at that point is marked as white, otherwise it is set to black.

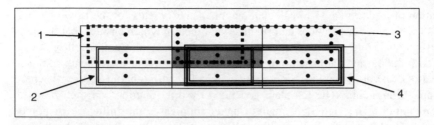

Figure 4.28 Regions for zero crossing detection

```
zerox(pic):= | newpic←zero(pic)
             | for x∈1.. cols(pic)-2
             |     for y∈1.. rows(pic)-2
             |
             |         int₀←  Σ      Σ      picy1,x1
             |             x1=x-1  y1= y-1
             |
             |         int₁←  Σ      Σ      picy1,x1
             |             x1=x-1   y1=y
             |
             |         int₂←  Σ      Σ      picy1,x1
             |             x1=x    y1=y-1
             |
             |         int₃←  Σ      Σ      picy1,x1
             |             x1=x    y1=y
             |         maxval←max(int)
             |         minval←min(int)
             |         newpicy,x ←255 if (maxval>0)·(minval<0)
             |
             | newpic
```

Code 4.15 Zero crossing detector

The action of the Marr–Hildreth operator is given in Figure 4.29, applied to the face image in Figure 4.21(a). The output of the LoG operator is hard to interpret visually and is not shown here (remember that it is the zero-crossings which mark the edge points and it is hard to see them). The detected zero-crossings (for a 3×3 neighbourhood) are shown in Figure 4.29(b) and (c) for LoG operators of size and variance 11×11 with $\sigma = 0.8$ and 15×15 with $\sigma = 1.8$, respectively. These show that the selection of window size and variance can be used to provide edges at differing scales. Some of the smaller regions in Figure 4.29(b) join to form larger regions in Figure 4.29(c). Note that one virtue of the Marr–Hildreth operator is its ability to provide *closed* edge borders, which the Canny operator cannot. Another virtue is that it avoids the recursion associated with hysteresis thresholding that can require a massive stack size for large images.

The Fourier transform of a LoG operator is shown in relief in Figure 4.30(a) and as a contour plot in Figure 4.30(b). The transform is circular–symmetric, as expected. Since the transform reveals that the LoG operator omits low and high frequencies (those close to the origin, and

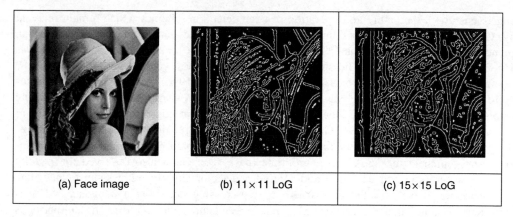

| (a) Face image | (b) 11×11 LoG | (c) 15×15 LoG |

Figure 4.29 Marr–Hildreth edge detection

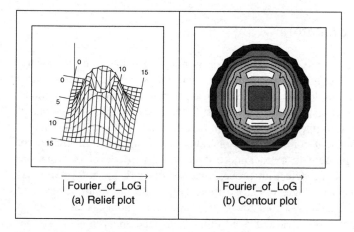

| Fourier_of_LoG | | Fourier_of_LoG |
| (a) Relief plot | (b) Contour plot |

Figure 4.30 Fourier transform of LoG operator

those far away from the origin) it is equivalent to a *band-pass filter*. Choice of the value of σ controls the spread of the operator in the spatial domain and the 'width' of the band in the frequency domain: setting σ to a high value gives low-pass filtering, as expected. This differs from first order edge detection templates which offer a *high-pass* (differencing) filter along one axis with a *low-pass* (smoothing) action along the other axis.

The Marr–Hildreth operator has stimulated much attention, perhaps in part because it has an appealing relationship to human vision, and its ability for multiresolution analysis (the ability to detect edges at differing scales). In fact, it has been suggested that the original image can be reconstructed from the zero-crossings at different scales. One early study (Haralick, 1984) concluded that the Marr–Hildreth could give good performance. Unfortunately, the implementation appeared to be different from the original LoG operator (and has appeared in some texts in this form), as noted by one of the Marr–Hildreth study's originators (Grimson and Hildreth, 1985). This lead to a somewhat spirited reply (Haralick, 1985) clarifying concern, but also raising issues about the nature and operation of edge detection schemes which remain relevant today. Given the requirement for convolution of large templates, attention quickly focused on frequency domain implementation (Huertas and Medioni, 1986), and speed improvement was later considered in some detail (Forshaw, 1988). Later, schemes were developed to refine the edges produced via the LoG approach (Ulupinar and Medioni, 1990). Although speed and accuracy are major concerns with the Marr–Hildreth approach, it is also possible for zero-crossing detectors to mark as edge points ones that have no significant contrast, motivating study of their authentication (Clark, 1989). Gunn (1999) studied the relationship between mask size of the LoG operator and its error rate. Essentially, an acceptable error rate defines a truncation error, which in turn gives an appropriate mask size. Gunn also observed the paucity of studies on zero-crossing detection and offered a detector slightly more sophisticated than the one here (as it includes the case where a zero-crossing occurs at a boundary, whereas the one here assumes that the zero-crossing can only occur at the centre). The similarity is not coincidental: Mark developed the one here after conversations with Steve Gunn, who he works with!

4.4 Other edge detection operators

There have been many approaches to edge detection. This is not surprising, since it is often the first stage in a vision process. The most popular operators are the Sobel, Canny and Marr–Hildreth operators. Clearly, in any implementation, there is a *compromise* between (computational) cost and efficiency. In some cases, it is difficult to justify the extra complexity associated with the Canny and the Marr–Hildreth operators. This is in part due to the images: few images contain the adverse noisy situations that complex edge operators are designed to handle. Also, when finding shapes, it is often prudent to extract more than enough low-level information, and to let the more sophisticated shape detection process use, or discard, the information as appropriate. For these reasons, we will study only two more edge detection approaches, and only briefly. These operators are the *Spacek* and the *Petrou* operators: both are designed to be optimal and both have different properties and a different basis (the *smoothing* functional in particular) to the Canny and Marr–Hildreth approaches. The Spacek and Petrou Operators will be reviewed briefly, by virtue of their optimality. Essentially, while Canny maximized the ratio of the signal-to-noise ratio with the localization, Spacek (1986) maximized the ratio of the product of the signal-to-noise ratio and the peak separation with the localization. In Spacek's work, since the edge was again modelled as a step function, the ideal filter appeared to be of the same form as Canny's. Spacek's operator can give better performance than Canny's formulation (Jia and Nixon, 1985), as such challenging the optimality of the Gaussian operator for noise smoothing (in step edge detection), although such advantage should be explored in application.

Petrou questioned the validity of the step-edge model for real images (Petrou and Kittler, 1991). Given that the composite performance of an image acquisition system can be considered to be that of a low-pass filter, any step changes in the image will be smoothed to become a ramp. As such, a more plausible model of the edge is a ramp, rather than a step. Since the process is based on ramp edges, and because of limits imposed by its formulation, the Petrou operator uses templates that much wider to preserve optimal properties. As such, the operator can impose greater computational complexity, but is a natural candidate for applications with the conditions for which its properties were formulated.

Of the other approaches, Korn (1988) developed a unifying operator for symbolic representation of grey level change. The *Susan* operator (Smith and Brady, 1997) derives from an approach aimed to find more that just edges, since it can also be used to derive *corners* (where feature boundaries change direction sharply, as in *curvature* detection in Section 4.8) and structure-preserving image noise reduction. Essentially, SUSAN derives from smallest univalue segment assimilating nucleus, which concerns aggregating the difference between elements in a (circular) template centred on the nucleus. The USAN is essentially the number of pixels within the circular mask that have similar brightness to the nucleus. The edge strength is then derived by subtracting the USAN size from a geometric threshold, which is, say, three-quarters of the maximum USAN size. The method includes a way of calculating edge direction, which is essential if non-maximum suppression is to be applied. The advantages are in simplicity (and hence speed), since it is based on simple operations, and the possibility of extension to find other feature types.

4.5 Comparison of edge detection operators

The selection of an edge operator for a particular application depends on the application itself. As has been suggested, it is not usual to require the sophistication of the advanced operators in many applications. This is reflected in analysis of the performance of the edge operators on the eye image. To provide a different basis for comparison, we shall consider the difficulty of low-level feature extraction in ultrasound images. As has been seen earlier (Section 3.5.4), ultrasound images are very *noisy* and require *filtering* before analysis. Figure 4.31(a) is part of the ultrasound image which could have been filtered using the truncated median operator (Section 3.5.3). The image contains a feature called the pitus (the 'splodge' in the middle) and we shall see how different edge operators can be used to detect its perimeter, although without noise filtering. Earlier, in Section 3.5.4, we considered a comparison of statistical operators on an ultrasound image. The median is perhaps the most popular of these processes for general (i.e. non-ultrasound) applications. Accordingly, it is of interest that one study (Bovik et al., 1987) has suggested that the known advantages of median filtering (the removal of noise with the preservation of edges, especially for salt and pepper noise) are shown to good effect if it is used as a prefilter to first and second order approaches, although with the cost of the median filter. However, we will not consider median filtering here: its choice depends more on suitability to a particular application.

The results for all edge operators have been generated using hysteresis thresholding, where the thresholds were selected manually for best performance. The basic first order operator (Figure 4.31b) responds rather nicely to the noise and it is difficult to select a threshold that reveals a major part of the pitus border. Some is present in the Prewitt and Sobel operators' results (Figure 4.31c and d, respectively), but there is still much noise in the processed image, although there is less in the Sobel. The Laplacian operator (Figure 4.31e) gives very little information indeed, as to be expected with such noisy imagery. However, the more advanced operators can be used to good effect. The Marr–Hildreth approach improves matters (Figure 4.31f), but suggests that it is difficult to choose a LoG operator of appropriate size to detect a feature of these dimensions in such noisy imagery, illustrating the compromise between the size of operator needed for noise filtering and the size needed for the target feature. However, the Canny and Spacek operators can be used to good effect, as shown in Figure 4.31(g) and (h), respectively. These reveal much of the required information, together with data away from the

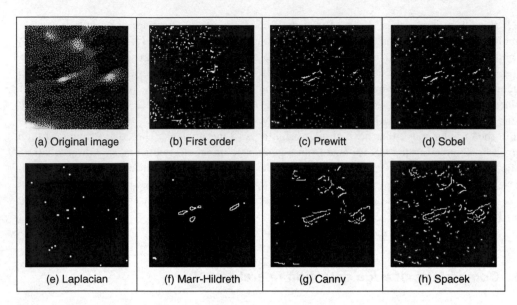

Figure 4.31 Comparison of edge detection operators

pitus itself. In an automated analysis system, for this application, the extra complexity of the more sophisticated operators would clearly be warranted.

4.6 Further reading on edge detection

Few computer vision and image processing texts omit detail concerning edge detection operators, although few give explicit details concerning implementation. Many of the earlier texts omit the more recent techniques. Parker (1994) only includes C code for some of the most basic edge detection techniques. Further information can be found in journal papers; Petrou's excellent study of edge detection (Petrou, 1994) highlights study of the performance factors involved in the optimality of the Canny, Spacek and Petrou operators with extensive tutorial support (although I suspect that Petrou junior may one day be embarrassed by the frequency with which his youthful mugshot is used: his teeth show up very well!). There have been a number of surveys of edge detection highlighting performance attributes in comparison. See, for example, Torre and Poggio (1986), which gives a theoretical study of edge detection and considers some popular edge detection techniques in light of this analysis. One survey (Heath et al., 1997) surveys many approaches, comparing them in particular with the Canny operator (and states where code for some of the techniques they compared can be found). This showed that best results can be achieved by tuning an edge detector for a particular application and highlighted good results by the Bergholm operator (Bergholm, 1987). Marr (1982) considers the Marr–Hildreth approach to edge detection in the light of human vision (and its influence on perception), with particular reference to scale in edge detection. More recently, Yitzhaky and Peli (2003) suggest 'a general tool to assist in practical implementations of parametric edge detectors where an automatic process is required' and use statistical tests to evaluate edge detector performance. Since edge detection is one of the most important vision techniques, it continues to be a focus of research interest. Accordingly, it is always worth looking at recent conference proceedings to see any new techniques, or perhaps more likely performance comparison or improvement, that may help you to solve a problem.

4.7 Phase congruency

The comparison of edge detectors highlights some of their innate problems: incomplete contours, the need for selective thresholding, and their response to noise. Further, the selection of a threshold is often inadequate for all the regions in an image, since there are many changes in local illumination. We shall find that some of these problems can be handled at a higher level, when shape extraction can be arranged to accommodate partial data and to reject spurious information. There is, however, interest in refining the low-level feature extraction techniques further.

Phase congruency is a feature detector with two main advantages: it can detect a *broad range* of features; and it is *invariant to* local (and smooth) *change in illumination*. As the name suggests, it is derived by frequency domain considerations operating on the considerations of phase (i.e. time). It is illustrated detecting some 1D features in Figure 4.32, where the features are the solid lines: a (noisy) *step* function in Figure 4.32(a), and a peak (or impulse) in Figure 4.32(b). By Fourier transform analysis, any function is made up from the controlled addition of sinewaves of differing frequencies. For the step function to occur (the solid line in Figure 4.32a), the constituent frequencies (the dotted lines in Figure 4.32a) must all change at the same time, so they add up to give the edge. Similarly, for the peak to occur, the constituent frequencies must all peak at the same time; in Figure 4.32(b) the solid line is the peak and the dotted lines are some of its constituent frequencies. This means that to find the feature in which we are interested, we can determine points where events happen at the same time: this is phase congruency. By way of generalization, a triangle wave is made of peaks and troughs: phase congruency implies that the peaks and troughs of the constituent signals should coincide.

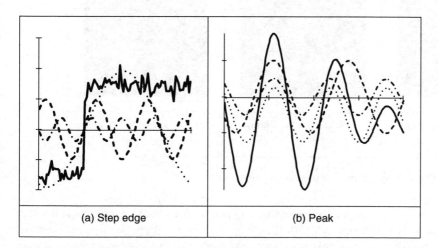

(a) Step edge   (b) Peak

Figure 4.32 Low-level feature extraction by phase congruency

The constituent sinewaves plotted in Figure 4.32(a) were derived by taking the Fourier transform of a step and then determining the sinewaves according to their magnitude and phase. The Fourier transform in Equation 2.15 delivers the complex Fourier components **Fp**. These can be used to show the constituent signals xc by

$$xc\,(t) = |\mathbf{Fp}_u|\, e^{j\left(\frac{2\pi}{N}ut + \phi(\mathbf{Fp}_u)\right)} \tag{4.29}$$

where $|\mathbf{Fp}_u|$ is again the magnitude of the uth Fourier component (Equation 2.7) and $\phi\,(\mathbf{Fp}_u) = \langle \mathbf{Fp}_u$ is the argument, the phase in Equation 2.8. The (dotted) frequencies displayed in Figure 4.32 are the first four odd components (the even components for this function are zero, as shown in the Fourier transform of the step in Figure 2.11). The addition of these components is indeed the inverse Fourier transform which reconstructs the step feature.

The advantages are that detection of congruency is invariant with local contrast: the sinewaves still add up so the changes are still in the same place, even if the magnitude of the step edge is much smaller. In images, this implies that we can change the contrast and still detect edges. This is illustrated in Figure 4.33. Here, a standard image processing image, the 'cameraman' image from the early UCSD dataset, has been changed between the left and right sides so that the contrast changes in the two halves of the image (Figure 4.33a). Edges detected by Canny are shown in Figure 4.33(b) and by phase congruency in 4.33(c). The basic structure of the edges detected by phase congruency is very similar to that structure detected by Canny, and the phase congruency edges appear somewhat cleaner (there is a single line associated with the tripod control in phase congruency); both detect the change in brightness between the two halves. There is a major difference though: the building in the lower right side of the image is barely detected in the Canny image, whereas it can clearly be seen by phase congruency. Its absence is due to the parameter settings used in the Canny operator. These can be changed, but if the contrast were to change again, then the parameters would need to be reoptimized for the new arrangement. This is not the case for phase congruency.

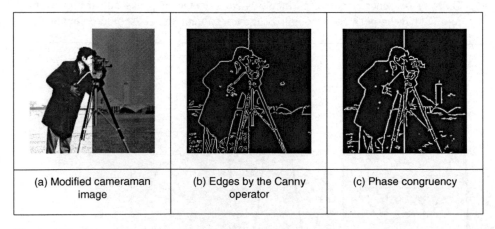

| (a) Modified cameraman image | (b) Edges by the Canny operator | (c) Phase congruency |

Figure 4.33 Edge detection by Canny and by phase congruency

Such a change in brightness might appear unlikely in practical application, but this is not the case with moving objects which interact with illumination or in fixed applications where illumination changes. In studies aimed to extract spinal information from digital videofluoroscopic X-ray images to provide guidance for surgeons (Zheng et al., 2004), phase congruency was found to be immune to the changes in contrast caused by slippage of the shield used to protect the patient while acquiring the image information. One such image is shown in Figure 4.34. The lack of shielding is apparent in the bloom at the side of the images. This changes as the subject is moved, so it proved difficult to optimize the parameters for Canny over the whole sequence (Figure 4.34b), but the detail of a section of the phase congruency result (Figure 4.34c) shows that the vertebrae information is readily available for later high-level feature extraction.

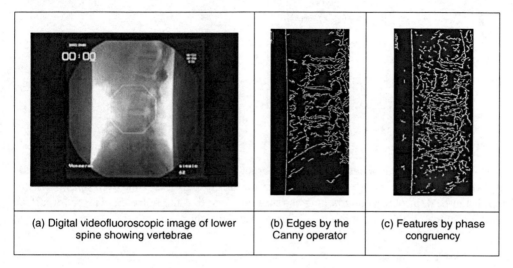

| (a) Digital videofluoroscopic image of lower spine showing vertebrae | (b) Edges by the Canny operator | (c) Features by phase congruency |

Figure 4.34 Spinal contour by phase congruency (Zheng et al., 2004)

The original notions of phase congruency are the concepts of *local energy* (Morrone and Owens, 1987), with links to the human visual system (Morrone and Burr, 1988). One of the most sophisticated implementations was by Kovesi (1999), with added advantage that his Matlab implementation is available on the web (http://www.csse.uwa.edu.au/~pk/Research/research.html), as well as much more information. Essentially, we seek to determine features by detection of points at which Fourier components are maximally in phase. By extension of the Fourier reconstruction functions in Equation 4.29, Morrone and Owens defined a measure of phase congruency PC as

$$PC(x) = \max_{\bar{\phi}(x) \in 0, 2\pi} \left(\frac{\sum_u |\mathbf{Fp}_u| \cos\left(\phi_u(x) - \bar{\phi}(x)\right)}{\sum_u |\mathbf{Fp}_u|} \right) \tag{4.30}$$

where $\phi_u(x)$ represents the local phase of the component $\mathbf{Fp}_u$ at position x. Essentially, this computes the ratio of the sum of projections onto a vector (the sum in the numerator) to the total vector length (the sum in the denominator). The value of $\bar{\phi}(x)$ that maximizes this equation is the amplitude weighted mean local phase angle of all the Fourier terms at the point being considered. In Figure 4.35 the resulting vector is made up of four components, illustrating the projection of the second onto the resulting vector. Clearly, the value of PC ranges from 0 to 1, the maximum occurring when all elements point along the resulting vector. As such, the resulting phase congruency is a *dimensionless normalized measure* which is thresholded for image analysis.

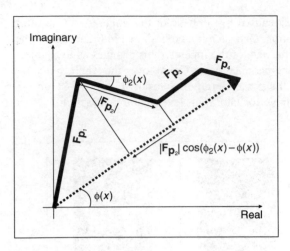

Figure 4.35 Summation in phase congruency

In this way, we have calculated the phase congruency for the step function in Figure 4.36(a), which is shown in Figure 4.36(b). Here, the position of the step is at time step 40; this is the position of the peak in phase congruency, as required. Note that the noise can be seen to affect the result, although the phase congruency is largest at the right place.

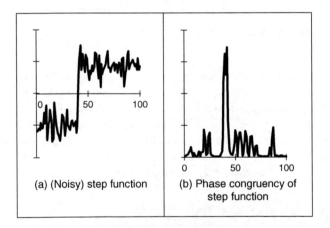

(a) (Noisy) step function

(b) Phase congruency of step function

Figure 4.36 One-dimensional phase congruency

One interpretation of the measure is that since for small angles, $\cos \theta = 1 - \theta^2$, then Equation 4.30 expresses the ratio of the magnitudes weighted by the variance of the difference to the summed magnitude of the components. There is certainly difficulty with this measure, apart from difficulty in implementation: it is sensitive to *noise*, as is any phase measure; it is not *conditioned* by the magnitude of a response (small responses are not discounted); and it is not well *localized* (the measure varies with the cosine of the difference in phase, not with the difference itself, although it does avoid discontinuity problems with direct use of angles). In fact, the phase congruency is directly proportional to the local energy (Venkatesh and Owens, 1989), so an alternative approach is to search for maxima in the local energy. The notion of local energy allows us to compensate for the sensitivity to the detection of phase in noisy situations.

For these reasons, Kovesi developed a wavelet-based measure which improved performance, while accommodating noise. In basic form, phase congruency can be determined by convolving a set of wavelet filters with an image, and calculating the difference between the average filter response and the individual filter responses. The response of a (1D) signal I to a set of wavelets at scale n is derived from the convolution of the cosine and sine wavelets (discussed in Section 2.7.3), denoted M_n^e and M_n^o, respectively

$$[e_n(x), o_n(x)] = [I(x) * M_n^e, I(x) * M_n^o] \tag{4.31}$$

to deliver the even and odd components at the nth scale $e_n(x)$ and $o_n(x)$, respectively. The amplitude of the transform result at this scale is the local energy

$$A_n(x) = \sqrt{e_n(x)^2 + o_n(x)^2} \tag{4.32}$$

At each point x we will have an array of vectors which correspond to each scale of the filter. Given that we are only interested in phase congruency that occurs over a wide range of frequencies (rather than just at a couple of scales), the set of wavelet filters needs to be designed so that adjacent components overlap. By summing the even and odd components we obtain

$$F(x) = \sum_n e_n(x)$$

$$H(x) = \sum_n o_n(x) \tag{4.33}$$

and a measure of the total energy A as

$$\sum_n A_n(x) \approx \sum_n \sqrt{e_n(x)^2 + o_n(x)^2} \tag{4.34}$$

Then a measure of phase congruency is

$$PC(x) = \frac{\sqrt{F(x)^2 + H(x)^2}}{\sum_n A_n(x) + \varepsilon} \tag{4.35}$$

where the addition of a small factor ε in the denominator avoids division by zero and any potential result when values of the numerator are very small. This gives a measure of phase congruency, which is essentially a measure of the local energy. Kovesi improved on this, improving on the response to noise, developing a measure which reflects the confidence that the signal is significant relative to the noise. Further, he considers in detail the frequency domain considerations, and its extension to two dimensions (Kovesi, 1999). For 2D (image) analysis, phase congruency can be determined by convolving a set of wavelet filters with an image, and calculating the difference between the average filter response and the individual filter responses. The filters are constructed in the frequency domain by using complementary spreading functions; the filters must be constructed in the Fourier domain because the log-Gabor function has a singularity at $\omega = 0$. To construct a filter with appropriate properties, a filter is constructed in a manner similar to the Gabor wavelet, but here in the frequency domain and using different functions. Following Kovesi's implementation, the first filter is a low-pass filter, here a Gaussian filter g with L different orientations

$$g(\theta, \theta_l) = \frac{1}{\sqrt{2\pi}\sigma_s} e^{-\frac{(\theta - \theta_l)^2}{2\sigma_s^2}} \tag{4.36}$$

where θ is the orientation, σ_s controls the spread about that orientation and θ_l is the angle is local orientation focus. The other spreading function is a band-pass filter, here a log-Gabor filter lg with M different scales.

$$lg\,(\omega, \omega_m) = \begin{cases} 0 & \omega = 0 \\ \dfrac{1}{\sqrt{2\pi}\sigma_\beta} e^{-\frac{\left(\log\left(\omega/\omega_m\right)\right)^2}{2(\log(\beta))^2}} & \omega \neq 0 \end{cases} \tag{4.37}$$

where ω is the scale, β controls bandwidth at that scale and ω_m is the centre frequency at that scale. The combination of these functions provides a 2D filter $l2Dg$ which can act at different scales and orientations.

$$l2Dg\,(\omega, \omega_m, \theta, \theta_l) = g\,(\theta, \theta_l) \times lg\,(\omega, \omega_m) \tag{4.38}$$

One measure of phase congruency based on the convolution of this filter with the image $\mathbf{P}$ is derived by inverse Fourier transformation $\mathfrak{I}^{-1}$ of the filter $l2Dg$ (to yield a spatial domain operator) which is convolved as

$$S\,(m)_{x,y} = \mathfrak{I}^{-1}\,(l2Dg\,(\omega, \omega_m, \theta, \theta_l))_{x,y} * \mathbf{P}_{x,y} \tag{4.39}$$

to deliver the convolution result S at the mth scale. The measure of phase congruency over the M scales is then

$$PC_{x,y} = \frac{\left| \sum\limits_{m=1}^{M} S\,(m)_{x,y} \right|}{\sum\limits_{m=1}^{M} \left| S\,(m)_{x,y} \right| + \varepsilon} \tag{4.40}$$

where the addition of a small factor ε numerator again avoids division by zero and any potential result when values of S are very small. This gives a measure of phase congruency, but is certainly a bit of an ouch, especially as it still needs refinement.

Note that keywords recur within phase congruency: frequency domain, wavelets and convolution. By its nature, we are operating in the frequency domain and there is not enough room in this text, and it is inappropriate to the scope here, to expand further. Despite this, the performance of phase congruency certainly encourages its consideration, especially if local illumination is likely to vary and if a range of features is to be considered. It is derived by an alternative conceptual basis, and this gives different insight, as well as performance. Even better, there is a Matlab implementation available, for application to images, allowing you to replicate its excellent results. There has been further research, noting especially its extension in ultrasound image analysis (Mulet-Parada and Noble, 2000) and its extension to spatiotemporal form (Myerscough and Nixon, 2004).

4.8 Localized feature extraction

Two main areas are covered here. The traditional approaches aim to derive local features by measuring specific image properties. The main target has been to estimate curvature: peaks of local curvature are corners, and analysing an image by its corners is especially suited to images of artificial objects. The second area includes more modern approaches that improve performance by using region or patch-based analysis. We shall start with the more established curvature-based operators, before moving to the patch or region-based analysis.

4.8.1 Detecting image curvature (corner extraction)

4.8.1.1 Definition of curvature

Edges are perhaps the low-level image features that are most obvious to human vision. They preserve significant features, so we can usually recognize what an image contains from its edge-detected version. However, there are other low-level features that can be used in computer vision. One important feature is *curvature*. Intuitively, we can consider curvature as the rate of change in edge direction. This rate of change characterizes the points in a curve; points where the edge direction changes rapidly are *corners*, whereas points where there is little change in edge direction correspond to straight lines. Such extreme points are very useful for shape description and matching, since they represent significant information with reduced data.

Curvature is normally defined by considering a parametric form of a planar curve. The parametric contour $v(t) = x(t) U_x + y(t) U_y$ describes the points in a continuous curve as the endpoints of the position vector. Here, the values of t define an arbitrary parameterization, the unit vectors are again $U_x = [1, 0]$ and $U_y = [0, 1]$. Changes in the position vector are given by the tangent vector function of the curve $v(t)$. That is, $\dot{v}(t) = \dot{x}(t) U_x + \dot{y}(t) U_y$. This vectorial expression has a simple intuitive meaning. If we think of the trace of the curve as the motion of a point and t is related to time, the tangent vector defines the instantaneous motion. At any moment, the point moves with a speed given by $|\dot{v}(t)| = \sqrt{\dot{x}^2(t) + \dot{y}^2(t)}$ in the direction $\varphi(t) = \tan^{-1}(\dot{y}(t)/\dot{x}(t))$. The curvature at a point $v(t)$ describes the changes in the direction $\varphi(t)$ with respect to changes in arc length. That is,

$$\kappa(t) = \frac{d\varphi(t)}{ds} \tag{4.41}$$

where s is arc length, along the edge itself. Here φ is the angle of the tangent to the curve. That is, $\varphi = \theta \pm 90°$, where θ is the gradient direction defined in Equation 4.13. That is, if we apply an edge detector operator to an image, we have for each pixel a gradient direction value that represents the normal direction to each point in a curve. The tangent to a curve is given by an orthogonal vector. Curvature is given with respect to arc length because a curve parameterized by arc length maintains a constant speed of motion. Thus, curvature represents changes in direction for constant displacements along the curve. By considering the chain rule, we have

$$\kappa(t) = \frac{d\varphi(t)}{dt} \frac{dt}{ds} \tag{4.42}$$

The differential ds/dt defines the change in arc length with respect to the parameter t. If we again consider the curve as the motion of a point, this differential defines the instantaneous change in distance with respect to time. That is, the instantaneous speed. Thus,

$$ds/dt = |\dot{v}(t)| = \sqrt{\dot{x}^2(t) + \dot{y}^2(t)} \tag{4.43}$$

and

$$dt/ds = 1 / \sqrt{\dot{x}^2(t) + \dot{y}^2(t)} \tag{4.44}$$

By considering that $\varphi(t) = \tan^{-1}(\dot{y}(t)/\dot{x}(t))$, then the curvature at a point $v(t)$ in Equation 4.42 is given by

$$\kappa(t) = \frac{\dot{x}(t)\ddot{y}(t) - \dot{y}(t)\ddot{x}(t)}{[\dot{x}^2(t) + \dot{y}^2(t)]^{3/2}} \tag{4.45}$$

This relationship is called the *curvature function* and it is the standard measure of curvature for *planar* curves (Apostol, 1966). An important feature of curvature is that it relates the derivative of a tangential vector to a normal vector. This can be explained by the simplified Serret–Frenet equations (Goetz, 1970) as follows. We can express the tangential vector in polar form as

$$\dot{v}(t) = |\dot{v}(t)| \left(\cos(\varphi(t)) + j \sin(\varphi(t)) \right) \qquad (4.46)$$

If the curve is parameterized by arc length, then $|\dot{v}(t)|$ is constant. Thus, the derivative of a tangential vector is simply given by

$$\ddot{v}(t) = |\dot{v}(t)| \left(-\sin(\varphi(t)) + j \cos(\varphi(t)) \right) \left(d\varphi(t)/dt \right) \qquad (4.47)$$

Since we are using a normal parameterization, then $d\varphi(t)/dt = d\varphi(t)/ds$. Thus, the tangential vector can be written as

$$\ddot{v}(t) = \kappa(t)\mathbf{n}(t) \qquad (4.48)$$

where $\mathbf{n}(t) = |v(t)| \left(-\sin(\varphi(t)) + j \cos(\varphi(t)) \right)$ defines the direction of $\ddot{v}(t)$, while the curvature $\kappa(t)$ defines its modulus. The derivative of the normal vector is given by $\dot{\mathbf{n}}(t) = |\dot{v}(t)| \left(-\cos(\varphi(t)) - i \sin(\varphi(t)) \right) (d\varphi(t)/ds)$, which can be written as

$$\dot{\mathbf{n}}(t) = -\kappa(t)\dot{v}(t) \qquad (4.49)$$

Clearly, $\mathbf{n}(t)$ is normal to $\dot{v}(t)$. Therefore, for each point in the curve, there is a pair of orthogonal vectors $\dot{v}(t)$ and $\mathbf{n}(t)$ whose moduli are proportionally related by the curvature.

In general, the curvature of a parametric curve is computed by evaluating Equation 4.45. For a straight *line*, for example, the second derivatives $\ddot{x}(t)$ and $\ddot{y}(t)$ are *zero*, so the curvature function is *nil*. For a *circle* of radius r, we have that $\dot{x}(t) = r\cos(t)$ and $\dot{y}(t) = -r\sin(t)$. Thus, $\ddot{y}(t) = -r\cos(t)$, $\ddot{x}(t) = -r\sin(t)$ and $\kappa(t) = 1/r$. However, for curves in digital images, the derivatives must be computed from discrete data. This can be done in three main ways. The most obvious approach is to calculate curvature by directly computing the difference between angular direction of successive edge pixels in a curve. A second approach is to derive a measure of curvature changes in image intensity. Finally, a measure of curvature can be obtained by correlation.

4.8.1.2 Computing differences in edge direction

Perhaps the easier way to compute curvature in digital images is to measure the *angular change* along the curve's path. This approach was considered in early corner detection techniques (Bennett and MacDonald, 1975; Groan and Verbeek, 1978; Kitchen and Rosenfeld, 1982) and it merely computes the *difference* in edge *direction* between connected pixels forming a discrete curve. That is, it approximates the derivative in Equation 4.41 as the difference between neighbouring pixels. As such, curvature is simply given by

$$k(t) = \varphi_{t+1} - \varphi_{t-1} \qquad (4.50)$$

where the sequence $\ldots \varphi_{t-1}, \varphi_t, \varphi_{t+1}, \varphi_{t+2} \ldots$ represents the gradient direction of a sequence of pixels defining a curve segment. Gradient direction can be obtained as the angle given by an edge detector operator. Alternatively, it can be computed by considering the position of pixels in the sequence. That is, by defining $\varphi_t = (y_{t-1} - y_{t+1})/(x_{t-1} - x_{t+1})$, where (x_t, y_t) denotes pixel t in the sequence. Since edge points are only defined at discrete points, this angle can only take eight values, so the computed curvature is very ragged. This can be smoothed out by considering the difference in mean angular direction of n pixels on the leading and trailing curve segment. That is,

$$k_n(t) = \frac{1}{n}\sum_{i=1}^{n}\varphi_{t+i} - \frac{1}{n}\sum_{i=-n}^{-1}\varphi_{t+i} \qquad (4.51)$$

The average also gives some immunity to noise and it can be replaced by a weighted average if Gaussian smoothing is required. The number of pixels considered, the value of n, defines a compromise between accuracy and noise sensitivity. Notice that filtering techniques may also be used to reduce the quantization effect when angles are obtained by an edge detection operator. As we have already discussed, the level of filtering the filtering is related to the size of the template (as in Section 3.4.3).

To compute angular differences, we need to determine connected edges. This can easily be implemented with the code already developed for hysteresis thresholding in the Canny edge operator. To compute the difference of points in a curve, the connect routine (Code 4.12) only needs to be arranged to store the difference in edge direction between connected points. Code 4.16 shows an implementation for curvature detection. First, edges and magnitudes are determined. Curvature is only detected at edge points. As such, we apply maximal suppression. The function Cont returns a matrix containing the connected neighbour pixels of each edge. Each edge pixel is connected to one or two neighbours. The matrix Next stores only the direction of consecutive pixels in an edge. We use a value of -1 to indicate that there is no connected neighbour. The function NextPixel obtains the position of a neighbouring pixel

```
%Curvature detection
function outputimage=CurvConnect(inputimage)

         [rows,columns]=size(inputimage);    %Image size
         outputimage=zeros(rows,columns);    %Result image
         [Mag,Ang]=Edges(inputimage);        %Edge Detection
         Mag=MaxSupr(Mag,Ang);               %Maximal Suppression
         Next=Cont(Mag,Ang);                 %Next connected pixels


         %Compute curvature in each pixel
         for x=1:columns-1
           for y=1:rows-1
             if Mag(y,x)~=0
               n=Next(y,x,1); m=Next(y,x,2);
                 if(n~=-1 & m~=-1)
                   [px,py]=NextPixel(x,y,n);
                   [qx,qy]=NextPixel(x,y,m);
                 outputimage(y,x)=abs(Ang(py,px)-Ang(qy,qx));
             end
           end
         end
       end
```

Code 4.16   Curvature by differences

by taking the position of a pixel and the direction of its neighbour. The curvature is computed as the difference in gradient direction of connected neighbour pixels.

The result of applying this form of curvature detection to an image is shown in Figure 4.37. Figure 4.37(a) contains the silhouette of an object; Figure 4.39(b) is the curvature obtained by computing the rate of change of edge direction. In this figure, curvature is defined only at the edge points. Here, by its formulation the measurement of curvature κ gives just a thin line of differences in edge direction which can be seen to track the perimeter points of the shapes (at points where there is measured curvature). The brightest points are those with greatest curvature. To show the results, we have scaled the curvature values to use 256 intensity values. The estimates of corner points could be obtained by a uniformly thresholded version of Figure 4.37(b), well in theory anyway!

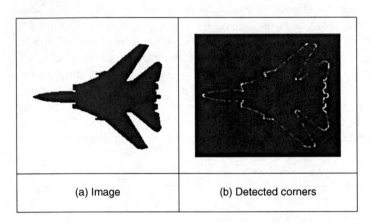

| (a) Image | (b) Detected corners |

Figure 4.37   Curvature detection by difference

Unfortunately, as can be seen, this approach does not provide reliable results. It is essentially a reformulation of a first order edge detection process and presupposes that the corner information lies within the threshold data (and uses no corner structure in detection). One of the major difficulties with this approach is that measurements of angle can be severely affected by *quantization error* and accuracy is *limited* (Bennett and MacDonald, 1975), a factor which will return to plague us later when we study methods for describing shapes.

4.8.1.3   *Measuring curvature by changes in intensity (differentiation)*

As an alternative way of measuring curvature, we can derive the curvature as a function of *changes* in *image intensity*. This derivation can be based on the measure of angular changes in the discrete image. We can represent the direction at each image point as the function $\varphi'(x, y)$. Thus, according to the definition of curvature, we should compute the change in these direction values normal to the image edge (i.e. along the curves in an image). The curve at an edge can be locally approximated by the points given by the parametric line defined by $x(t) = x + t\cos(\varphi'(x, y))$ and $y(t) = y + t\sin(\varphi'(x, y))$. Thus, the curvature is given by the change in the function $\varphi'(x, y)$ with respect to t. That is,

$$\kappa_{\varphi'}(x, y) = \frac{\partial \varphi'(x, y)}{\partial t} = \frac{\partial \varphi'(x, y)}{\partial x}\frac{\partial x(t)}{\partial t} + \frac{\partial \varphi'(x, y)}{\partial y}\frac{\partial y(t)}{\partial t} \tag{4.52}$$

where $\partial x(t)/\partial t = \cos(\varphi')$ and $\partial y(t)/\partial t = \sin(\varphi')$. By considering the definition of the gradient angle, we have that the normal tangent direction at a point in a line is given by $\varphi'(x,y) = \tan^{-1}\left(Mx/(-My)\right)$. From this geometry we can observe that

$$\cos(\varphi') = -My\left/\sqrt{Mx^2+My^2}\right. \quad \text{and} \quad \sin(\varphi') = Mx\left/\sqrt{Mx^2+My^2}\right. \tag{4.53}$$

By differentiation of $\varphi'(x,y)$ and by considering these definitions we obtain

$$\kappa_{\varphi'}(x,y) = \frac{1}{(Mx^2+My^2)^{\frac{3}{2}}}\left\{My^2\frac{\partial Mx}{\partial x} - MxMy\frac{\partial My}{\partial x} + Mx^2\frac{\partial My}{\partial y} - MxMy\frac{\partial Mx}{\partial y}\right\} \tag{4.54}$$

This defines a *forward* measure of curvature along the edge direction. We can use an alternative direction to measure of curvature. We can differentiate *backwards* (in the direction of $-\varphi'(x,y)$) giving $\kappa_{-\varphi'}(x,y)$. In this case we consider that the curve is given by $x(t) = x+t\cos(-\varphi'(x,y))$ and $y(t) = y+t\sin(-\varphi'(x,y))$. Thus,

$$\kappa_{-\varphi'}(x,y) = \frac{1}{(Mx^2+My^2)^{\frac{3}{2}}}\left\{My^2\frac{\partial Mx}{\partial x} - MxMy\frac{\partial My}{\partial x} - Mx^2\frac{\partial My}{\partial y} + MxMy\frac{\partial Mx}{\partial y}\right\} \tag{4.55}$$

Two *further* measures can be obtained by considering the forward and a backward differential along the *normal*. These differentials cannot be related to the actual definition of curvature, but can be explained intuitively. If we consider that curves are more than one pixel wide, differentiation along the edge will measure the difference between the gradient angle between interior and exterior borders of a wide curve. In theory, the tangent angle should be the same. However, in discrete images there is a change due to the measures in a window. If the curve is a straight line, then the interior and exterior borders are the same. Thus, gradient direction normal to the edge does not change locally. As we bend a straight line, we increase the difference between the curves defining the interior and exterior borders. Thus, we expect the measure of gradient direction to change. That is, if we differentiate along the normal direction, we maximize detection of gross curvature. The value $\kappa_{\perp\varphi'}(x,y)$ is obtained when $x(t) = x+t\sin(\varphi'(x,y))$ and $y(t) = y+t\cos(\varphi'(x,y))$. In this case,

$$\kappa_{\perp\varphi'}(x,y) = \frac{1}{(Mx^2+My^2)^{\frac{3}{2}}}\left\{Mx^2\frac{\partial My}{\partial x} - MxMy\frac{\partial My}{\partial x} - MxMy\frac{\partial My}{\partial y} + MyMy\frac{\partial Mx}{\partial y}\right\} \tag{4.56}$$

In a *backward* formulation along a *normal* direction to the edge, we obtain:

$$\kappa_{-\perp\varphi'}(x,y) = \frac{1}{(Mx^2+My^2)^{\frac{3}{2}}}\left\{-Mx^2\frac{\partial My}{\partial x} + MxMy\frac{\partial Mx}{\partial x} - MxMy\frac{\partial My}{\partial y} + My^2\frac{\partial Mx}{\partial y}\right\} \tag{4.57}$$

This was originally used by Kass et al. (1988) as a means to detect *line terminations*, as part of a feature extraction scheme called snakes (active contours), which are covered in Chapter 6. Code 4.17 shows an implementation of the four measures of curvature. The function Gradient is used to obtain the gradient of the image and to obtain its derivatives. The output image is obtained by applying the function according to the selection of parameter op.

Low-level feature extraction (including edge detection) 157

```
%Gradient Corner Detector
%op=T  tangent direction
%op=TI tangent inverse
%op=N  normal direction
%op=NI normal  inverse

function outputimage=GradCorner(inputimage,op)
         [rows,columns]=size(inputimage);    %Image size
         outputimage=zeros(rows,columns);    %Result image
         [Mx,My]=Gradient(inputimage);       %Gradient images
         [M,A]=Edges(inputimage);            %Edge Suppression
         M=MaxSupr(M,A);
         [Mxx,Mxy]=Gradient(Mx);             %Derivatives of the
         [Myx,Myy]=Gradient(My);             %gradient image


         %compute curvature
         for x=1:columns
           for y=1:rows
             if(M(y,x)~=0)
                   My2=My(y,x)^2; Mx2=Mx(y,x)^2; MxMy=Mx(y,x)*My(y,x);
                   if((Mx2+My2)~=0)
                       if(op=='TI')
                           outputimage(y,x)=(1/(Mx2+My2)^1.5)*(My2*Mxx(y,x)
                                             -MxMy*Myx(y,x)-Mx2*Myy(y,x)
                                             +MxMy*Mxy(y,x));
                       elseif (op=='N')
                           outputimage(y,x)=(1/(Mx2+My2)^1.5)*(Mx2*Myx(y,x)
                                             -MxMy*Mxx(y,x)-MxMy*Myy(y,x)
                                             +My2*Mxy(y,x));
                       elseif (op=='NI')
                           outputimage(y,x)=(1/(Mx2+My2)^1.5)*(-Mx2*Myx(y,x)
                                             +MxMy*Mxx(y,x)-MxMy*Myy(y,x)
                                             +My2*Mxy(y,x));
                       else %tangential as default
                           outputimage(y,x)=(1/(Mx2+My2)^1.5)*(My2*Mxx(y,x)
                                             -MxMy*Myx(y,x)+Mx2*Myy(y,x)
                                             -MxMy*Mxy(y,x));
                       end
                   end
               end
           end
         end
```

Code 4.17 Curvature by measuring changes in intensity

Let us see how the four functions for estimating curvature from image intensity perform for the image given in Figure 4.37(a). In general, points where the curvature is large are highlighted by each function. Different measures of curvature (Figure 4.38) highlight differing points on the feature boundary. All measures appear to offer better performance than that derived by reformulating hysteresis thresholding (Figure 4.37b), although there is little discernible performance advantage between the direction of differentiation. As the results in Figure 4.38 suggest, detecting curvature directly from an image is not a totally reliable way of determining curvature, and

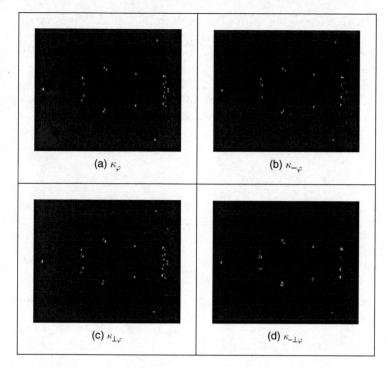

(a) κ_φ

(b) $\kappa_{-\varphi}$

(c) $\kappa_{\perp\varphi}$

(d) $\kappa_{-\perp\varphi}$

Figure 4.38 Comparing image curvature detection operators

hence corner information. This is in part due to the higher order of the differentiation process. (Also, scale has not been included within the analysis.)

4.8.1.4 Moravec and Harris detectors

In the previous section, we measured curvature as the derivative of the function $\varphi(x, y)$ along a particular direction. Alternatively, a measure of curvature can be obtained by considering changes along a particular direction in the image $\mathbf{P}$ itself. This is the basic idea of *Moravec's corner* detection operator. This operator computes the average change in image intensity when a window is shifted in several directions. That is, for a pixel with coordinates (x, y), and a window size of $2w + 1$ we have:

$$\mathbf{E}_{u,v}(x, y) = \sum_{i=-w}^{w} \sum_{j=-w}^{w} \left[\mathbf{P}_{x+i,y+j} - \mathbf{P}_{x+i+u,y+j+v} \right]^2 \tag{4.58}$$

This equation approximates the *autocorrelation* function in the direction (u, v). A measure of curvature is given by the minimum value of $\mathbf{E}_{u,v}(x, y)$ obtained by considering the shifts (u, v) in the four main directions. That is, by $(1,0)$, $(0,-1)$, $(0,1)$ and $(-1,0)$. The minimum is chosen because it agrees with the following two observations. First, if the pixel is in an edge defining a straight line, $\mathbf{E}_{u,v}(x, y)$ is small for a shift along the edge and large for a shift perpendicular to the edge. In this case, we should choose the small value since the curvature of the edge is small. Secondly, if the edge defines a corner, then all the shifts produce a large value. Thus, if we also chose the minimum, this value indicates high curvature. The main problem with this approach is that it considers only a small set of possible shifts. This problem is solved in the *Harris corner detector* (Harris and Stephens, 1988) by defining an analytic expression for the autocorrelation. This expression can be obtained by considering the local approximation of intensity changes.

We can consider that the points $\mathbf{P}_{x+i,y+j}$ and $\mathbf{P}_{x+i+u,y+j+v}$ define a vector (u, v) in the image. Thus, in a similar fashion to the development given in Equation 4.58, the increment in the image function between the points can be approximated by the directional derivative $u\partial\mathbf{P}_{x+i,y+j}/\partial x + v\partial\mathbf{P}_{x+i,y+j}/\partial y$. Thus, the intensity at $\mathbf{P}_{x+i+u,y+j+v}$ can be approximated as

$$\mathbf{P}_{x+i+u,y+j+v} = \mathbf{P}_{x+i,y+j} + \frac{\partial\mathbf{P}_{x+i,y+j}}{\partial x}u + \frac{\partial\mathbf{P}_{x+i,y+j}}{\partial y}v \tag{4.59}$$

This expression corresponds to the three first terms of the Taylor expansion around $\mathbf{P}_{x+i,y+j}$ (an expansion to first order). If we consider the approximation in Equation 4.58 we have:

$$\mathbf{E}_{u,v}(x, y) = \sum_{i=-w}^{w}\sum_{j=-w}^{w}\left[\frac{\partial\mathbf{P}_{x+i,y+j}}{\partial x}u + \frac{\partial\mathbf{P}_{x+i,y+j}}{\partial y}v\right]^2 \tag{4.60}$$

By expansion of the squared term (and since u and v are independent of the summations), we obtain:

$$\mathbf{E}_{u,v}(x, y) = A(x, y)u^2 + 2C(x, y)uv + B(x, y)v^2 \tag{4.61}$$

where

$$A(x, y) = \sum_{i=-w}^{w}\sum_{j=-w}^{w}\left(\frac{\partial\mathbf{P}_{x+i,y+j}}{\partial x}\right)^2 \quad B(x, y) = \sum_{i=-w}^{w}\sum_{j=-w}^{w}\left(\frac{\partial\mathbf{P}_{x+i,y+j}}{\partial y}\right)^2$$

$$C(x, y) = \sum_{i=-w}^{w}\sum_{j=-w}^{w}\left(\frac{\partial\mathbf{P}_{x+i,y+j}}{\partial x}\right)\left(\frac{\partial\mathbf{P}_{x+i,y+j}}{\partial y}\right) \tag{4.62}$$

That is, the summation of the squared components of the gradient direction for all the pixels in the window. In practice, this average can be weighted by a Gaussian function to make the measure less sensitive to noise (i.e. by filtering the image data). To measure the curvature at a point (x, y), it is necessary to find the vector (u, v) that minimizes $\mathbf{E}_{u,v}(x, y)$ given in Equation 4.61. In a basic approach, we can recall that the minimum is obtained when the window is displaced in the direction of the edge. Thus, we can consider that $u = \cos(\varphi(x, y))$ and $v = \sin(\varphi(x, y))$. These values were defined in Equation 4.53. Accordingly, the minima values that define curvature are given by

$$\kappa_{u,v}(x, y) = \min \mathbf{E}_{u,v}(x, y) = \frac{A(x, y)M_y^2 + 2C(x, y)M_xM_y + B(x, y)M_x^2}{M_x^2 + M_y^2} \tag{4.63}$$

In a more sophisticated approach, we can consider the form of the function $\mathbf{E}_{u,v}(x, y)$. We can observe that this is a quadratic function, so it has two principal axes. We can rotate the function such that its axes have the same direction as the axes of the coordinate system. That is, we rotate the function $\mathbf{E}_{u,v}(x, y)$ to obtain

$$\mathbf{F}_{u,v}(x, y) = \alpha(x, y)^2 u^2 + \beta(x, y)^2 v^2 \tag{4.64}$$

The values of α and β are proportional to the *autocorrelation* function along the principal axes. Accordingly, if the point (x, y) is in a region of constant intensity, both values are small. If the point defines a straight border in the image, then one value is large and the other is small. If the point defines an edge with high curvature, both values are large. Based on these observations a measure of curvature is defined as

$$\kappa_k(x, y) = \alpha\beta - k(\alpha + \beta)^2 \tag{4.65}$$

The first term in this equation makes the measure large when the values of α and β increase. The second term is included to decrease the values in flat borders. The parameter k must be selected to control the sensitivity of the detector. The higher the value, the more sensitive the computed curvature will be to changes in the image (and therefore to noise).

In practice, to compute $\kappa_k(x, y)$ it is not necessary to compute explicitly the values of α and β, but the curvature can be measured from the coefficient of the quadratic expression in Equation 4.61. This can be derived by considering the matrix forms of Equations 4.61 and 4.64. If we define the vector $D^T = [u, v]$, then Equations 4.60 and 4.63 can be written as

$$\mathbf{E}_{u,v}(x, y) = \mathbf{D}^T \mathbf{M} \mathbf{D} \quad \text{and} \quad \mathbf{F}_{u,v}(x, y) = \mathbf{D}^T \mathbf{Q} \mathbf{D} \tag{4.66}$$

where ^{T} denotes the transpose and where

$$\mathbf{M} = \begin{bmatrix} A(x, y) & C(x, y) \\ C(x, y) & B(x, y) \end{bmatrix} \quad \text{and} \quad \mathbf{Q} = \begin{bmatrix} \alpha & 0 \\ 0 & \beta \end{bmatrix} \tag{4.67}$$

To relate Equations 4.60 and 4.63, we consider that $F_{u,v}(x, y)$ is obtained by rotating $E_{u,v}(x, y)$ by a transformation R that rotates the axis defined by D. That is,

$$\mathbf{F}_{u,v}(x, y) = (\mathbf{R}\mathbf{D})^T \mathbf{M} \mathbf{R} \mathbf{D} \tag{4.68}$$

This can be arranged as

$$\mathbf{F}_{u,v}(x, y) = \mathbf{D}^T \mathbf{R}^T \mathbf{M} \mathbf{R} \mathbf{D} \tag{4.69}$$

By comparison with Equation 4.66, we have:

$$\mathbf{Q} = \mathbf{R}^T \mathbf{M} \mathbf{R} \tag{4.70}$$

This defines a well-known equation of linear algebra and it means that $\mathbf{Q}$ is an orthogonal decomposition of $\mathbf{M}$. The diagonal elements of $\mathbf{Q}$ are called the eigenvalues. We can use Equation 4.70 to obtain the value of $\alpha\beta$, which defines the first term in Equation 4.65 by considering the determinant of the matrices. That is, $\det(\mathbf{Q}) = \det(\mathbf{R}^T) \det(\mathbf{M}) \det(\mathbf{R})$. Since $\mathbf{R}$ is a rotation matrix $\det(\mathbf{R}^T) \det(\mathbf{R}) = 1$, thus

$$\alpha\beta = A(x, y) B(x, y) - C(x, y)^2 \tag{4.71}$$

which defines the first term in Equation 4.65. The second term can be obtained by taking the trace of the matrices on each side of this equation. Thus, we have:

$$\alpha + \beta = A(x, y) + B(x, y) \tag{4.72}$$

We can also use Equation 4.70 to obtain the value of $\alpha + \beta$, which defines the first term in Equation 4.65. By taking the trace of the matrices in each side of this equation, we have:

$$\kappa_k(x, y) = A(x, y) B(x, y) - C(x, y)^2 - k(A(x, y) + B(x, y))^2 \tag{4.73}$$

Code 4.18 shows an implementation for Equations 4.64 and 4.73. The equation to be used is selected by the op parameter. Curvature is only computed at edge points; that is, at pixels whose edge magnitude is different of zero after applying maximal suppression. The first part of the code computes the coefficients of the matrix **M**. Then, these values are used in the curvature computation.

```
%Harris Corner Detector
%op=H  Harris
%op=M  Minimum direction
function outputimage=Harris(inputimage,op)

  w=4;                                    %Window size=2w+1
  k=0.1;                                  %Second term constant
  [rows,columns]=size(inputimage);       %Image size
  outputimage=zeros(rows,columns);       %Result image
  [difx,dify]=Gradient(inputimage);      %Differential
  [M,A]=Edges(inputimage);               %Edge Suppression
  M=MaxSupr(M,A);

  %compute correlation
  for x=w+1:columns-w                %pixel (x,y)
    for y=w+1:rows-w
      if M(y,x)~=0
        %compute window average
        A=0;B=0;C=0;
      for i=-w:w
        for j=-w:w
          A=A+difx(y+i,x+j)^2;
          B=B+dify(y+i,x+j)^2;
          C=C+difx(y+i,x+j)*dify(y+i,x+j);
        end
      end

      if(op=='H')
          outputimage(y,x)=A*B-C^2-k*((A+B)^2);;
      else
          dx=difx(y,x);
          dy=dify(y,x);

          if dx*dx+dy*dy~=0
            outputimage(y,x)=((A*dy*dy-
            2*C*dx*dy+B*dx*dx)/(dx*dx+dy*dy));
          end
        end
      end
    end
  end
end
```

Code 4.18 Harris corner detector

Figure 4.39 shows the results of computing curvature using this implementation. The results are capable of showing the different curvature in the border. We can observe that $\kappa_k(x, y)$

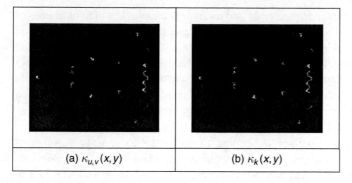

| (a) $\kappa_{u,v}(x,y)$ | (b) $\kappa_k(x,y)$ |

Figure 4.39 Curvature via the Harris operator

produces more contrast between lines with low and high curvature than $\kappa_{u,v}(x, y)$. The reason is the inclusion of the second term in Equation 4.73. In general, the measure of correlation is not only useful to compute curvature; this technique has much wider application in finding points for matching pairs of images.

4.8.1.5 *Further reading on curvature*

Many of the arguments earlier advanced on extensions to edge detection in Section 4.4 apply to corner detection as well, so the same advice applies. There is much less attention paid by established textbooks to corner detection through Davis (2005) devotes a chapter to the topic. Van Otterloo's fine book on shape analysis (van Otterloo, 1991) contains a detailed analysis of measurement of (planar) curvature.

There are other important issues in corner detection. It has been suggested that corner extraction can be augmented by local knowledge to improve performance (Rosin, 1996). There are many other corner detection schemes, each offering different attributes, although with differing penalties. Important work has focused on characterizing shapes using corners. In a scheme analogous to the **primal sketch** introduced earlier, there is a *curvature primal sketch* (Asada and Brady, 1986), which includes a set of primitive parameterized curvature discontinuities (such as termination and joining points). There are many other approaches: one suggestion is to define a corner as the intersection between two lines; this requires a process to find the lines. Other techniques use methods that describe shape variation to find corners. We commented that filtering techniques can be included to improve the detection process; however, filtering can also be used to obtain a multiple detail representation. This representation is very useful to shape characterization. A *curvature scale space* has been developed (Mokhtarian and Mackworth, 1986; Mokhtarian and Bober, 2003) to give a compact way of representing shapes, and at different scales, from coarse (low-level) to fine (detail), and with the ability to handle appearance transformations.

4.8.2 Modern approaches: region/patch analysis

4.8.2.1 *Scale invariant feature transform*

The *scale invariant feature transform* (SIFT) (Lowe, 1999, 2004) aims to resolve many of the practical problems in low-level feature extraction and their use in matching images. The earlier

Harris operator is sensitive to changes in image scale and as such is unsuited to matching images of differing size. SIFT involves two stages: feature extraction and description. The description stage concerns use of the low-level features in object matching, and this will be considered later. Low-level feature extraction within the SIFT approach selects salient features in a manner invariant to image scale (feature size) and rotation, and with partial invariance to change in illumination. Further, the formulation reduces the probability of poor extraction due to occlusion clutter and noise. It also shows how many of the techniques considered previously can be combined and capitalized on, to good effect.

First, the difference of Gaussians operator is applied to an image to identify features of potential interest. The formulation aims to ensure that feature selection does not depend on feature size (scale) or orientation. The features are then analysed to determine location and scale before the orientation is determined by local gradient direction. Finally, the features are transformed into a representation that can handle variation in illumination and local shape distortion. Essentially, the operator uses local information to refine the information delivered by standard operators. The detail of the operations is best left to the source material (Lowe, 1999, 2004), for it is beyond the level or purpose here. As such, we shall concentrate on principle only.

The features detected for the Lena image are illustrated in Figure 4.40. Here, the major features detected are shown by white lines, where the length reflects magnitude and the direction reflects the feature's orientation. These are the major features, which include the rim of the hat, face features and the boa. The minor features are the smaller white lines: the ones shown here are concentrated around a background feature. In the full set of features detected at all scales in this image, there are many more of the minor features, concentrated particularly in the textured regions of the image (Figure 4.43). Later, we shall see how this can be used within shape extraction, but the purpose here is the basic low-level features extracted by this new technique.

(a) Original image   (b) Output points with magnitude and direction

Figure 4.40   Detecting features with the SIFT operator

In the first stage, the difference of Gaussians for an image **P** is computed in the manner of Equation 4.28 as

$$D(x, y, \sigma) = \frac{(g(x, y, k\sigma) - g(x, y, \sigma)) * \mathbf{P}}{k\sigma - \sigma}$$

$$= L(x, y, k\sigma) - L(x, y, k)$$

(4.74)

The function L is a scale-space function which can be used to define smoothed images at different scales. Note again the influence of scale-space in the more modern techniques. Rather than any difficulty in locating zero-crossing points, the features are the maxima and minima of the function. Candidate keypoints are then determined by comparing each point in the function with its immediate neighbours. The process then proceeds to analysis between the levels of scale, given appropriate sampling of the scale-space. This then implies comparing a point with its eight neighbours at that scale and with the nine neighbours in each of the adjacent scales, to determine whether it is a minimum or a maximum, as well as image resampling to ensure comparison between the different scales.

To filter the candidate points to reject those which are the result of low local contrast (low edge strength) or which are poorly localized along an edge, a function is derived by local curve fitting, which indicates local edge strength and stability as well as location. Uniform thresholding then removes the keypoints with low contrast. Those that have poor localization, i.e. their position is likely to be influenced by noise, can be filtered by considering the ratio of curvature along an edge to that perpendicular to it, in a manner following the Harris operator in Section 4.8.1.4, by thresholding the ratio of Equations 4.71 and 4.72.

To characterize the filtered keypoint features at each scale, the gradient magnitude is calculated in exactly the manner of Equations 4.12 and 4.13 as

$$M_{\text{SIFT}}(x, y) = \sqrt{(L(x+1, y) - L(x-1, y))^2 + (L(x, y+1) - L(x, y-1))^2}$$

(4.75)

$$\theta_{\text{SIFT}}(x, y) = \tan^{-1}\left(\frac{L(x, y+1) - L(x, y-1)}{(L(x+1, y) - L(x-1, y))}\right)$$

(4.76)

The peak of the histogram of the orientations around a keypoint is then selected as the local direction of the feature. This can be used to derive a canonical orientation, so that the resulting descriptors are invariant with rotation. As such, this contributes to the process which aims to reduce sensitivity to camera viewpoint and to non-linear change in image brightness (linear changes are removed by the gradient operations) by analysing regions in the locality of the selected viewpoint. The main description (Lowe, 2004) considers the technique's basis in much greater detail, and outlines factors important to its performance, such as the need for sampling and performance in noise.

As shown in Figure 4.41, the technique can certainly operate well, and scale is illustrated by applying the operator to the original image and to one at half the resolution. In all, 601 keypoints are determined in the original resolution image and 320 keypoints at half the resolution. By inspection, the major features are retained across scales (a lot of minor regions in the leaves disappear at lower resolution), as expected. Alternatively, the features can be filtered further by magnitude, or even direction (if appropriate). If you want more than results to convince you, implementations are available for Windows and Linux (http://www.cs.ubc.ca/spider/lowe/research.html): a feast for a developer. These images were derived using siftWin32, version 4.

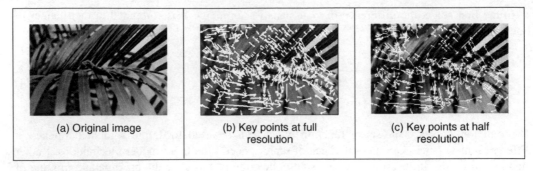

(a) Original image

(b) Key points at full resolution

(c) Key points at half resolution

Figure 4.41 SIFT feature detection at different scales

4.8.2.2 *Saliency*

The new *saliency* operator (Kadir and Brady, 2001) was also motivated by the need to extract robust and relevant features. In the approach, regions are considered salient if they are simultaneously unpredictable both in some feature and scale–space. Unpredictability (rarity) is determined in a statistical sense, generating a space of saliency values over position and scale, as a basis for later understanding. The technique aims to be a generic approach to scale and saliency compared to conventional methods, because both are defined independent of a particular basis morphology–meaning that it is not based on a particular geometric feature like a blob, edge or corner. The technique operates by determining the *entropy* (a measure of rarity) within patches at scales of interest and the saliency is a weighted summation of where the entropy peaks. The new method has practical capability in that it can be made invariant to rotation, translation, non-uniform scaling and uniform intensity variations and robust to small changes in viewpoint. An example result of processing the image in Fig. 4.42(a) is shown in Figure 4.42(b) where the 200 most salient points are shown circled, and the radius of the circle is indicative of the scale. Many of the points are around the walking subject and others highlight significant features in the background, such as the waste bins, the tree or the time index. An example use of saliency was within an approach to learn and recognize object class models (such as faces, cars or animals)

(a) Original image

(b) Top 200 saliency matches circled

Figure 4.42 Detecting features by saliency

from unlabelled and unsegmented cluttered scenes, irrespective of their overall size (Fergus et al., 2003). For further study and application, descriptions and Matlab binaries are available from Kadir's website (http://www.robots.ox.ac.uk/~timork/).

4.8.2.3 Other techniques and performance issues

There has been a recent comprehensive performance review (Mikolajczyk and Schmid, 2005), comparing established and new patch-based operators. The techniques that were compared included SIFT, differential derivatives by differentiation, cross-correlation for matching, and a gradient location and orientation-based histogram (an extension to SIFT, which performed well); the saliency approach was not included. The criterion used for evaluation concerned the number of correct matches, and the number of false matches, between feature points selected by the techniques. The matching process was between an original image and one of the same scene when subject to one of six image transformations. The image transformations covered practical effects that can change image appearance, and were: rotation, scale change, viewpoint change, image blur, JPEG compression, and illumination. For some of these there were two scene types available, which allowed for separation of understanding of scene type and transformation. The study observed that, within its analysis, 'the SIFT-based descriptors perform best', but this is a complex topic and selection of technique is often application dependent. Note that there is further interest in performance evaluation, and in invariance to higher order changes in viewing geometry, such as invariance to affine and projective transformation.

4.9 Describing image motion

We have looked at the main low-level features that we can extract from a single image. In the case of motion, we must consider more than one image. If we have two images obtained at different times, the simplest way in which we can detect *motion* is by image *differencing*. That is, changes or motion can be located by subtracting the intensity values; when there is not motion, the subtraction will give a zero value and when an object in the image moves their pixel's intensity changes, so the subtraction will give a value different of zero.

To denote a sequence of images, we include a time index in our previous notation. That is, $\mathbf{P}(t)_{x,y}$. Thus, the image at the origin of our time is $\mathbf{P}(0)_{x,y}$ and the next image is $\mathbf{P}(1)_{x,y}$. As such, the image differencing operation which delivered the difference image $\mathbf{D}$ is given by

$$\mathbf{D}(t) = \mathbf{P}(t) - \mathbf{P}(t-1) \tag{4.77}$$

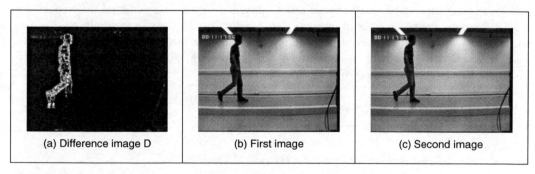

| (a) Difference image D | (b) First image | (c) Second image |

Figure 4.43 Detecting motion by differencing

Figure 4.43 shows an example of this operation. The image in 4.43(a) is the result of subtracting the image in Figure 4.43(b) from the one in Figure 4.43(c). This shows rather more than just the bits that are moving; we have not just highlighted the moving subject, we have also highlighted bits above the subject's head and around his feet. This is due mainly to change in the lighting (the shadows around the feet are to do with the subject's interaction with the lighting). However, perceived change can also be due to motion of the camera and to the motion of other objects in the field of view. In addition to these inaccuracies, perhaps the most important limitation of differencing is the lack of information about the movement itself. That is, we cannot see exactly *how* image points have moved. To describe the way in which the points in an image move, we should study how the pixels' position changes in each image frame.

4.9.1 Area-based approach

When a scene is captured at different times, 3D elements are mapped into corresponding pixels in the images. Thus, if image features are not occluded, they can be related to each other and motion can be characterized as a collection of displacements in the image plane. The displacement corresponds to the projection of movement of the objects in the scene and it is referred to as the *optical flow*. If you were to take an image, and its optical flow, you should be able to construct the *next* frame in the image sequence. So optical flow is like a measurement of velocity, the movement in pixels/unit of time, more simply pixels/frame. Optical flow can be found by looking for corresponding features in images. We can consider alternative features such as points, pixels, curves or complex descriptions of objects.

The problem of finding correspondences in images has motivated the development of many techniques that can be distinguished by the features, the constraints imposed and the optimization or searching strategy (Dhond and Aggarwal, 1989). When features are pixels, the correspondence can be found by observing the similarities between intensities in image regions (local neighbourhood). This approach is known as area-based matching and it is one of the most common techniques used in computer vision (Barnard and Fichler, 1987). In general, pixels in non-occluded regions can be related to each other by means of a general transformation of the form

$$\mathbf{P}(t+1)_{x+\delta x, y+\delta y} = \mathbf{P}(t)_{x,y} + \mathbf{H}(t)_{x,y} \qquad (4.78)$$

where the function $\mathbf{H}(t)_{x,y}$ compensates for intensity differences between the images, and $(\delta x, \delta y)$ defines the displacement vector of the pixel at time $t+1$. That is, the intensity of the pixel in the frame at time $t+1$ is equal to the intensity of the pixel in the position (x, y) in the previous frame plus some small change due to physical factors and temporal differences that induce the photometric changes in images. These factors can be due, for example, to shadows, specular reflections, differences in illumination or changes in observation angles. In a general case, it is extremely difficult to account for the photometric differences, thus the model in Equation 4.78 is generally simplified by assuming that

- the *brightness* of a point in an image is *constant*
- that *neighbouring* points move with *similar* velocity.

According to the first assumption, we have $\mathbf{H}(x) \approx 0$. Thus,

$$\mathbf{P}(t+1)_{x+\delta x, y+\delta y} = \mathbf{P}(t)_{x,y} \qquad (4.79)$$

Many techniques have used this relationship to express the matching process as an optimization or variational problem (Jordan and Bovik, 1992). The objective is to find the vector $(\delta x, \delta y)$ that minimizes the error given by

$$e_{x,y} = S\left(\mathbf{P}(t+1)_{x+\delta x, y+\delta y}, \mathbf{P}(t)_{x,y}\right) \qquad (4.80)$$

where $S()$ represents a function that measures the similarity between pixels. As such, the optimum is given by the displacements that minimizes the image differences. Alternative measures of similarity can be used to define the matching cost (Jordan and Bovik, 1992). For example, we can measure the difference by taking the absolute of the arithmetic difference. Alternatively, we can consider the correlation or the squared values of the difference or an equivalent normalized form. In practice, it is difficult to try to establish a conclusive advantage of a particular measure, since they will perform differently depending on the kind of image, the kind of noise and the nature of the motion we are observing. As such, one is free to use any measure as long as it can be justified based on particular practical or theoretical observations. The correlation and the squared difference will be explained in more detail in the next chapter when we consider how a template can be located in an image. We shall see that if we want to make the estimation problem in Equation 4.80 equivalent to maximum likelihood estimation then we should minimize the squared error. That is,

$$e_{x,y} = \left(\mathbf{P}(t+1)_{x+\delta x, y+\delta y}, \mathbf{P}(t)_{x,y}\right)^2 \qquad (4.81)$$

In practice, the implementation of the minimization is extremely prone to error since the displacement is obtained by comparing intensities of single pixels; it is very likely that the intensity changes or that a pixel can be confused with other pixels. To improve the performance, the optimization includes the second assumption presented above. If neighbouring points move with similar velocity, we can determine the displacement by considering not just a single pixel, but pixels in a neighbourhood. Thus,

$$e_{x,y} = \sum_{(x',y') \in W} \left(\mathbf{P}(t+1)_{x'+\delta x, y'+\delta y}, \mathbf{P}(t)_{x',y'}\right)^2 \qquad (4.82)$$

That is, the error in the pixel at position (x, y) is measured by comparing all the pixels (x', y') in a window W. This makes the measure more stable by introducing an implicit smoothing factor. The size of the window is a compromise between noise and accuracy. The automatic selection of the window parameter has attracted some interest (Kanade and Okutomi, 1994). Another important problem is the amount of computation involved in the minimization when the displacement between frames is large. This has motivated the development of hierarchical implementations. Other extensions have considered more elaborate assumptions about the speed of neighbouring pixels.

A straightforward implementation of the minimization of the square error is presented in Code 4.19. This function has a pair of parameters that define the maximum displacement and the window size. The optimum displacement for each pixel is obtained by comparing the error for all the potential integer displacements. In a more complex implementation, it is possible to obtain displacements with subpixel accuracy (Lawton, 1983). This is normally achieved by a postprocessing step based on subpixel interpolation or by matching surfaces obtained by fitting the data at the integer positions. The effect of the selection of different window parameters can be seen in the example shown in Figure 4.44. Figure 4.44(a) and (b) show an object moving up into a static background (at least for the two frames we are considering). Figure 4.44(c)–(e) shows the displacements obtained by considering windows of increasing size. Here, we can observe that as the size of the window increases, the result is smoother, but detail has been about

```
%Optical flow by correlation
%d: max displacement., w:window size 2w+1
function FlowCorr(inputimage1,inputimage2,d,w)

%Load images
L1=double(imread(inputimage1, 'bmp'));
L2=double(imread(inputimage2,'bmp'));

%image size
[rows,columns]=size(L1); %L2 must have the same size

%result image
u=zeros(rows,columns);
v=zeros(rows,columns);

%correlation for each pixel
for x1=w+d+1:columns-w-d
      for y1=w+d+1:rows-w-d
             min=99999; dx=0; dy=0;
             %displacement position
             for x2=x1-d:x1+d
                   for y2=y1-d:y1+d
                   sum=0;
                   for i=-w:w% window
                         for j=-w:w
                               sum=sum+(double(L1(y1+j,x1+i))-
                               double(L2(y2+j,x2+i)))^2;
                         end
                   end
                   if (sum<min)
                         min=sum;
                         dx=x2-x1; dy=y2-y1;
                   end
             end
      end
      u(y1,x1)=dx;
      v(y1,x1)=dy;
      end
end

%display result
quiver(u,v,.1);
```

Code 4.19 Implementation of area-based motion computation

the boundary of the object. We can also observe that when the window is small, the are noisy displacements near the object's border. This can be explained by considering that Equation 4.78 supposes that pixels appear in both images, but this is not true near the border since pixels appear and disappear (i.e. occlusion) from and behind the moving object. In addition, there are problems in regions that lack of intensity variations (texture). This is because the minimization

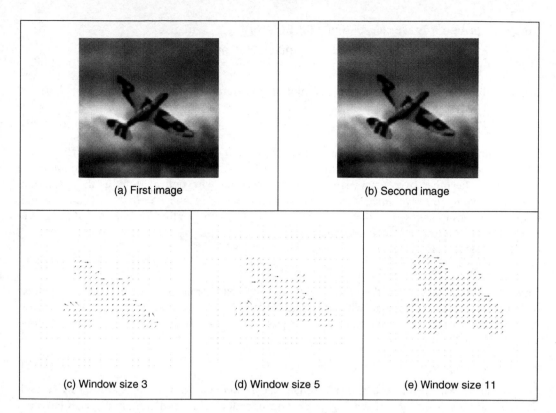

(a) First image   (b) Second image

(c) Window size 3   (d) Window size 5   (e) Window size 11

Figure 4.44 Example of area-based motion computation

function in Equation 4.81 is almost flat and there is no clear evidence of the motion. In general, there is not a very effective way of handling these problems since they are caused by the lack of information in the image.

4.9.2 Differential approach

Another popular way to estimate motion focuses on the observation of the differential changes in the pixel values. There are many ways of calculating the optical flow by this approach (Nagel, 1987; Barron et al., 1994). We shall discuss one of the more popular techniques (Horn and Schunk, 1981). We start by considering the intensity equity in Equation 4.79. According to this, the brightness at the point in the *new* position should be the same as the brightness at the *old* position. Like Equation 4.5, we can expand $\mathbf{P}(t+\delta t)_{x+\delta x, y+\delta y}$ by using a Taylor series as

$$\mathbf{P}(t+\delta t)_{x+\delta x, y+\delta y} = \mathbf{P}(t)_{x,y} + \delta x \frac{\partial \mathbf{P}(t)_{x,y}}{\partial x} + \delta y \frac{\partial \mathbf{P}(t)_{x,y}}{\partial y} + \delta t \frac{\partial \mathbf{P}(t)_{x,y}}{\partial t} + \xi \tag{4.83}$$

where ξ contains higher order terms. If we take the limit as $\delta t \to 0$ then we can ignore ξ as it also tends to zero, which leaves

$$\mathbf{P}(t+\delta t)_{x+\delta x, y+\delta y} = \mathbf{P}(t)_{x,y} + \delta x \frac{\partial \mathbf{P}(t)_{x,y}}{\partial x} + \delta y \frac{\partial \mathbf{P}(t)_{x,y}}{\partial y} + \delta t \frac{\partial \mathbf{P}(t)_{x,y}}{\partial t} \tag{4.84}$$

Low-level feature extraction (including edge detection)   171

Now by Equation 4.79 we can substitute for $\mathbf{P}(t + \delta t)_{x+\delta x, y+\delta y}$ to give

$$\mathbf{P}(t)_{x,y} = \mathbf{P}(t)_{x,y} + \delta x \frac{\partial \mathbf{P}(t)_{x,y}}{\partial x} + \delta y \frac{\partial \mathbf{P}(t)_{x,y}}{\partial y} + \delta t \frac{\partial \mathbf{P}(t)_{x,y}}{\partial t} \tag{4.85}$$

which with some rearrangement gives the motion constraint equation

$$\frac{\delta x}{\delta t} \frac{\partial \mathbf{P}}{\partial x} + \frac{\delta y}{\delta t} \frac{\partial \mathbf{P}}{\partial y} = -\frac{\partial \mathbf{P}}{\partial t} \tag{4.86}$$

We can recognize some terms in this equation. $\partial \mathbf{P}/\partial x$ and $\partial \mathbf{P}/\partial y$ are the first order differentials of the image intensity along the two image axes. $\partial \mathbf{P}/\partial t$ is the rate of change of image intensity with time. The other two factors are the ones concerned with optical flow, as they describe movement along the two image axes. Let us call $u = \delta x/\delta t$ and $v = \delta y/\delta t$. These are the optical flow components: u is the *horizontal optical flow* and v is the *vertical optical flow*. We can write these into our equation to give

$$u \frac{\partial \mathbf{P}}{\partial x} + v \frac{\partial \mathbf{P}}{\partial y} = -\frac{\partial \mathbf{P}}{\partial t} \tag{4.87}$$

This equation suggests that the optical flow and the spatial rate of intensity change together describe how an image changes with time. The equation can be expressed more simply in vector form in terms of the intensity change $\nabla \mathbf{P} = [\nabla x \; \nabla y] = [\partial \mathbf{P}/\partial x \; \partial \mathbf{P}/\partial y]$ and the optical flow $\mathbf{v} = [u \; v]^T$, as the dot product

$$\nabla \mathbf{P} \cdot \mathbf{v} = -\dot{\mathbf{P}} \tag{4.88}$$

We already have operators that can estimate the spatial intensity change, $\nabla x = \partial \mathbf{P}/\partial x$ and $\nabla y = \partial \mathbf{P}/\partial y$, by using one of the edge detection operators described earlier. We also have an operator which can estimate the rate of change of image intensity, $\nabla t = \partial \mathbf{P}/\partial t$, as given by Equation 4.77. Unfortunately, we cannot determine the optical flow components from Equation 4.87 since we have one equation in two unknowns (there are many possible pairs of values for u and v that satisfy the equation). This is called the *aperture problem* and makes the problem *ill-posed*. Essentially, we seek estimates of u and v that minimize the error in Equation 4.90 over the entire image. By expressing Equation 4.87 as

$$u \nabla x + v \nabla y + \nabla t = 0 \tag{4.89}$$

we then seek estimates of u and v that minimize the error ec for all the pixels in an image

$$ec = \iint (u \nabla x + v \nabla y + \nabla t)^2 \mathrm{d}x \mathrm{d}y \tag{4.90}$$

We can approach the solution (equations to determine u and v) by considering the second assumption we made earlier, namely that neighbouring points move with similar velocity. This is called the *smoothness constraint* as it suggests that the velocity field of the brightness varies in a smooth manner without abrupt change (or discontinuity). If we add this in to the formulation, we turn a problem that is ill-posed, without unique solution, to one that is well-posed. Properly, we define the smoothness constraint as an integral over the area of interest, as in Equation 4.90. Since we want to maximize smoothness, we seek to minimize the rate of change of the optical flow. Accordingly, we seek to minimize an integral of the rate of change of flow along both axes. This is an error es, as

$$es = \iint \left(\left(\frac{\partial u}{\partial x} \right)^2 + \left(\frac{\partial u}{\partial y} \right)^2 + \left(\frac{\partial v}{\partial x} \right)^2 + \left(\frac{\partial v}{\partial y} \right)^2 \right) \mathrm{d}x \mathrm{d}y \tag{4.91}$$

The total error is the compromise between the importance of the assumption of constant brightness and the assumption of smooth velocity. If this compromise is controlled by a *regularization parameter* λ, then the total error e is

$$e = \lambda \times ec + es$$

$$= \iint \left(\lambda \times \left(u \frac{\partial \mathbf{P}}{\partial x} + v \frac{\partial \mathbf{P}}{\partial y} + \frac{\partial \mathbf{P}}{\partial t} \right)^2 + \left(\left(\frac{\partial u}{\partial x} \right)^2 + \left(\frac{\partial u}{\partial y} \right)^2 + \left(\frac{\partial v}{\partial x} \right)^2 + \left(\frac{\partial v}{\partial y} \right)^2 \right) \right) dxdy$$

(4.92)

There are several ways to approach the solution (Horn, 1986), but the most appealing is perhaps also the most direct. We are concerned with providing estimates of optical flow at image points. So we are interested in computing the values for $u_{x,y}$ and $v_{x,y}$. We can form the error at image points, like $es_{x,y}$. Since we are concerned with image points, we can form $es_{x,y}$ by using first order differences, just like Equation 4.1 at the start of this chapter. Equation 4.90 can be implemented in discrete form as

$$es_{x,y} = \sum_x \sum_y \frac{1}{4} \left(\left(u_{x+1,y} - u_{x,y} \right)^2 + \left(u_{x,y+1} - u_{x,y} \right)^2 + \left(v_{x+1,y} - v_{x,y} \right)^2 + \left(v_{x,y+1} - v_{x,y} \right)^2 \right)$$ (4.93)

The discrete form of the smoothness constraint is then that the average rate of change of flow should be minimized. To obtain the discrete form of Equation 4.92 we then add in the discrete of ec (the discrete form of Equation 4.90) to give

$$ec_{x,y} = \sum_x \sum_y \left(u_{x,y} \nabla x_{x,y} + v_{x,y} \nabla y_{x,y} + \nabla t_{x,y} \right)^2$$

(4.94)

where $\nabla x_{x,y} = \partial \mathbf{P}_{x,y} / \partial x$, $\nabla y_{x,y} = \partial \mathbf{P}_{x,y} / \partial y$ and $\nabla t_{x,y} = \partial \mathbf{P}_{x,y} / \partial t$ are local estimates, at the point with coordinates x,y of the rate of change of the picture with horizontal direction, vertical direction and time, respectively. Accordingly, we seek values for $u_{x,y}$ and $v_{x,y}$ that minimize the total error e as given by

$$e_{x,y} = \sum_x \sum_y \left(\lambda \times ec_{x,y} + es_{x,y} \right)$$

$$= \sum_x \sum_y \left(\begin{array}{l} \lambda \times \left(u_{x,y} \nabla x_{x,y} + v_{x,y} \nabla y_{x,y} + \nabla t_{x,y} \right)^2 + \\ \frac{1}{4} \left(\left(u_{x+1,y} - u_{x,y} \right)^2 + \left(u_{x,y+1} - u_{x,y} \right)^2 + \left(v_{x+1,y} - v_{x,y} \right)^2 + \left(v_{x,y+1} - v_{x,y} \right)^2 \right) \end{array} \right)$$

(4.95)

Since we seek to minimize this equation with respect to $u_{x,y}$ and $v_{x,y}$ we differentiate it separately, with respect to the two parameters of interest, and the resulting equations when equated to zero should yield the equations we seek. As such,

$$\frac{\partial e_{x,y}}{\partial u_{x,y}} = \left(\lambda \times 2 \left(u_{x,y} \nabla x_{x,y} + v_{x,y} \nabla y_{x,y} + \nabla t_{x,y} \right) \nabla x_{x,y} + 2 \left(u_{x,y} - \overline{u}_{x,y} \right) \right) = 0$$ (4.96)

and

$$\frac{\partial e_{x,y}}{\partial v_{x,y}} = \left(\lambda \times 2 \left(u_{x,y} \nabla x_{x,y} + v_{x,y} \nabla y_{x,y} + \nabla t_{x,y} \right) \nabla y_{x,y} + 2 \left(v_{x,y} - \overline{v}_{x,y} \right) \right) = 0$$ (4.97)

This gives a pair of equations in $u_{x,y}$ and $v_{x,y}$

$$\begin{array}{l} \left(1 + \lambda \left(\nabla x_{x,y} \right)^2 \right) u_{x,y} + \lambda \nabla x_{x,y} \nabla y_{x,y} v_{x,y} = \overline{u}_{x,y} - \lambda \nabla x_{x,y} \nabla t_{x,y} \\ \lambda \nabla x_{x,y} \nabla y_{x,y} u_{x,y} + \left(1 + \lambda \left(\nabla y_{x,y} \right)^2 \right) v_{x,y} = \overline{v}_{x,y} - \lambda \nabla x_{x,y} \nabla t_{x,y} \end{array}$$

(4.98)

This is a pair of equations in u and v with solution

$$\left(1+\lambda\left(\left(\nabla x_{x,y}\right)^2+\left(\nabla y_{x,y}\right)^2\right)\right)u_{x,y}=\left(1+\lambda\left(\nabla y_{x,y}\right)^2\right)\bar{u}_{x,y}-\lambda\nabla x_{x,y}\nabla y_{x,y}\bar{v}_{x,y}-\lambda\nabla x_{x,y}\nabla t_{x,y}$$
$$\left(1+\lambda\left(\left(\nabla x_{x,y}\right)^2+\left(\nabla y_{x,y}\right)^2\right)\right)v_{x,y}=-\lambda\nabla x_{x,y}\nabla y_{x,y}\bar{u}_{x,y}+\left(1+\lambda\left(\nabla x_{x,y}\right)^2\right)\bar{v}_{x,y}-\lambda\nabla y_{x,y}\nabla t_{x,y}$$

$$(4.99)$$

The solution to these equations is in iterative form, where we shall denote the estimate of u at iteration n as $u^{<n>}$, so each iteration calculates new values for the flow at each point according to

$$u_{x,y}^{<n+1>}=\bar{u}_{x,y}^{<n>}-\lambda\left(\frac{\nabla x_{x,y}\bar{u}_{x,y}+\nabla y_{x,y}\bar{v}_{x,y}+\nabla t_{x,y}}{\left(1+\lambda\left(\nabla x_{x,y}^2+\nabla y_{x,y}^2\right)\right)}\right)\left(\nabla x_{x,y}\right)$$

$$(4.100)$$

$$v_{x,y}^{<n+1>}=\bar{v}_{x,y}^{<n>}-\lambda\left(\frac{\nabla x_{x,y}\bar{u}_{x,y}+\nabla y_{x,y}\bar{v}_{x,y}+\nabla t_{x,y}}{\left(1+\lambda\left(\nabla x_{x,y}^2+\nabla y_{x,y}^2\right)\right)}\right)\left(\nabla y_{x,y}\right)$$

Now we have it, the pair of equations gives iterative means for calculating the images of optical flow based on differentials. To estimate the first order differentials, rather than use our earlier equations, we can consider neighbouring points in quadrants in successive images. This gives approximate estimates of the gradient based on the two frames. That is,

$$\nabla x_{x,y}=\frac{\left(\mathbf{P}(0)_{x+1,y}+\mathbf{P}(1)_{x+1,y}+\mathbf{P}(0)_{x+1,y+1}+\mathbf{P}(1)_{x+1,y+1}\right)-}{8}\frac{\left(\mathbf{P}(0)_{x,y}+\mathbf{P}(1)_{x,y}+\mathbf{P}(0)_{x,y+1}+\mathbf{P}(1)_{x,y+1}\right)}{8}$$

$$(4.101)$$

$$\nabla y_{x,y}=\frac{\left(\mathbf{P}(0)_{x,y+1}+\mathbf{P}(1)_{x,y+1}+\mathbf{P}(0)_{x+1,y+1}+\mathbf{P}(1)_{x+1,y+1}\right)-}{8}\frac{\left(\mathbf{P}(0)_{x,y}+\mathbf{P}(1)_{x,y}+\mathbf{P}(0)_{x+1,y}+\mathbf{P}(1)_{x+1,y}\right)}{8}$$

In fact, in a later reflection (Horn and Schunk, 1993) on the earlier presentation, Horn noted with rancour that some difficulty experienced with the original technique had been caused by use of simpler methods of edge detection which are not appropriate here, as the simpler versions do not deliver a correctly positioned result between two images. The time differential is given by the difference between the two pixels along the two faces of the cube, as

$$\nabla t_{x,y}=\frac{\left(\mathbf{P}(1)_{x,y}+\mathbf{P}(1)_{x+1,y}+\mathbf{P}(1)_{x,y+1}+\mathbf{P}(1)_{x+1,y+1}\right)-}{8}\frac{\left(\mathbf{P}(0)_{x,y}+\mathbf{P}(0)_{x+1,y}+\mathbf{P}(0)_{x,y+1}+\mathbf{P}(0)_{x+1,y+1}\right)}{8}$$

$$(4.102)$$

Note that if the spacing between the images is other than one unit, this will change the denominator in Equations 4.101 and 4.102, but this is a constant scale factor. We also need means to calculate the averages. These can be computed as

$$\bar{u}_{x,y}=\frac{u_{x-1,y}+u_{x,y-1}+u_{x+1,y}+u_{x,y+1}}{2}+\frac{u_{x-1,y-1}+u_{x-1,y+1}+u_{x+1,y-1}+u_{x+1,y+1}}{4}$$

$$\bar{v}_{x,y}=\frac{v_{x-1,y}+v_{x,y-1}+v_{x+1,y}+v_{x,y+1}}{2}+\frac{v_{x-1,y-1}+v_{x-1,y+1}+v_{x+1,y-1}+v_{x+1,y+1}}{4}$$

$$(4.103)$$

The implementation of the computation of optical flow by the iterative solution in Equation 4.100 is presented in Code 4.20. This function has two parameters that define the smoothing parameter

```
%Optical flow by gradient method
%s = smoothing parameter
%n = number of iterations
function OpticalFlow(inputimage1,inputimage2,s,n)

%Load images
L1=double(imread(inputimage1, 'bmp'));
L2=double(imread(inputimage2, 'bmp'));

%Image size
[rows,columns]=size(I1); %I2 must have the same size

%Result flow
u=zeros(rows,columns);
v=zeros(rows,columns);

%Temporal flow
tu=zeros(rows,columns);
tv=zeros(rows,columns);

%Flow computation
for k=1:n    %iterations
   for x=2:columns-1
       for y=2:rows-1
       %derivatives
         Ex=(L1(y,x+1)-L1(y,x)+L2(y,x+1)-L2(y,x)+L1(y+1,x+1)
            -L1(y+1,x)+L2(y+1,x+1)-L2(y+1,x))/4;
         Ey=(L1(y+1,x)-L1(y,x)+L2(y+1,x)-L2(y,x)+L1(y+1,x+1)
            -L1(y,x+1)+L2(y+1,x+1)-L2(y,x+1))/4;
         Et=(L2(y,x)-L1(y,x)+L2(y+1,x)-L1(y+1,x)+L2(y,x+1)
            -L1(y,x+1)+L2(y+1,x+1)-L1(y+1,x+1))/4;
       %average
       AU=(u(y,x-1)+u(y,x+1)+u(y-1,x)+u(y+1,x))/4;
       AV=(v(y,x-1)+v(y,x+1)+v(y-1,x)+v(y+1,x))/4;
       %update estimates
       A=(Ex*AU+Ey*AV+Et);
       B=(1+s*(Ex*Ex+Ey*Ey));
       tu(y,x)= AU-(Ex*s*A/B);
       tv(y,x)= AV-(Ey*s*A/B);
       end%for (x,y)
end
%update
for x=2:columns-1
    for y=2:rows-1
      u(y,x)=tu(y,x); v(y,x)=tv(y,x);
    end %for (x,y)
    end
end %iterations

%display result
quiver(u,v,1);
```

Code 4.20 Implementation of gradient-based motion

and the number of iterations. In the implementation, we use the matrices u, v, tu and tv to store the old and new estimates in each iteration. The values are updated according to Equation 4.100. Derivatives and averages are computed by using simplified forms of Equations 4.101–4.103. In a more elaborate implementation, it is convenient to include averages, as we discussed in the case of single image feature operators. This will improve the accuracy and reduce noise. In addition, since derivatives can only be computed for small displacements, generally, gradient algorithms are implemented with a hierarchical structure. This will enable the computation of displacements larger than one pixel.

Figure 4.45 shows some examples of optical flow computation. In these examples, we used the same images as in Figure 4.44. The first row in the figure shows three results obtained by different number of iterations and fixed smoothing parameter. In this case, the estimates converged quite quickly. Note that at the start, the estimates of flow in are quite noisy, but they quickly improve; as the algorithm progresses the results are refined and a more smooth and accurate motion is obtained. The second row in Figure 4.45 shows the results for a fixed number of iterations and a variable smoothing parameter. The regularization parameter controls the compromise between the detail and the smoothness. A *large* value of λ will enforce the *smoothness* constraint, whereas a *small* value will make the *brightness* constraint dominate the result. In the results we can observe that the largest vectors point in the expected direction, upwards, while some of the smaller vectors are not exactly correct. This is because there are occlusions and some regions have similar textures. We could select the brightest of these points

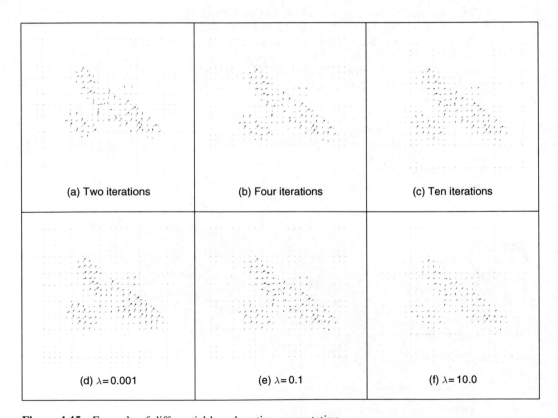

Figure 4.45 Example of differential-based motion computation

by thresholding according to magnitude. That would leave the largest vectors (the ones that point in exactly the right direction).

Optical flow has been used in automatic gait recognition (Little and Boyd, 1998; Huang et al., 1999), among other applications, partly because the displacements can be large between successive images of a walking subject, which makes the correlation approach suitable (note that fast versions of area-based correspondence are possible; Zabir and Woodfill, 1994). Figure 4.46 shows the result for a walking subject where brightness depicts magnitude (direction is not shown). Figure 4.46(a) shows the result for the differential approach, where the flow is clearly more uncertain than that produced by the correlation approach shown in Figure 4.46(b). Another reason for using the correlation approach is that we are not concerned with rotation as people (generally!) walk along flat surfaces. If 360° rotation is to be considered then you have to match regions for every rotation value and this can make the correlation-based techniques computationally very demanding indeed.

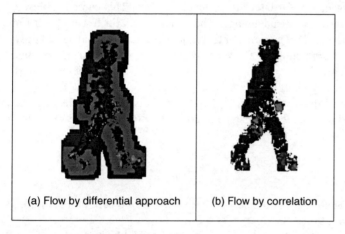

(a) Flow by differential approach   (b) Flow by correlation

Figure 4.46   Optical flow of walking subject

4.9.3   Further reading on optical flow

Determining optical flow does not get much of a mention in the established textbooks, even though it is a major low-level feature description. Rather naturally, it is to be found in depth in one of its early proponent's textbooks (Horn, 1986). One approach to motion estimation has considered the frequency domain (Adelson and Bergen, 1985) (yes, Fourier transforms get everywhere!). For a further overview of dense optical flow, see (Bulthoff et al. 1989) and for implementation, see (Little et al. 1988). The major survey (Beauchemin and Barron, 1995) of the approaches to optical flow is rather dated now, as is their performance appraisal (Barron et al., 1994). Such an (accuracy) appraisal is particularly useful in view of the number of ways there are to estimate it. The nine techniques studied included the differential approach we have studied here, a Fourier technique and a correlation-based method. Their conclusion was that a local differential method and a phase-based method (Fleet and Jepson, 1990) offered the most consistent performance on the datasets studied. However, there are many variables, not only in the data but also in implementation, that might lead to preference for a particular technique. Clearly, there are many impediments to the successful calculation of optical flow such as change in illumination or occlusion (and by other moving objects). In fact, there have been a number of studies on performance, e.g. of affine flow in Grossmann and Santos-Victor (1997). A thorough analysis of correlation techniques has been developed (Giachetti, 2000) with new algorithms for sub-pixel estimation. One study (Liu et al., 1998) notes how developments have been made for fast or for accurate techniques, without consideration of the trade-off between these two factors. The study compared the techniques mentioned previously with two newer approaches (one fast and one accurate), and also surveys real-time implementations that include implementation via parallel computers and special purpose VLSI chips.

4.10 Conclusions

This chapter has covered the main ways to extract low-level feature information. In some cases this can prove sufficient for understanding the image. Often, however, the function of low-level feature extraction is to provide information for later higher level analysis. This can be achieved in a variety of ways, with advantages and disadvantages and quickly or at a lower speed (or requiring a faster processor/more memory!). The range of techniques presented here has certainly proved sufficient for the majority of applications. There are other, more minor techniques, but the main approaches to boundary, corner and motion extraction have proved sufficiently robust and with requisite performance that they shall endure for some time.

Next, we move on to using this information at a higher level. This means collecting the information so as to find shapes and objects, the next stage in understanding the image's content.

4.11 References

Adelson, E. H. and Bergen, J. R., Spatiotemporal Energy Models for the Perception of Motion, *J. Opt. Soc. Am.*, **A2**(2), pp. 284–299, 1985

Apostol, T. M., *Calculus*, 2nd edn, Vol. 1, Xerox College Publishing, Waltham, 1966

Asada, H. and Brady, M., The Curvature Primal Sketch, *IEEE Trans. PAMI*, **8**(1), pp. 2–14, 1986

Barnard, S. T. and Fichler, M. A., Stereo Vision, In: *Encyclopedia of Artificial Intelligence*, John Wiley, New York, pp. 1083–2090, 1987

Barron, J. L., Fleet, D. J. and Beauchemin, S. S., Performance of Optical Flow Techniques, *Int. J. Comput. Vision*, **12**(1), pp. 43–77, 1994

Beauchemin, S. S. and Barron, J. L., The Computation of Optical Flow, *Communs ACM*, pp. 433–467, 1995

Bennet, J. R. and MacDonald, J. S., On the Measurement of Curvature in a Quantized Environment, *IEEE Trans. Comput.*, **C-24**(8), pp. 803–820, 1975

Bergholm, F., Edge Focussing, *IEEE Trans. PAMI*, **9**(6), pp. 726–741, 1987

Bovik, A. C., Huang, T. S. and Munson, D. C., The Effect of Median Filtering on Edge Estimation and Detection, *IEEE Trans. PAMI*, **9**(2), pp. 181–194, 1987.

Bulthoff, H., Little, J. and Poggio, T., A Parallel Algorithm for Real-Time Computation of Optical Flow, *Nature*, **337**(9), pp. 549–553, 1989

Canny, J., A Computational Approach to Edge Detection, *IEEE Trans. PAMI*, **8**(6), pp. 679–698, 1986

Clark, J. J., Authenticating Edges Produced by Zero-Crossing Algorithms, *IEEE Trans. PAMI*, **11**(1), pp. 43–57, 1989

Davies, E. R., *Machine Vision: Theory, Algorithms and Practicalities*, 3rd edn, Morgan Kaufmann (Elsevier), 2005

Deriche, R., Using Canny's Criteria to Derive a Recursively Implemented Optimal Edge Detector, *Int. J. Comput. Vision*, **1**, pp. 167–187, 1987

Dhond, U. R. and Aggarwal, J. K., Structure From Stereo – A Review, *IEEE Trans. SMC*, **19**(6), pp. 1489–1510, 1989

Fergus, R., Perona, P. and Zisserman, A., Object Class Recognition by Unsupervised Scale-Invariant Learning, *Proc. CVPR 2003*, II, pp. 264–271, 2003

Fleet, D. J. and Jepson, A. D., Computation of Component Image Velocity from Local Phase Information, *Int. J. Comput. Vision*, 5(1), pp. 77–104, 1990

Forshaw, M. R. B., Speeding Up the Marr–Hildreth Edge Operator, *CVGIP*, **41**, pp. 172–185, 1988

Giachetti, A., Matching Techniques to Compute Image Motion, *Image Vision Comput.*, **18**(3), pp. 247–260, 2000

Goetz, A., *Introduction to Differential Geometry*, Addison-Wesley, Reading, MA, 1970

Grimson, W. E. L. and Hildreth, E. C., Comments on 'Digital Step Edges from Zero Crossings of Second Directional Derivatives', *IEEE Trans. PAMI*, **7**(1), pp. 121–127, 1985

Groan, F. and Verbeek, P., Freeman-Code Probabilities of Object Boundary Quantized Contours, *CVGIP*, **7**, pp. 391–402, 1978

Grossmann, E. and Santos-Victor, J., Performance Evaluation of Optical Flow: Assessment of a New Affine Flow Method, *Robotics Auton. Syst.*, **21**, pp. 69–82, 1997

Gunn, S. R., On the Discrete Representation of the Laplacian of Gaussian, *Pattern Recog.*, **32**(8), pp. 1463–1472, 1999

Haddon, J. F., Generalized Threshold Selection for Edge Detection, *Pattern Recog.*, **21**(3), pp. 195–203, 1988

Haralick, R. M., Digital Step Edges from Zero-Crossings of Second Directional Derivatives, *IEEE Trans. PAMI*, **6**(1), pp. 58–68, 1984

Haralick, R. M., Author's Reply, *IEEE Trans. PAMI*, **7**(1), pp. 127–129, 1985

Harris, C. and Stephens, M., A Combined Corner and Edge Detector, *Proc. 4th Alvey Vision Conference*, pp. 147–151, 1988

Heath, M. D., Sarkar, S., Sanocki, T. and Bowyer, K. W., A Robust Visual Method of Assessing the Relative Performance of Edge Detection Algorithms, *IEEE Trans. PAMI*, **19**(12), pp. 1338–1359, 1997

Horn, B. K. P., *Robot Vision*, MIT Press, Cambridge, MA, 1986

Horn, B. K. P. and Schunk, B. G., Determining Optical Flow, *Artif. Intell.*, **17**, pp. 185–203, 1981

Horn, B. K. P. and Schunk, B. G., Determining Optical Flow: A Retrospective, *Artif. Intell.*, **59**, pp. 81–87, 1993

Huang, P. S., Harris, C. J. and Nixon, M. S., Human Gait Recognition in Canonical Space using Temporal Templates, *IEE Proc. Vision Image Signal Process.*, **146**(2), pp. 93–100, 1999

Huertas, A. and Medioni, G., Detection of Intensity Changes with Subpixel Accuracy using Laplacian–Gaussian Masks, *IEEE Trans. PAMI*, **8**(1), pp. 651–664, 1986

Jia, X. and Nixon, M. S., Extending the Feature Vector for Automatic Face Recognition, *IEEE Trans. PAMI*, **17**(12), pp. 1167–1176, 1995

Jordan, J. R. III and Bovik, A. C. M. S., Using Chromatic Information in Dense Stereo Correspondence, *Pattern Recog.*, **25**, pp. 367–383, 1992

Kadir, T. and Brady, M., Scale, Saliency and Image Description, *Int. J. Comput. Vision*, **45**(2), pp. 83–105, 2001

Kanade, T. and Okutomi, M., A Stereo Matching Algorithm with an Adaptive Window: Theory and Experiment, *IEEE Trans. PAMI*, **16**, pp. 920–932, 1994

Kass, M., Witkin, A. and Terzopoulos, D., Snakes: Active Contour Models, *Int. J. Comput. Vis.*, **1**(4), 321–331, 1988

Kitchen, L. and Rosenfeld, A., Gray-Level Corner Detection, *Pattern Recog. Lett.*, **1**(2), pp. 95–102, 1982

Korn, A. F., Toward a Symbolic Representation of Intensity Changes in Images, *IEEE Trans. PAMI*, **10**(5), pp. 610–625, 1988

Kovesi, P., Image Features from Phase Congruency. *Videre: J. Comput. Vision Res.*, **1**(3), pp. 1–27, 1999

Lawton, D. T., Processing Translational Motion Sequences, *CVGIP*, **22**, pp. 116–144, 1983

Lee, C. K., Haralick, M. and Deguchi, K., Estimation of Curvature from Sampled Noisy Data, *ICVPR'93*, pp. 536–541, 1993

Lindeberg, T., Scale-Space Theory: A Basic Tool for Analysing Structures at Different Scales, *J. Appl. Statist.*, **21**(2), pp. 224–270, 1994

Little, J. J. and Boyd, J. E., Recognizing People By Their Gait: The Shape of Motion, *Videre*, **1**(2), pp. 2–32, 1998, online at http://mitpress.mit.edu/e-journals/VIDE/001/v12.html

Little, J. J., Bulthoff, H. H. and Poggio, T., Parallel Optical Flow using Local Voting, *Proc. Int. Conf. Comput. Vision*, pp. 454–457, 1988

Liu, H., Hong, T.-S., Herman, M., Camus, T. and Chellappa, R., Accuracy vs Efficiency Trade-offs in Optical Flow Algorithms, *Comput. Vision Image Understand.*, **72**(3), pp. 271–286, 1998

Lowe, D. G., Object Recognition from Local Scale-Invariant Features, *Proc. Int. Conf. Comput. Vision*, pp. 1150–1157, 1999

Lowe, D. G., Distinctive Image Features from Scale-Invariant Key Points, *Int. J. Comput. Vision*, **60**(2), pp. 91–110, 2004

Lucas, B. and Kanade, T., An Iterative Image Registration Technique with an Application to Stereo Vision, *Proc DARPA Image Understanding Workshop*, pp. 121–130, 1981

Marr, D., *Vision*, W. H. Freeman and Co., New York, 1982

Marr, D. C. and Hildreth, E., Theory of Edge Detection, *Proc. R. Soc. Lond.*, **B207**, pp. 187–217, 1980

Mikolajczyk, K. and Schmid, C., A Performance Evaluation of Local Descriptors, *IEEE Trans. PAMI*, **27**(10), pp. 1615–1630, 2005

Mokhtarian, F. and Bober, M., *Curvature Scale Space Representation: Theory, Applications and MPEG-7 Standardization*, Kluwer Academic, Dordrecht, 2003

Mokhtarian, F. and Mackworth, A. K., Scale-Space Description and Recognition of Planar Curves and Two-Dimensional Shapes, *IEEE Trans. PAMI*, **8**(1), pp. 34–43, 1986

Morrone, M. C. and Burr, D. C., Feature Detection in Human Vision: A Phase-Dependent Energy Model, *Proc. R. Soc. Lond. B, Biol. Sci.*, **235**(1280), pp. 221–245, 1988

Morrone, M. C. and Owens, R. A., Feature Detection from Local Energy, *Pattern Recog. Lett.*, **6**, pp. 303–313, 1987

Mulet-Parada, M. and Noble, J. A., 2D+T Acoustic Boundary Detection in Echocardiography, *Med. Image Analysis*, **4**, 21–30, 2000

Myerscough, P.J. and Nixon, M. S., Temporal Phase Congruency, *Proc. IEEE Southwest Symposium on Image Analysis and Interpretation SSIAI '04*, pp. 76–79, 2004

Nagel, H. H., On the Estimation of Optical Flow: Relations between Different Approaches and Some New Results, *Artif. Intell.*, **33**, pp. 299–324, 1987 van Otterloo, P. J., *A Contour-Oriented Approach to Shape Analysis*, Prentice Hall International (UK), Hemel Hempstead, 1991

Parker, J. R., *Practical Computer Vision using C*, Wiley & Sons, New York, 1994

Petrou, M., The Differentiating Filter Approach to Edge Detection, *Adv. Electron. Electron Phys.*, **88**, pp. 297–345, 1994

Petrou, M. and Kittler, J., Optimal Edge Detectors for Ramp Edges, *IEEE Trans. PAMI*, **13**(5), pp. 483–491, 1991

Prewitt, J. M. S. and Mendelsohn, M. L., The Analysis of Cell Images, *Ann. N. Y. Acad. Sci.*, **128**, pp. 1035–1053, 1966

Roberts, L. G., Machine Perception of Three-Dimensional Solids, In: *Optical and Electro-Optical Information Processing*, MIT Press, Cambridge, MA, pp. 159–197, 1965

Rosin, P. L., Augmenting Corner Descriptors, *Graphical Models Image Process.*, **58**(3), pp. 286–294, 1996

Smith, S. M. and Brady, J. M., SUSAN – A New Approach to Low Level Image Processing. *Int. J. Comput. Vision*, **23**(1), pp. 45–78, May 1997

Sobel, I. E., *Camera Models and Machine Perception*, PhD Thesis, Stanford University, 1970

Spacek, L. A., Edge Detection and Motion Detection, *Image Vision Comput.*, **4**(1), pp. 43–56, 1986

Torre, V. and Poggio, T. A., On Edge Detection, *IEEE Trans. PAMI*, **8**(2), pp. 147–163, 1986

Ulupinar, F. and Medioni, G., Refining Edges Detected by a LoG Operator, *CVGIP*, **51**, pp. 275–298, 1990

Venkatesh, S. and Owens, R. A., An Energy Feature Detection Scheme. *Proc. Int. Conf. Image Process.*, Singapore, pp. 553–557, 1989

Venkatesh, S. and Rosin, P. L., Dynamic Threshold Determination by Local and Global Edge Evaluation, *Graphical Models Image Process.*, **57**(2), pp. 146–160, 1995

Vliet, L. J. and Young, I. T., A Nonlinear Laplacian Operator as Edge Detector in Noisy Images, *CVGIP*, **45**, pp. 167–195, 1989

Yitzhaky, Y. and Peli, E., A Method for Objective Edge Detection Evaluation and Detector Parameter Selection, *IEEE Trans. PAMI*, **25**(8), pp. 1027–1033, 2003

Zabir, R. and Woodfill, J., Non-Parametric Local Transforms for Computing Visual Correspondence, *Proc. Eur. Conf.Comput. Vision*, pp. 151–158, 1994

Zheng, Y., Nixon, M. S. and Allen, R., Automatic Segmentation of Lumbar Vertebrae in Digital Videofluoroscopic Imaging, *IEEE Trans. Med. Imaging*, **23**(1), pp. 45–52. 2004

5

Feature extraction by shape matching

5.1 Overview

High-level *feature extraction* concerns finding shapes in computer images. To be able to recognize faces automatically, for example, one approach is to extract the component features. This requires extraction of, say, the eyes, the ears and the nose, which are the major facial features. To find them, we can use their shape: the white part of the eyes is ellipsoidal; the mouth can appear as two lines, as do the eyebrows. Shape extraction implies finding their position, their orientation and their size. This feature extraction process can be viewed as similar to the way in which we perceive the world: many books for babies describe basic geometric shapes such as triangles, circles and squares. More complex pictures can be decomposed into a structure of simple shapes. In many applications, analysis can be guided by the way in which the shapes are arranged. For the example of face image analysis, we expect to find the eyes above (and either side of) the nose, and we expect to find the mouth below the nose.

In feature extraction, we generally seek *invariance properties* so that the extraction process does not vary according to chosen (or specified) conditions. That is, techniques should find shapes reliably and robustly whatever the value of any parameter that can control the appearance of a shape. As a basic *invariant*, we seek immunity to changes in the *illumination* level: we seek to find a shape whether it is light or dark. In principle, as long as there is contrast between a shape and its background, the shape can be said to exist, and can then be detected. (Clearly, any computer vision technique will fail in extreme lighting conditions; you cannot see anything when it is completely dark.) Following illumination, the next most important parameter is *position*: we seek to find a shape wherever it appears. This is usually called *position*, *location* or *translation invariance*. Then, we often seek to find a shape irrespective of its *rotation* (assuming that the object or the camera has an unknown orientation); this is usually called *rotation* or *orientation invariance*. Then, we might seek to determine the object at whatever *size* it appears, which might be due to physical change, or to how close the object has been placed to the camera. This requires *size* or *scale invariance*. These are the main invariance properties we shall seek from our shape extraction techniques. However, nature (as usual) tends to roll balls under our feet: there is always *noise* in images. In addition, since we are concerned with shapes, there may be more than one in the image. If one is on top of the other it will *occlude*, or hide, the other, so not all of the shape of one object will be visible.

But before we can develop image analysis techniques, we need techniques to extract the shapes. Extraction is more complex than *detection*, since extraction implies that we have a description of a shape, such as its position and size, whereas detection of a shape merely implies knowledge of its existence within an image.

The techniques presented in this chapter are outlined in Table 5.1. To extract a shape from an image, it is necessary to identify it from the background elements. This can be done by considering the intensity information or by comparing the pixels against a given template. In the first approach, if the brightness of the shape is known, then the pixels that form the shape can be extracted by classifying the pixels according to a fixed intensity threshold. Alternatively, if the background image is known, this can be subtracted to obtain the pixels that define the shape of an object superimposed on the background. Template matching is a model-based approach in which the shape is extracted by searching for the best correlation between a known model and the pixels in an image. There are alternative ways in which to compute the correlation between the template and the image. Correlation can be implemented by considering the image or frequency domains. In addition, the template can be defined by considering intensity values or a binary shape. The *Hough transform* defines an efficient implementation of template matching for binary templates. This technique is capable of extracting simple shapes such as lines and quadratic forms, as well as arbitrary shapes. In any case, the complexity of the implementation can be reduced by considering invariant features of the shapes.

Table 5.1 Overview of Chapter 5

Main topic	Sub topics	Main points
Pixel operations	How we detect features at a *pixel* level. Moving object detection. *Limitations* and *advantages* of this approach. Need for *shape* information.	Thresholding. Background subtraction.
Template matching	Shape extraction by *matching*. Advantages and disadvantages. Need for *efficient* implementation.	Template matching. Direct and Fourier implementations. Noise and occlusion.
Hough transform	Feature extraction by *matching*. Hough transforms for *conic sections*. Hough transform for *arbitrary shapes*. *Invariant* formulations. Advantages in *speed* and *efficacy*.	Feature extraction by evidence gathering. Hough transforms for lines, circles and ellipses. Generalized and Invariant Hough transforms.

5.2 Thresholding and subtraction

Thresholding is a simple shape extraction technique, as shown in Section 3.3.4, where the images could be viewed as the result of trying to separate the eye from the background. If it can be assumed that the shape to be extracted is defined by its brightness, then thresholding an image at that brightness level should find the shape. Thresholding is clearly sensitive to change in illumination: if the image illumination changes then so will the perceived brightness of the target shape. Unless the threshold level can be arranged to adapt to the change in brightness level, any thresholding technique will fail. Its attraction is *simplicity*: thresholding does not require much computational effort. If the illumination level changes in a linear fashion, using histogram equalization will result in an image that does not vary. Unfortunately, the result of histogram equalization is sensitive to noise, shadows and variant illumination; noise can affect the resulting image quite dramatically and this will again render a thresholding technique useless.

Thresholding after *intensity normalization* (Section 3.3.2) is less sensitive to noise, since the noise is stretched with the original image and cannot affect the stretching process by much. It is, however, still sensitive to shadows and variant illumination. Again, it can only find application where the illumination can be carefully controlled. This requirement is germane to any application that uses basic thresholding. If the overall illumination level cannot be controlled, it is possible to threshold edge magnitude data since this is insensitive to overall brightness level, by virtue of the implicit differencing process. However, edge data is rarely continuous and there can be gaps in the detected perimeter of a shape. Another major difficulty, which applies to thresholding the brightness data as well, is that there are often more shapes than one. If the shapes are on top of each other, one occludes the other and the shapes need to be separated.

An alternative approach is to *subtract* an image from a known background before thresholding. (We saw how we can estimate the background in Section 3.4.2.) This assumes that the background is known precisely, otherwise many more details than just the target feature will appear in the resulting image; clearly, the subtraction will be unfeasible if there is *noise* on either image, and especially on both. In this approach, there is no implicit shape description, but if the thresholding process is sufficient, it is simple to estimate basic shape parameters, such as position.

The subtraction approach is illustrated in Figure 5.1. Here, we seek to separate or extract the walking subject from the background. We saw earlier, in Figure 3.22, how the median filter can be used to provide an estimate of the background to the sequence of images that Figure 5.1(a) comes from. When we subtract the background of Figure 3.22(i) from the image of Figure 5.1(a), we obtain most of the subject with some extra background just behind the subject's head. This is due to the effect of the moving subject on *lighting*. Also, removing the background removes some of the subject: the horizontal bars in the background have been removed from the subject by the subtraction process. These aspects are highlighted in the thresholded image (Figure 5.1c). It is not a particularly poor way of separating the subject from the background (we have the subject, but we have chopped out his midriff), but it is not especially good either.

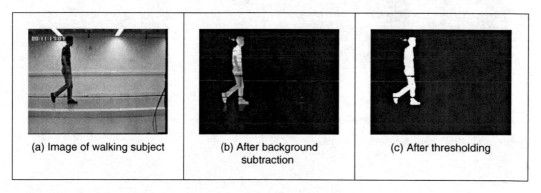

(a) Image of walking subject   (b) After background subtraction   (c) After thresholding

Figure 5.1 Shape extraction by subtraction and thresholding

Even though thresholding and subtraction are attractive (because of simplicity and hence their speed), the performance of both techniques is sensitive to partial shape data, noise, variation in illumination and occlusion of the target shape by other objects. Accordingly, many approaches to image interpretation use higher level information in shape extraction, namely how the pixels are connected within the shape. This can resolve these factors.

5.3 Template matching

5.3.1 Definition

Template matching is conceptually a simple process. We need to match a *template* to an image, where the template is a subimage that contains the shape we are trying to find. Accordingly, we centre the template on an image point and count up how many points in the template *matched* those in the image. The procedure is repeated for the entire image, and the point that led to the best match, the maximum count, is deemed to be the point where the shape (given by the template) lies within the image.

Consider that we want to find the template of Figure 5.2(b) in the image of Figure 5.2(a). The template is first positioned at the origin and then matched with the image to give a count which reflects how well the template matched that part of the image at that position. The count of matching pixels is increased by one for each point where the brightness of the template matches the brightness of the image. This is similar to the process of template convolution, illustrated earlier in Figure 3.11. The difference here is that points in the image are matched with those in the template, and the sum is of the number of matching points as opposed to the weighted sum of image data. The best match is when the template is placed at the position where the rectangle is matched to itself. This process can be generalized to find, for example, templates of different *size* or *orientation*. In these cases, we have to try all the templates (at expected rotation and size) to determine the best match.

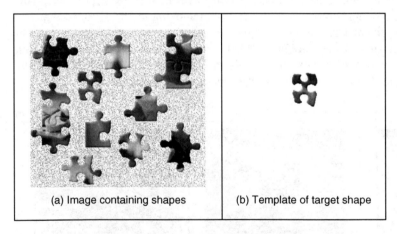

(a) Image containing shapes   (b) Template of target shape

Figure 5.2   Illustrating template matching

Formally, template matching can be defined as a method of parameter estimation. The parameters define the position (and pose) of the template. We can define a template as a discrete function $\mathbf{T}x, y$. This function takes values in a window. That is, the coordinates of the points $(x, y) \in \mathbf{W}$. For example, for a 2×2 template we have that the set of points $\mathbf{W} = \{(0,0), (0,1), (1,0), (1,1)\}$.

Let us consider that each pixel in the image $\mathbf{I}x, y$ is corrupted by additive Gaussian noise. The noise has a mean value of zero and the (unknown) standard deviation is σ. Thus, the probability that a point in the template placed at coordinates (i, j) matches the corresponding pixel at position $(x, y) \in \mathbf{W}$ is given by the normal distribution

$$p_{i,j}(x, y) = \frac{1}{\sqrt{2\pi}\sigma} e^{-\frac{1}{2}\left(\frac{\mathbf{I}_{x+i,y+j}-\mathbf{T}_{x,y}}{\sigma}\right)^2} \tag{5.1}$$

Since the noise affecting each pixel is independent, the probability that the template is at position (i, j) is the combined probability of each pixel that the template covers. That is,

$$L_{i,j} = \prod_{(x,y)\in\mathbf{W}} p_{i,j}(x, y) \tag{5.2}$$

By substitution of Equation 5.1, we have:

$$L_{i,j} = \left(\frac{1}{\sqrt{2\pi}\sigma}\right)^n e^{-\frac{1}{2}\sum_{(x,y)\in\mathbf{W}}\left(\frac{\mathbf{I}_{x+i,y+j}-\mathbf{T}_{x,y}}{\sigma}\right)^2} \tag{5.3}$$

where n is the number of pixels in the template. This function is called the *likelihood* function. Generally, it is expressed in logarithmic form to simplify the analysis. Notice that the logarithm scales the function, but it does not change the position of the maximum. Thus, by taking the logarithm, the likelihood function is redefined as

$$\ln(L_{i,j}) = n \ln\left(\frac{1}{\sqrt{2\pi}\sigma}\right) - \frac{1}{2}\sum_{(x,y)\in\mathbf{W}}\left(\frac{\mathbf{I}_{x+i,y+j}-\mathbf{T}_{x,y}}{\sigma}\right)^2 \tag{5.4}$$

In *maximum likelihood estimation*, we have to choose the parameter that maximizes the likelihood function. That is, the positions that minimize the rate of change of the objective function

$$\frac{\partial \ln(L_{i,j})}{\partial i} = 0 \quad \text{and} \quad \frac{\partial \ln(L_{i,j})}{\partial j} = 0 \tag{5.5}$$

That is,

$$\sum_{(x,y)\in\mathbf{W}} (\mathbf{I}_{x+i,y+j} - \mathbf{T}_{x,y}) \frac{\partial \mathbf{I}_{x+i,y+j}}{\partial i} = 0$$

$$\sum_{(x,y)\in\mathbf{W}} (\mathbf{I}_{x+i,y+j} - \mathbf{T}_{x,y}) \frac{\partial \mathbf{I}_{x+i,y+j}}{\partial j} = 0 \tag{5.6}$$

We can observe that these equations are also the solution of the minimization problem given by

$$\min e = \sum_{(x,y)\in\mathbf{W}} (\mathbf{I}_{x+i,y+j} - \mathbf{T}_{x,y})^2 \tag{5.7}$$

That is, maximum likelihood estimation is equivalent to choosing the template position that minimizes the squared error (the squared values of the differences between the template points and the corresponding image points). The position where the template best matches the image is the estimated position of the template within the image. Thus, if you measure the match using the squared error criterion, then you will be choosing the *maximum likelihood* solution. This implies that the result achieved by template matching is optimal for images corrupted by Gaussian noise. A more detailed examination of the method of least squares is given in Appendix 3, Section 11.2. (Note that the *central limit theorem* suggests that practically experienced noise can be assumed to be Gaussian distributed, although many images appear to contradict this assumption.) You can use other error criteria, such as the absolute difference, rather than the squared difference or, if you feel more adventurous, you might consider robust measures such as M-estimators.

We can derive alternative forms of the squared error criterion by considering that Equation 5.7 can be written as

$$\min \ e = \sum_{(x,y)\in W} \mathbf{I}_{x+i,y+j}^2 - 2\mathbf{I}_{x+i,y+j}\mathbf{T}_{x,y} + \mathbf{T}_{x,y}^2 \tag{5.8}$$

The last term does not depend on the template position (i, j). As such, it is constant and cannot be minimized. Thus, the optimum in this equation can be obtained by minimizing

$$\min \ e = \sum_{(x,y)\in W} \mathbf{I}_{x+i,y+j}^2 - 2 \sum_{(x,y)\in W} \mathbf{I}_{x+i,y+j}\mathbf{T}_{x,y} \tag{5.9}$$

If the first term

$$\sum_{(x,y)\in W} \mathbf{I}_{x+i,y+j}^2 \tag{5.10}$$

is approximately constant, then the remaining term gives a measure of the similarity between the image and the template. That is, we can maximize the *cross-correlation* between the template and the image. Thus, the best position can be computed by

$$\max \ e = \sum_{(x,y)\in W} \mathbf{I}_{x+i,y+j}\mathbf{T}_{x,y} \tag{5.11}$$

However, the squared term in Equation 5.10 can vary with position, so the match defined by Equation 5.11 can be poor. In addition, the range of the cross-correlation is dependent on the size of the template and it is non-invariant to changes in image lighting conditions. Thus, in an implementation it is more convenient to use either Equation 5.7 or Equation 5.9 (in spite of being computationally more demanding than the cross-correlation in Equation 5.11).

Alternatively, cross-correlation can be *normalized* as follows. We can rewrite Equation 5.8 as

$$\min \ e = 1 - 2\frac{\sum_{(x,y)\in W} \mathbf{I}_{x+i,y+j}\mathbf{T}_{x,y}}{\sum_{(x,y)\in W} \mathbf{I}_{x+i,y+j}^2} \tag{5.12}$$

Here, the first term is constant and, thus, the optimum value can be obtained by

$$\max \ e = \frac{\sum_{(x,y)\in W} \mathbf{I}_{x+i,y+j}\mathbf{T}_{x,y}}{\sum_{(x,y)\in W} \mathbf{I}_{x+i,y+j}^2} \tag{5.13}$$

In general, it is convenient to normalize the grey level of each image window under the template. That is,

$$\max \ e = \frac{\sum_{(x,y)\in W} \left(\mathbf{I}_{x+i,y+j} - \overline{\mathbf{I}}_{i,j}\right)\left(\mathbf{T}_{x,y} - \overline{\mathbf{T}}\right)}{\sum_{(x,y)\in W} \left(\mathbf{I}_{x+i,y+j} - \overline{\mathbf{I}}_{i,j}\right)^2} \tag{5.14}$$

where $\overline{\mathbf{I}}_{i,j}$ is the mean of the pixels $\mathbf{I}_{x+i,y+j}$ for points within the window (i.e. $(x, y) \in W$) and $\overline{\mathbf{T}}$ is the mean of the pixels of the template. An alternative form of Equation 5.14 is given by *normalizing* the cross-correlation. This does not change the position of the optimum and gives an interpretation as the normalization of the cross-correlation vector. That is, the cross-correlation is divided by its modulus. Thus,

$$\max \ e = \frac{\sum\limits_{(x,y)\in W} \left(I_{x+i,y+j} - \bar{I}_{i,j}\right)\left(T_{x,y} - \bar{T}\right)}{\sqrt{\sum\limits_{(x,y)\in W} \left(I_{x+i,y+j} - \bar{I}_{i,j}\right)^2 \left(T_{x,y} - \bar{T}\right)^2}} \tag{5.15}$$

However, this equation has a similar computational complexity to the original formulation in Equation 5.7.

A particular implementation of template matching is when the image and the template are binary. In this case, the binary image can represent regions in the image or it can contain the edges. These two cases are illustrated in the example in Figure 5.3. The advantage of using binary images is that the amount of *computation* can be *reduced*. That is, each term in Equation 5.7 will take only two values: it will be one when $I_{x+i,y+j} = T_{x,y}$, and zero otherwise. Thus, Equation 5.7 can be implemented as

$$\max \ e = \sum\limits_{(x,y)\in W} \overline{I_{x+i,y+j} \oplus T_{x,y}} \tag{5.16}$$

where the symbol $\overline{\oplus}$ denotes the exclusive NOR operator. This equation can be easily implemented and requires significantly less resource than the original matching function.

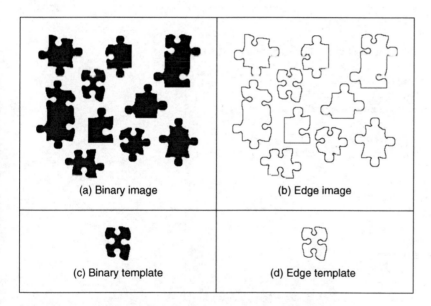

(a) Binary image	(b) Edge image
(c) Binary template	(d) Edge template

Figure 5.3   Example of binary and edge template matching

Template matching develops an *accumulator space* that stores the match of the template to the image at different locations; this corresponds to an implementation of Equation 5.7. It is called an accumulator, since the match is *accumulated* during application. Essentially, the accumulator is a two-dimensional (2D) array that holds the difference between the template and the image at different positions. The position in the image gives the same position of match in the accumulator. Alternatively, Equation 5.11 suggests that the peaks in the accumulator resulting from template correlation give the location of the template in an image: the coordinates of the point of best match. Accordingly, template correlation and template matching can be viewed as similar processes. The location of a template can be determined by either process. The binary implementation of template matching (Equation 5.16) is usually concerned with thresholded edge data. This equation will be reconsidered in the definition of the Hough transform, the topic of the following section.

The Matlab code to implement template matching is the function TMatching given in Code 5.1. This function first clears an accumulator array, accum, then searches the whole picture, using pointers i and j, and then searches the whole template for matches, using pointers x and y. Notice that the position of the template is given by its centre. The accumulator elements are incremented according to Equation 5.7. The accumulator array is delivered as the result. The match for each position is stored in the array. After computing all the matches, the minimum element in the array defines the position where most pixels in the template matched those in the image. As such, the minimum is deemed to be the coordinates of the point where the template's shape is most likely to lie within the original image. It is possible to implement a version of template matching without the accumulator array, by storing the location of the minimum alone. This will give the same result and it requires little storage. However, this implementation will provide a result that cannot support later image interpretation that may require knowledge of more than just the best match.

```
%Template Matching Implementation

function accum=TMatching(inputimage,template)

%Image size & template size
  [rows,columns]=size(inputimage);
  [rowsT,columnsT]=size(template);

  %Centre of the template
cx=floor(columnsT/2)+1;    cy=floor(rowsT/2)+1;

%Accumulator
accum=zeros(rows,columns);
%Template Position
for i=cx:columns-cx
   for j=cy:rows-cy
      %Template elements
        for x=1-cx:cx-1
          for y=1-cy:cy-1
                err=(double(inputimage(j+y,i+x))
                -double(template(y+cy,x+cx)))^2;
              accum(j,i)=accum(j,i)+err;
          end
        end
   end
end
```

Code 5.1 Implementation of template matching

The results of applying the template matching procedure are illustrated in Figure 5.4. This example shows the accumulator arrays for matching the images shown in Figures 5.2(a), 5.3(a) and 5.3(b) with their respective templates. The dark points in each image are at the coordinates of the origin of the position where the template best matched the image (the minimum). Note that there is a border where the template has not been matched to the image data. At these border points, the template extended beyond the image data, so no matching has been performed. This is the same border as experienced with template convolution (Section 3.4.1). We can observe that a clearer minimum is obtained (Figure 5.4c) from the edge images of Figure 5.3. This is because for grey-level and binary images, there is some match when the template is not exactly in the best position. In the case of edges, the count of matching pixels is less.

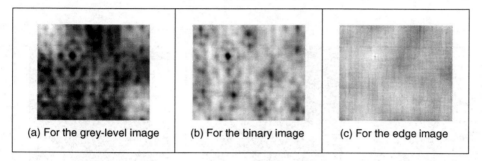

| (a) For the grey-level image | (b) For the binary image | (c) For the edge image |

Figure 5.4   Accumulator arrays from template matching

Most applications require further degrees of freedom such as rotation (orientation), scale (size) or perspective deformations. Rotation can be handled by rotating the template or by using polar coordinates; scale invariance can be achieved using templates of differing size. Having more parameters of interest implies that the accumulator space becomes larger; its dimensions increase by one for each extra parameter of interest. *Position*-invariant template matching, as considered here, implies a 2D parameter space, whereas the extension to *scale*- and *position*-invariant template matching requires a 3D parameter space.

The computational cost of template matching is *large*. If the template is square and of size $m \times m$ and is matched to an image of size $N \times N$, since the m^2 pixels are matched at all image points (except for the border) the computational cost is $O(N^2 m^2)$. This is the cost for position-invariant template matching. Any further parameters of interest *increase* the computational cost in proportion to the number of values of the extra parameters. This is clearly a large penalty and so a direct digital implementation of template matching is slow. Accordingly, this guarantees interest in techniques that can deliver the same result, but faster, such as using a Fourier implementation based on fast transform calculus.

The main *advantages* of template matching are its *insensitivity* to *noise* and *occlusion*. Noise can occur in any image, on any signal, just like on a telephone line. In digital photographs, the noise might appear low, but in computer vision it is made worse by edge detection by virtue of the differencing (differentiation) processes. Likewise, shapes can easily be occluded or *hidden*: a person can walk behind a lamppost, or illumination can cause occlusion. The *averaging* inherent in template matching reduces the susceptibility to noise; the *maximization* process reduces susceptibility to occlusion.

These advantages are illustrated in Figure 5.5, which shows detection in the presence of increasing noise. Here, we will use template matching to locate the vertical rectangle near the top of the image (so we are matching a binary template of a black template on a white background to the binary image). The lowest noise level is in Figure 5.5(a) and the highest is in Figure 5.5(c); the position of the origin of the detected rectangle is shown as a black cross in a white square. The position of the origin is detected correctly in Figure 5.5(a) and (b) but incorrectly in the noisiest image, Figure 5.5(c). Clearly, template matching can handle quite high noise corruption. (Admittedly this is somewhat artificial: the noise would usually be filtered out by one of the techniques described in Chapter 3, but we are illustrating basic properties here.) The ability to handle noise is shown by correct determination of the position of the target shape, until the noise becomes too much and there are more points due to noise than there are due to the shape itself. When this occurs, the votes resulting from the noise exceed those occurring from the shape, and so the maximum is not found where the shape exists.

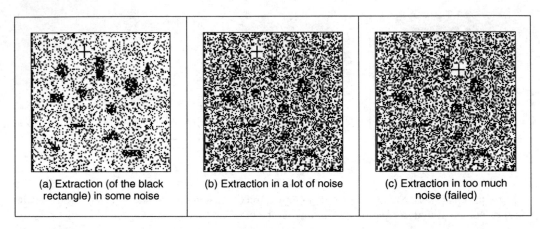

| (a) Extraction (of the black rectangle) in some noise | (b) Extraction in a lot of noise | (c) Extraction in too much noise (failed) |

Figure 5.5   Template matching in noisy images

Occlusion is shown by placing a grey bar across the image; in Figure 5.6(a) the bar does not occlude the target rectangle, whereas in Figure 5.6(c) the rectangle is completely obscured. As with performance in the presence of noise, detection of the shape fails when the votes occurring from the shape exceed those from the rest of the image, and the cross indicating the

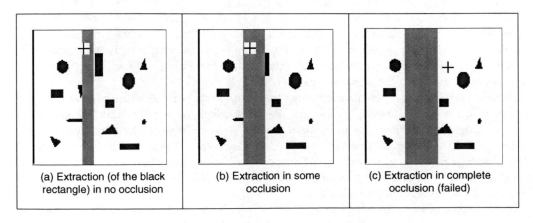

| (a) Extraction (of the black rectangle) in no occlusion | (b) Extraction in some occlusion | (c) Extraction in complete occlusion (failed) |

Figure 5.6   Template matching in occluded images

position of the origin of the rectangle is drawn in completely the wrong place. This is what happens when the rectangle is completely obscured in Figure 5.6(c).

So it can operate well, with practical advantage. We can include edge detection to concentrate on a shape's borders. Its main problem is *speed*: a direct implementation is slow, especially when handling shapes that are rotated or scaled (and there are other implementation difficulties too). Recalling that from Section 3.4.2 that template matching can be speeded up by using the Fourier transform, let us see whether that can be used here too.

5.3.2 Fourier transform implementation

We can implement template matching via the Fourier transform by using the *duality* between convolution and multiplication. This duality establishes that a multiplication in the space domain corresponds to a convolution in the frequency domain and vice versa. This can be exploited for faster computation by using the frequency domain, given the fast Fourier transform (FFT) algorithm. Thus, to find a shape we can compute the cross-correlation as a multiplication in the frequency domain. However, the matching process in Equation 5.11 is *correlation* (Section 2.3), *not* convolution. Thus, we need to express the correlation in terms of a convolution. This can be done as follows. First, we can rewrite the *correlation* in Equation 5.11 as

$$\mathbf{I} \otimes \mathbf{T} = \sum_{(x,y) \in W} \mathbf{I}_{x',y'} \mathbf{T}_{x'-i,y'-j} \tag{5.17}$$

where $x' = x + i$ and $y' = y + j$. *Convolution* is defined as

$$\mathbf{I} * \mathbf{T} = \sum_{(x,y) \in W} \mathbf{I}_{x',y'} \mathbf{T}_{i-x',j-y'} \tag{5.18}$$

Thus, to implement template matching in the frequency domain, we need to express Equation 5.17 in terms of Equation 5.18. This can be achieved by considering that

$$\mathbf{I} \otimes \mathbf{T} = \mathbf{I} * \mathbf{T}' = \sum_{(x,y) \in W} \mathbf{I}_{x',y'} \mathbf{T}'_{i-x',j-y'} \tag{5.19}$$

where

$$\mathbf{T}' = \mathbf{T}_{-x,-y} \tag{5.20}$$

That is, correlation is equivalent to convolution when the template is changed according to Equation 5.20. This equation reverses the coordinate axes and it corresponds to a horizontal and a vertical flip.

In the frequency domain, convolution corresponds to *multiplication*. As such, Equation 5.19 can be implemented by

$$\mathbf{I} * \mathbf{T}' = \Im^{-1}\big(\Im(\mathbf{I})\Im(\mathbf{T}')\big) \tag{5.21}$$

where $\Im$ denotes Fourier transformation as in Chapter 2 (and calculated by the FFT) and $\Im^{-1}$ denotes the inverse FFT. This is computationally faster than its direct implementation, given the speed advantage of the FFT. There are two ways of implementing this equation. In the first approach, we can compute $\mathbf{T}'$ by flipping the template and then computing its Fourier transform, $\Im(\mathbf{T}')$. In the second approach, we compute the transform of $\Im(\mathbf{T})$ and then we compute the complex *conjugate*. That is,

$$\Im(\mathbf{T}') = [\Im(\mathbf{T})]^* \tag{5.22}$$

where [ ]* denotes the complex conjugate of the transform data (yes, it is an unfortunate symbol clash with convolution, but both are standard symbols). So conjugation of the transform of the template implies that the product of the two transforms leads to correlation. That is,

$$\mathbf{I} * \mathbf{T}' = \Im^{-1}\left(\Im(\mathbf{I})\big[\Im(\mathbf{T})\big]^{*}\right) \tag{5.23}$$

For both implementations, Equations 5.21 and 5.23 will evaluate the match, and more quickly for large templates than by direct implementation of template matching. Note that one assumption is that the transforms are of the same size, even though the template's shape is usually much smaller than the image. There is a selection of approaches; a simple solution is to include extra zero values (*zero-padding*) to make the image of the template the same size as the image.

The code to implement template matching by Fourier, FTConv, is given in Code 5.2. The implementation takes the image and the flipped template. The template is zero-padded and then transforms are evaluated. The required convolution is obtained by multiplying the transforms and then applying the inverse. The resulting image is the magnitude of the inverse transform. This could be invoked as a single function, rather than as procedure, but the implementation is less clear. This process can be formulated using brightness or edge data, as appropriate. Should we seek *scale* invariance, to find the position of a template irrespective of its size, then we need to formulate a set of templates that range in size between the maximum expected variation. Each of the templates of differing size is then matched by frequency domain multiplication. The maximum frequency domain value, for all sizes of template, indicates the position of the template and gives a value for its size. This can be a rather lengthy procedure when the template ranges considerably in size.

```
%Fourier Transform Convolution

function FTConv(inputimage,template)

%image size
[rows,columns]=size(inputimage);

%FT
Fimage=fft2(inputimage,rows,columns);
Ftemplate=fft2(template,rows,columns);

%Convolution
G=Fimage.*Ftemplate;

%Modulus
Z=log(abs(fftshift(G)));

%Inverse
R=real(ifft2(G));
```

Code 5.2 Implementation of convolution by the frequency domain

Figure 5.7 illustrates the results of template matching in the Fourier domain. This example uses the image and template shown in Figure 5.2. Figure 5.7(a) shows the flipped and padded template. The Fourier transforms of the image and of the flipped template are given in Figure 5.7(b)

and (c), respectively. These transforms are multiplied, point by point, to achieve the image in Figure 5.7(d). When this is inverse Fourier transformed, the result (Figure 5.7e) shows where the template best matched the image (the coordinates of the template's top left-hand corner). The resulting image contains several local maxima (in white). This can be explained by the fact that this implementation does not consider the term in Equation 5.10. In addition, the shape can partially match several patterns in the image. Figure 5.7(f) shows a zoom of the region where the peak is located. We can see that this peak is well defined. In contrast to template matching, the implementation in the frequency domain does not have any border. This is due to the fact that Fourier theory assumes picture replication to infinity. Note that in application, the Fourier transforms do not need to be rearranged (`fftshif`) so that the d.c. is at the centre, since this has been done here for display purposes only.

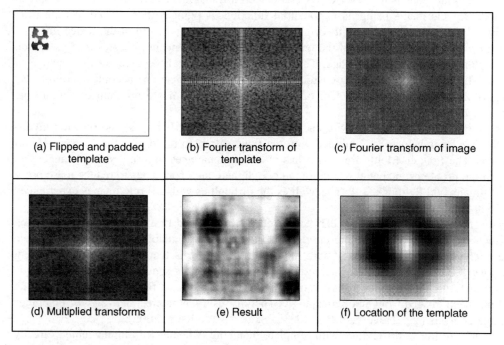

| (a) Flipped and padded template | (b) Fourier transform of template | (c) Fourier transform of image |
| (d) Multiplied transforms | (e) Result | (f) Location of the template |

Figure 5.7 Template matching by Fourier transformation

There are several further difficulties in using the transform domain for template matching in discrete images. If we seek rotation invariance, then an image can be expressed in terms of its polar coordinates. Discretization gives further difficulty since the points in a rotated discrete shape can map imperfectly to the original shape. This problem is better manifest when an image is scaled in size to become larger. In such a case, the spacing between points will increase in the enlarged image. The difficulty is how to allocate values for pixels in the enlarged image which are not defined in the enlargement process. There are several interpolation approaches, but it can often appear prudent to reformulate the original approach. Further difficulties can include the influence of the image borders: Fourier theory assumes that an image replicates spatially to infinity. Such difficulty can be reduced by using window operators, such as the Hamming or the Hanning windows. These difficulties do not arise for optical Fourier transforms and so using the Fourier transform for position-invariant template matching is often confined to optical implementations.

5.3.3 Discussion of template matching

The advantages associated with template matching are mainly theoretical since it can be very difficult to develop a template matching technique that operates satisfactorily. The results presented here have been for *position* invariance only. This can cause difficulty if invariance to *rotation* and *scale* is also required. This is because the template is stored as a discrete set of points. When these are rotated, *gaps* can appear owing to the discrete nature of the coordinate system. If the template is increased in size then there will be missing points in the scaled-up version. Again, there is a frequency domain version that can handle variation in size, since scale-invariant template matching can be achieved using the *Mellin transform* (Bracewell, 1986). This avoids using many templates to accommodate the variation in size by evaluating the scale-invariant match in a single pass. The Mellin transform essentially scales the spatial coordinates of the image using an exponential function. A point is then moved to a position given by a logarithmic function of its original coordinates. The transform of the scaled image is then multiplied by the transform of the template. The maximum again indicates the best match between the transform and the image. This can be considered to be equivalent to a change of variable. The logarithmic mapping ensures that scaling (multiplication) becomes addition. By the logarithmic mapping, the problem of scale invariance becomes a problem of finding the position of a match.

The Mellin transform only provides scale-invariant matching. For scale and position invariance, the Mellin transform is combined with the Fourier transform, to give the *Fourier–Mellin* transform. The Fourier–Mellin transform has many disadvantages in a digital implementation, owing to the problems in spatial resolution, although there are approaches to reduce these problems (Altmann and Reitbock, 1984), as well as the difficulties with discrete images experienced in Fourier transform approaches.

Again, the Mellin transform appears to be much better suited to an *optical* implementation (Casasent and Psaltis, 1977), where *continuous* functions are available, rather than to discrete image analysis. A further difficulty with the Mellin transform is that its result is independent of the *form factor* of the template. Accordingly, a rectangle and a square appear to be the same to this transform. This implies a loss of information since the form factor can indicate that an object has been imaged from an oblique angle. There is resurgent interest in *log-polar* mappings for image analysis (e.g. Traver and Pla, 2003; Zokai and Wollberg, 2005).

So, there are innate difficulties with template matching, whether it is implemented directly or by transform operations. For these reasons, and because many shape extraction techniques require more than just edge or brightness data, direct digital implementations of feature extraction are usually preferred. This is perhaps also influenced by the speed advantage that one popular technique can confer over template matching. This is the Hough transform, which is covered next.

5.4 Hough transform

5.4.1 Overview

The *Hough transform* (HT) (Hough, 1962) is a technique that locates shapes in images. In particular, it has been used to extract *lines*, *circles* and *ellipses* (or conic sections). In the case of lines, its mathematical definition is equivalent to the Radon transform (Deans, 1981). The HT was introduced by Hough (1962) and then used to find bubble tracks rather than shapes in images. However, Rosenfeld (1969) noted its potential advantages as an image processing algorithm. The HT was thus implemented to find lines in images (Duda and Hart, 1972) and it has been extended greatly, since it has many advantages and many potential routes for improvement. Its prime advantage is that it can deliver the *same* result as that for template matching, but *faster* (Stockman and Agrawala, 1977; Sklansky, 1978; Princen et al., 1992b). This is achieved by a reformulation of the template matching process, based on an *evidence-gathering* approach, where the evidence is the *votes* cast in an accumulator array. The HT implementation defines a *mapping* from the image points into an accumulator space (Hough space). The mapping is achieved in a computationally efficient manner, based on the function that describes the target shape. This mapping requires much fewer computational resources than template matching. However, it still requires significant storage and has high computational requirements. These problems are addressed later, since they give focus for the continuing development of the HT. However, the fact that the HT is equivalent to template matching has given sufficient impetus for the technique to be among the most popular of all existing shape extraction techniques.

5.4.2 Lines

We will first consider finding lines in an image. In a Cartesian parameterization, collinear points in an image with coordinates (x, y) are related by their slope m and an intercept c according to:

$$y = mx + c \tag{5.24}$$

This equation can be written in homogeneous form as

$$Ay + Bx + 1 = 0 \tag{5.25}$$

where $A = -1/c$ and $B = m/c$. Thus, a line is defined by giving a pair of values (A, B). However, we can observe a symmetry in the definition in Equation 5.25. This equation is symmetric since a pair of coordinates (x, y) also defines a line in the space with parameters (A, B). That is, Equation 5.25 can be seen as the equation of a line for fixed coordinates (x, y) or as the equation of a line for fixed parameters (A, B). Thus, pairs can be used to define points and lines simultaneously (Aguado et al., 2000a). The HT gathers evidence of the point (A, B) by considering that all the points (x, y) define the same line in the space (A, B). That is, if the set of collinear points $\{(x_i, y_i)\}$ defines the line (A, B), then

$$Ay_i + Bx_i + 1 = 0 \tag{5.26}$$

This equation can be seen as a system of equations and it can simply be rewritten in terms of the Cartesian parameterization as

$$c = -x_i m + y_i \tag{5.27}$$

Thus, to determine the line we must find the values of the parameters (m, c) [or (A, B) in homogeneous form] that satisfy Equation 5.27 (or 5.26, respectively). However, we must notice that the system is generally overdetermined. That is, we have more equations than unknowns. Thus, we must find the solution that comes close to satisfying all the equations simultaneously. This kind of problem can be solved, for example, using linear least squares techniques. The HT uses an evidence-gathering approach to provide the solution.

The relationship between a point (x_i, y_i) in an image and the line given in Equation 5.27 is illustrated in Figure 5.8. The points (x_i, y_i) and (x_j, y_j) in Figure 5.8(a) define the lines U_i and U_j in Figure 5.8(b), respectively. All the collinear elements in an image will define dual lines with the same concurrent point (A, B). This is independent of the line parameterization used. The HT solves it in an efficient way by simply counting the potential solutions in an accumulator

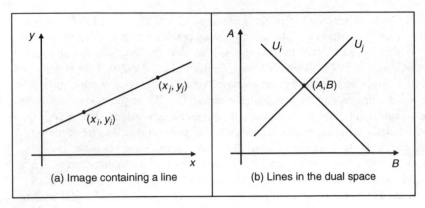

Figure 5.8 Illustrating the Hough transform for lines

array that stores the evidence, or votes. The count is made by tracing all the dual lines for each point (x_i, y_i). Each point in the trace increments an element in the array; thus, the problem of line extraction is transformed into the problem of locating a maximum in the accumulator space. This strategy is robust and has been demonstrated to be able to handle noise and occlusion.

The axes in the dual space represent the parameters of the line. In the case of the Cartesian parameterization m can take an *infinite* range of values, since lines can vary from horizontal to vertical. Since votes are gathered in a discrete array, this will produce *bias* errors. It is possible to consider a range of votes in the accumulator space that cover all possible values. This corresponds to techniques of antialiasing and can improve the gathering strategy (Brown, 1983; Kiryati and Bruckstein, 1991).

The implementation of the HT for lines, `HTLine`, is given in Code 5.3. It is important to observe that Equation 5.27 is not suitable for implementation since the parameters can take an infinite range of values. To handle the infinite range for c, we use two arrays in the implementation in Code 5.3. When the slope m is between $-45°$ and $45°$, c does not take a large value. For other values of m the intercept c can take a very large value. Thus, we consider an accumulator for each case. In the second case, we use an array that stores the intercept with the x-axis. This only solves the problem partially, since we cannot guarantee that the value of c will be small when the slope m is between $-45°$ and $45°$.

Figure 5.9 shows three examples of locating lines using the HT implemented in Code 5.3. In Figure 5.9(a) there is a single line which generates the peak seen in Figure 5.9(d). The magnitude of the peak is proportional to the number of pixels in the line from which it was generated. The edges of the wrench in Figure 5.9(b) and (c) define two main lines. Image 5.9(c) contains much more noise. This image was obtained by using a lower threshold value in the edge detector operator which gave rise to more noise. The accumulator results of the HT for the images in Figure 5.9(b) and (c) are shown in Figure 5.9(e) and (f), respectively. The two accumulator arrays are broadly similar in shape, and the peak in each is at the same place. The coordinates of the peaks are at combinations of parameters of the lines that best fit the image. The extra number of edge points in the noisy image of the wrench gives rise to more votes in the accumulator space, as can be seen by the increased number of votes in Figure 5.9(f) compared with Figure 5.9(e). Since the peak is in the same place, this shows that the HT can indeed tolerate noise. The results of extraction, when superimposed on the edge image, are shown in Figure 5.9(g)–(i). Only the two lines corresponding to significant peaks have been drawn for the image of the wrench. Here, we can see that the parameters describing the lines

```
%Hough Transform for Lines
function HTLine(inputimage)
%image size
[rows,columns]=size(inputimage);
%accumulator
acc1=zeros(rows,91);
acc2=zeros(columns,91);
%image
for x=1:columns
   for y=1:rows
      if(inputimage(y,x)==0)
        for m=-45:45
           b=round(y-tan((m*pi)/180)*x);
           if(b<rows & b>0)
              acc1(b,m+45+1)=acc1(b,m+45+1)+1;
           end
        end
        for m=45:135
           b=round(x-y/tan((m*pi)/180));
           if(b<columns & b>0)
              acc2(b,m-45+1)=acc2(b,m-45+1)+1;
           end
        end
      end
   end
end
```

Code 5.3 Implementation of the Hough transform for lines

have been extracted well. Note that the endpoints of the lines are not delivered by the HT, only the parameters that describe them. You have to go back to the image to obtain line length.

The HT delivers a correct response; that is, correct estimates of the parameters used to specify the line, so long as the number of collinear points along that line exceeds the number of collinear points on any other line in the image. As such, the HT has the same properties in respect of noise and occlusion as template matching. However, the non-linearity of the parameters and the discretization produce noisy accumulators. A major problem in implementing the basic HT for lines is the definition of an appropriate accumulator space. In application, Bresenham's line drawing algorithm (Bresenham, 1965) can be used to draw the lines of votes in the accumulator space. This ensures that lines of connected votes are drawn, as opposed to the use of Equation 5.27, which can lead to gaps in the drawn line. *Backmapping* (Gerig and Klein, 1986) can be used to determine exactly which edge points contributed to a particular peak. Backmapping is an *inverse* mapping from the accumulator space to the edge data and can allow for shape analysis of the image by removal of the edge points that contributed to particular peaks, and then by reaccumulation using the HT. Note that the computational cost of the HT depends on the number of edge points (n_e) and the length of the lines formed in the parameter space (l), giving a computational cost of $O(n_e l)$. This is considerably less than that for template matching, given earlier as $O(N^2 m^2)$.

One way to avoid the problems of the Cartesian parameterization in the HT is to base the mapping function on an alternative parameterization. One of the most proven techniques is

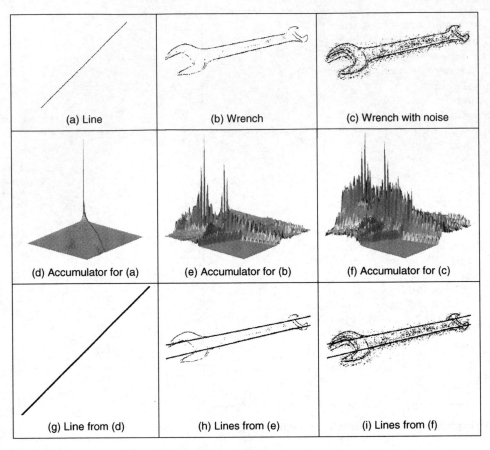

(a) Line	(b) Wrench	(c) Wrench with noise
(d) Accumulator for (a)	(e) Accumulator for (b)	(f) Accumulator for (c)
(g) Line from (d)	(h) Lines from (e)	(i) Lines from (f)

Figure 5.9 Applying the Hough transform for lines

called the *foot-of-normal* parameterization. This parameterizes a line by considering a point (x, y) as a function of an angle normal to the line, passing through the origin of the image. This gives a form of the HT for lines known as the *polar HT for lines* (Duda and Hart, 1972). The point where this line intersects the line in the image is given by

$$\rho = x\cos(\theta) + y\sin(\theta) \qquad (5.28)$$

where θ is the angle of the line normal to the line in an image and ρ is the length between the origin and the point where the lines intersect, as illustrated in Figure 5.10.

By recalling that two lines are perpendicular if the product of their slopes is -1, and by considering the geometry of the arrangement in Figure 5.10, we obtain

$$c = \frac{\rho}{\sin(\theta)} \qquad m = -\frac{1}{\tan(\theta)} \qquad (5.29)$$

By substitution in Equation 5.24 we obtain the polar form (Equation 5.28). This provides a different mapping function: votes are now cast in a sinusoidal manner, in a 2D accumulator array in terms of θ and ρ, the parameters of interest. The advantage of this alternative mapping is that the values of the parameters θ and ρ are now bounded to lie within a specific range. The range for θ is within 180°; the possible values of ρ are given by the image size, since

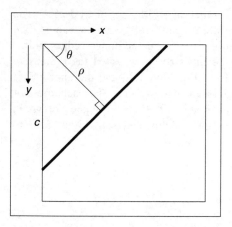

Figure 5.10 Polar consideration of a line

the maximum length of the line is $\sqrt{2} \times N$, where N is the (square) image size. The range of possible values is now fixed, so the technique is practicable.

As the voting function has now changed, we shall draw different loci in the accumulator space. In the conventional HT for lines, a straight line mapped to a straight line as in Figure 5.8. In the polar HT for lines, points map to curves in the accumulator space. This is illustrated in Figure 5.11(a)–(c), which shows the polar HT accumulator spaces for one, two and three points, respectively. For a single point in the upper row of Figure 5.11(a) we obtain a single curve shown in the lower row of Figure 5.11(a). For two points we obtain two curves, which intersect at a position which describes the parameters of the line joining them (Figure 5.11b). An additional curve is obtained for the third point and there is now a peak in the accumulator array containing three votes (Figure 5.11c).

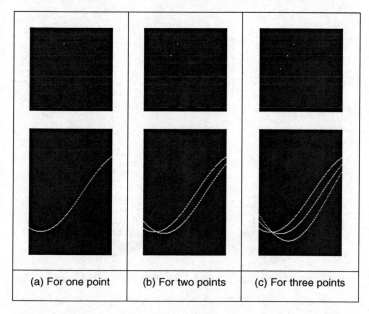

| (a) For one point | (b) For two points | (c) For three points |

Figure 5.11 Images and the accumulator space of the polar Hough transform

The implementation of the *polar HT for lines* is the function `HTPLine` in Code 5.4. The accumulator array is a set of 180 bins for value of θ in the range 0–180°, and for values of ρ in the range 0 to $\sqrt{N^2 + M^2}$, where $N \times M$ is the picture size. Then, for image (edge) points greater than a chosen threshold, the angle relating to the bin size is evaluated (as radians in the range 0 to π) and then the value of ρ is evaluated from Equation 5.28 and the appropriate accumulator cell is incremented so long as the parameters are within range. The accumulator arrays obtained by applying this implementation to the images in Figure 5.9 are shown in Figure 5.12. Figure 5.12(a) shows that a single line defines a well-delineated peak. Figure 5.12(b)

```
%Polar Hough Transform for Lines

function HTPLine(inputimage)

%image size
[rows,columns]=size(inputimage);

%accumulator
rmax=round(sqrt(rows^2+columns^2));
acc=zeros(rmax,180);

%image
for x=1:columns
   for y=1:rows
      if(inputimage(y,x)==0)
         for m=1:180
            r=round(x*cos((m*pi)/180)
               +y*sin((m*pi)/180));
            if(r<rmax & r>0)
               acc(r,m)= acc(r,m)+1; end
         end
      end
   end
end
```

Code 5.4   Implementation of the polar Hough transform for lines

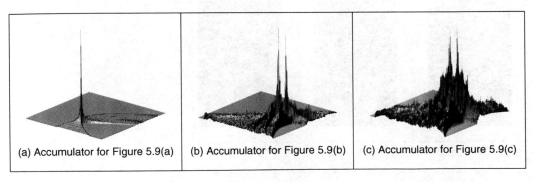

(a) Accumulator for Figure 5.9(a)   (b) Accumulator for Figure 5.9(b)   (c) Accumulator for Figure 5.9(c)

Figure 5.12   Applying the polar Hough transform for lines

and (c) show a clearer peak compared with the implementation of the Cartesian parameterization. This is because discretization effects are reduced in the polar parameterization. This feature makes the polar implementation far more practicable than the earlier, Cartesian, version.

5.4.3 Hough transform for circles

The HT can be extended by replacing the equation of the curve in the detection process. The equation of the curve can be given in *explicit* or *parametric* form. In explicit form, the HT can be defined by considering the equation for a circle given by

$$(x - x_0)^2 + (y - y_0)^2 = r^2 \tag{5.30}$$

This equation defines a locus of points (x, y) centred on an origin (x_0, y_0) and with radius r. This equation can again be visualized in two ways: as a locus of points (x, y) in an image, or as a locus of points (x_0, y_0) centred on (x, y) with radius r.

Figure 5.13 illustrates this dual definition. Each edge point in Figure 5.13(a) defines a set of circles in the accumulator space. These circles are defined by all possible values of the radius and they are centred on the coordinates of the edge point. Figure 5.13(b) shows three circles

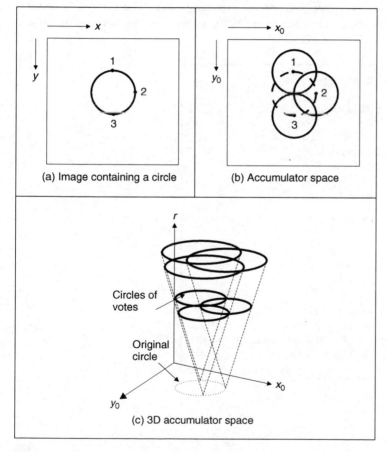

(a) Image containing a circle

(b) Accumulator space

(c) 3D accumulator space

Figure 5.13 Illustrating the Hough transform for circles

defined by three edge points. These circles are defined for a given radius value. Each edge point defines circles for the other values of the radius. This implies that the accumulator space is three dimensional (for the three parameters of interest) and that edge points map to a *cone* of votes in the accumulator space. Figure 5.13(c) illustrates this accumulator. After gathering evidence of all the edge points, the maximum in the accumulator space again corresponds to the parameters of the circle in the original image. The procedure of evidence gathering is the same as that for the HT for lines, but votes are generated in cones, according to Equation 5.30.

Equation 5.30 can be defined in *parametric* form as

$$x = x_0 + r\cos(\theta) \quad y = y_0 + r\sin(\theta) \tag{5.31}$$

The advantage of this representation is that it allows us to solve for the parameters. Thus, the HT mapping is defined by

$$x_0 = x - r\cos(\theta) \quad y_0 = y - r\sin(\theta) \tag{5.32}$$

These equations define the points in the accumulator space (Figure 5.13b) dependent on the radius r. Note that θ is not a free parameter, but defines the trace of the curve. The trace of the curve (or surface) is commonly referred to as the *point spread function*.

The implementation of the HT for circles, HTCircle, is shown in Code 5.5. This is similar to the HT for lines, except that the voting function corresponds to that in Equation 5.32 and the accumulator space is for circle data. The accumulator in the implementation is two dimensions,

```
%Hough Transform for Circles

function HTCircle(inputimage,r)

%image size
[rows,columns]=size(inputimage);

%accumulator
acc=zeros(rows,columns);

%image
for x=1:columns
   for y=1:rows
      if(inputimage(y,x)==0)
         for ang=0:360
            t=(ang*pi)/180;
            x0=round(x-r*cos(t));
            y0=round(y-r*sin(t));
            if(x0<columns & x0>0 & y0<rows & y0>0)
               acc(y0,x0)=acc(y0,x0)+1;
            end
         end
      end
   end
end
```

Code 5.5 Implementation of the Hough transform for circles

in terms of the centre parameters for a fixed value of the radius given as an argument to the function. This function should be called for all potential radii. A circle of votes is generated by varying t (i.e. θ, but Matlab does not allow Greek symbols!) from $0°$ to $360°$. The discretization of t controls the granularity of voting; too small an increment gives very fine coverage of the parameter space, too large a value results in very sparse coverage. The accumulator space, $\mathtt{acc}$ (initially zero), is incremented only for points whose coordinates lie within the specified range (in this case the centre cannot lie outside the original image).

The application of the HT for circles is illustrated in Figure 5.14. Figure 5.14(a) shows an image with a synthetic circle. In this figure, the edges are complete and well defined. The result of the HT process is shown in Figure 5.14(d). The peak of the accumulator space is at the centre of the circle. Note that votes exist away from the circle's centre, and rise towards

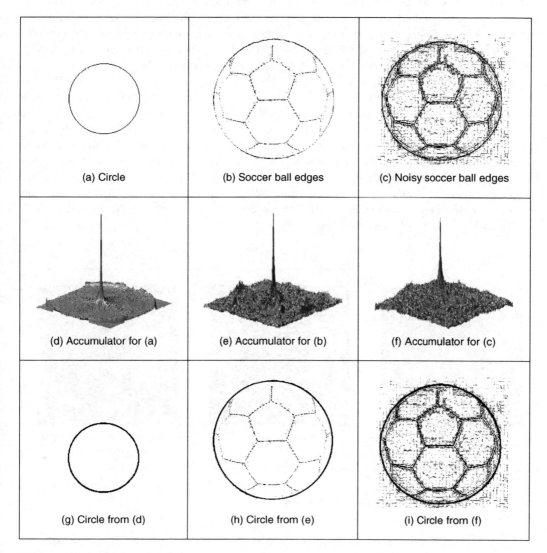

Figure 5.14 Applying the Hough transform for circles

the locus of the actual circle, although these background votes are much less than the actual peak. Figure 5.14(b) shows an example of data containing occlusion and noise. The image in Figure 5.14(c) corresponds to the same scene, but the noise level has been increased by changing the threshold value in the edge detection process. The accumulators for these two images are shown in Figure 5.14(e) and (f) and the circles related to the parameter space peaks are superimposed (in black) on the edge images in Figure 5.14(g)–(i). We can see that the HT has the ability to tolerate occlusion and noise. In Figure 5.14(c), there are many edge points, which implies that the amount of processing time increases. The HT will detect the circle (provide the right result) as long as more points are in a circular locus described by the parameters of the target circle than there are on any other circle. This is exactly the same performance as for the HT for lines, as expected, and is consistent with the result of template matching.

In application code, *Bresenham's algorithm* for discrete circles (Bresenham, 1977) can be used to draw the circle of votes, rather than use the polar implementation of Equation 5.32. This ensures that the complete locus of points is drawn and avoids the need to choose a value for increase in the angle used to trace the circle. Bresenham's algorithm can be used to generate the points in one octant, since the remaining points can be obtained by reflection. Backmapping can be used to determine which points contributed to the extracted circle.

An additional example of the circle HT extraction is shown in Figure 5.15. Figure 5.15(a) is a real image (albeit one with low resolution) which was processed by Sobel edge detection and thresholded to give the points in Figure 5.15(b). The circle detected by application of HTCircle with radius 5 pixels is shown in Figure 5.15(c) superimposed on the edge data. The extracted circle can be seen to *match* the edge data well. This highlights the two major advantages of the HT (and of template matching): its ability to handle *noise* and *occlusion*. Note that the HT merely finds the circle with the maximum number of points; it is possible to include other constraints to control the circle selection process, such as gradient direction for objects with known illumination profile. In the case of the human eye, the (circular) iris is usually darker than its white surroundings.

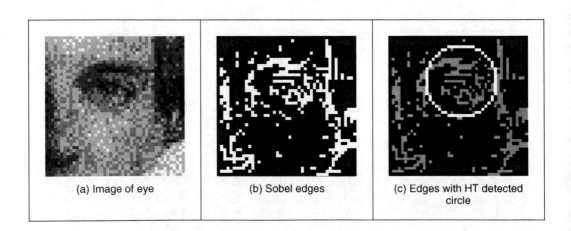

| (a) Image of eye | (b) Sobel edges | (c) Edges with HT detected circle |

Figure 5.15 Using the Hough transform for circles

Figure 5.15 also shows some of the difficulties with the HT, namely that it is essentially an implementation of template matching, and does not use some of the *richer* stock of information available in an image. For example, we might know constraints on *size*; the largest size and iris would be in an image like Figure 5.15. We also know some of the *topology*: the eye region contains two ellipsoidal structures with a circle in the middle. We might also know *brightness* information: the pupil is darker than the surrounding iris. These factors can be formulated as *constraints* on whether edge points can vote within the accumulator array. A simple modification is to make the votes proportional to edge magnitude; in this manner, points with high contrast will generate more votes and hence have more significance in the voting process. In this way, the feature extracted by the HT can be arranged to suit a particular application.

5.4.4 Hough transform for ellipses

Circles are very important in shape detection since many objects have a circular shape. However, because of the camera's viewpoint, circles do not always look like circles in images. Images are formed by mapping a shape in 3D space into a plane (the image plane). This mapping performs a perspective transformation. In this process, a circle is deformed to look like an ellipse. We can define the mapping between the circle and an ellipse by a similarity transformation. That is,

$$\begin{bmatrix} x \\ y \end{bmatrix} = \begin{bmatrix} \cos(\rho) & \sin(\rho) \\ -\sin(\rho) & \cos(\rho) \end{bmatrix} \begin{bmatrix} S_x \\ S_y \end{bmatrix} \begin{bmatrix} x' \\ y' \end{bmatrix} + \begin{bmatrix} t_x \\ t_y \end{bmatrix} \tag{5.33}$$

where (x', y') define the coordinates of the circle in Equation 5.31, ρ represents the orientation, (S_x, S_y) a scale factor and (t_x, t_y) a translation. If we define

$$a_0 = t_x \quad a_x = S_x \cos(\rho) \quad b_x = S_y \sin(\rho)$$
$$b_0 = t_y \quad a_y = -S_x \sin(\rho) \quad b_y = S_y \cos(\rho) \tag{5.34}$$

then we have that the circle is deformed into

$$x = a_0 + a_x \cos(\theta) + b_x \sin(\theta)$$
$$y = b_0 + a_y \cos(\theta) + b_y \sin(\theta) \tag{5.35}$$

This equation corresponds to the polar representation of an ellipse. This polar form contains six parameters $\left(a_0, b_0, a_x, b_x, a_y, b_y\right)$ that characterize the shape of the ellipse. θ is not a free parameter and it only addresses a particular point in the locus of the ellipse (just as it was used to trace the circle in Equation 5.32). However, one parameter is redundant since it can be computed by considering the orthogonality (independence) of the axes of the ellipse (the product $a_x b_x + a_y b_y = 0$, which is one of the known properties of an ellipse). Thus, an ellipse is defined by its centre (a_0, b_0) and three of the axis parameters (a_x, b_x, a_y, b_y). This gives five parameters, which is intuitively correct since an ellipse is defined by its centre (two parameters), it size along both axes (two more parameters) and its rotation (one parameter). In total, this states that five parameters describe an ellipse, so our three axis parameters must jointly describe size and rotation. In fact, the axis parameters can be related to the orientation and the length along the axes by

$$\tan(\rho) = \frac{a_y}{a_x} \quad a = \sqrt{a_x^2 + a_y^2} \quad b = \sqrt{b_x^2 + b_y^2} \tag{5.36}$$

where (a, b) are the axes of the ellipse, as illustrated in Figure 5.16.

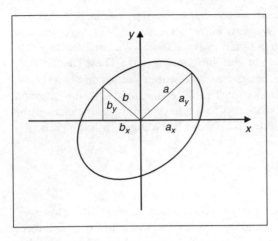

Figure 5.16   Definition of ellipse axes

In a similar way to Equation 5.31, Equation 5.35 can be used to generate the mapping function in the HT. In this case, the location of the centre of the ellipse is given by

$$a_0 = x - a_x \cos(\theta) + b_x \sin(\theta)$$

$$b_0 = y - a_y \cos(\theta) + b_y \sin(\theta)$$

$$(5.37)$$

The location is dependent on three parameters, thus the mapping defines the trace of a hyper-surface in a five-dimensional (5D) space. This space can be very large. For example, if there are 100 possible values for each of the five parameters, the 5D accumulator space contains 10^{10} values. This is 10 GB of storage, which is tiny nowadays (at least, when someone else pays!). Accordingly, there has been much interest in ellipse detection techniques which use much less space and operate much more quickly than direct implementation of Equation 5.37.

Code 5.6 shows the implementation of the HT mapping for ellipses. The function HTEllipse computes the centre parameters for an ellipse without rotation and with fixed axis length given as arguments. Thus, the implementation uses a 2D accumulator. In practice, to locate an ellipse, it is necessary to try all potential values of axis length. This is computationally impossible unless we limit the computation to a few values.

Figure 5.17 shows three examples of the application of the ellipse extraction process described in the Code 5.6. The first example (Figure 5.17a) illustrates the case of a perfect ellipse in a synthetic image. The array in Figure 5.17(d) shows a prominent peak whose position corresponds to the centre of the ellipse. The examples in Figure 5.17(b) and (c) illustrate the use of the HT to locate a circular form when the image has an oblique view. Each example was obtained by using a different threshold in the edge detection process. Figure 5.17(c) contains more noise data, which in turn gives rise to more noise in the accumulator. We can observe that there is more than one ellipse to be located in these two figures. This gives rise to the other high values in the accumulator space. As with the earlier examples for line and circle extraction, there is scope for interpreting the accumulator space, to discover which structures produced particular parameter combinations.

```
%Hough Transform for Ellipses

function HTEllipse(inputimage,a,b)

%image size
[rows,columns]=size(inputimage);

%accumulator
acc=zeros(rows,columns);

%image
for x=1:columns
  for y=1:rows
    if(inputimage(y,x)==0)
      for ang=0:360
        t=(ang*pi)/180;
        x0=round(x-a*cos(t));
        y0=round(y-b*sin(t));
        if(x0<columns & x0>0 & y0<rows & y0>0)
          acc(y0,x0)=acc(y0,x0)+1;
        end
      end
    end
  end
end
end
```

Code 5.6 Implementation of the Hough transform for ellipses

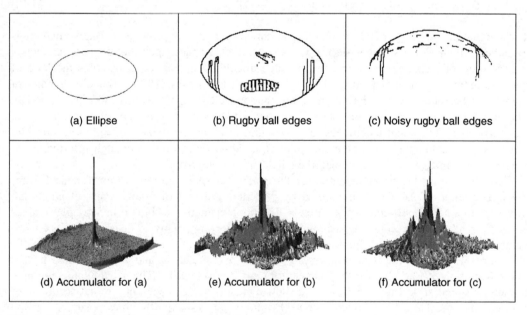

(a) Ellipse   (b) Rugby ball edges   (c) Noisy rugby ball edges

(d) Accumulator for (a)   (e) Accumulator for (b)   (f) Accumulator for (c)

Figure 5.17 Applying the Hough transform for ellipses

5.4.5 Parameter space decomposition

The HT gives the same (optimal) result as template matching and even though it is faster, it still requires significant computational resources. The previous sections showed that as we increase the complexity of the curve under detection, the computational requirements increase in an exponential way. Thus, the HT becomes less practical. For this reason, most of the research in the HT has focused on the development of techniques aimed to reduce its computational complexity (Illingworth and Kittler, 1988; Leavers, 1993). One important way to reduce the computation has been the use of geometric properties of shapes to decompose the parameter space. Several techniques have used different geometric properties. These geometric properties are generally defined by the relationship between points and derivatives.

5.4.5.1 Parameter space reduction for lines

For a line, the accumulator space can be reduced from two dimensions to one dimension by considering that we can compute the slope from the information of the image. The slope can be computed either by using the *gradient direction* at a point or by considering a pair of points. That is,

$$m = \varphi \quad \text{or} \quad m = \frac{y_2 - y_1}{x_2 - x_1} \tag{5.38}$$

where φ is the gradient direction at the point. In the case of two points, by considering Equation 5.24 we have:

$$c = \frac{x_2 y_1 - x_1 y_2}{x_2 - x_1} \tag{5.39}$$

Thus, according to Equation 5.29, one of the parameters of the polar representation for lines, θ, is now given by

$$\theta = -\tan^{-1}\left[\frac{1}{\varphi}\right] \quad \text{or} \quad \theta = \tan^{-1}\left[\frac{x_1 - x_2}{y_2 - y_1}\right] \tag{5.40}$$

These equations do not depend on the other parameter ρ and they provide alternative mappings to gather evidence. That is, they decompose the parametric space, such that the two parameters θ and ρ are now *independent*. The use of edge direction information constitutes the base of the line extraction method presented by O'Gorman and Clowes (1976). The use of pairs of points can be related to the definition of the randomized Hough transform (Xu et al., 1990). The number of feature points considered corresponds to all the combinations of points that form pairs. By using statistical techniques, it is possible to reduce the space of points to consider a representative sample of the elements; that is, a subset that provides enough information to obtain the parameters with predefined and small estimation errors.

Code 5.7 shows the implementation of the parameter space decomposition for the HT for lines. The slope of the line is computed by considering a pair of points. Pairs of points are restricted to a neighbourhood of 5×5 pixels. The implementation of Equation 5.40 gives values between $-90°$ and $90°$. Since the accumulators can only store positive values, we add $90°$ to all values. To compute ρ we use Equation 5.28, given the value of θ computed by Equation 5.40.

Figure 5.18 shows the accumulators for the two parameters θ and ρ as obtained by the implementation of Code 5.7 for the images in Figure 5.9(a) and (b). The accumulators are now one-dimensional, as in Figure 5.18(a), and show a clear peak. The peak in the first accumulator is close to 135°. Thus, by subtracting the 90° introduced to make all values positive, we find that the slope of the line $\theta = -45°$. The peaks in the accumulators in Figure 5.18(b) define

```
%Parameter Decomposition for the Hough Transform for Lines

function HTDLine(inputimage)

%image size
[rows,columns]=size(inputimage);

%accumulator
rmax=round(sqrt(rows^2+columns^2));
accro=zeros(rmax,1);
acct=zeros(180,1);

%image
for x=1:columns
    for y=1:rows
        if(inputimage(y,x)==0)
            for Nx=x-2:x+2
                for Ny=y-2:y+2
                    if(x~=Nx | y~=Ny)
                        if(Nx>0 & Ny>0 & Nx<columns & Ny<rows)
                            if(inputimage(Ny,Nx)==0)
                                if(Ny-y~=0)
                                    t=atan((x-Nx)/(Ny-y)); %Equation (5.40)
                                 else t=pi/2;
                                end
                                r=round(x*cos(t)+y*sin(t)); %Equation (5.28)

                                t=round((t+pi/2)*180/pi);
                                acct(t)=acct(t)+1;

                                if(r<rmax & r>0)
                                    accro(r)=accro(r)+1;
                                end
                            end
                        end
                    end
                end
            end
        end
    end
end
end
```

Code 5.7   Implementation of the parameter space reduction for the Hough transform for lines

two lines with similar slopes. The peak in the first accumulator represents the value of θ, while the two peaks in the second accumulator represent the location of the two lines. In general, when implementing parameter space decomposition it is necessary to follow a two-step process. First, it is necessary to gather data in one accumulator and search for the maximum. Secondly, the location of the maximum value is used as parameter value to gather data on the remaining accumulator.

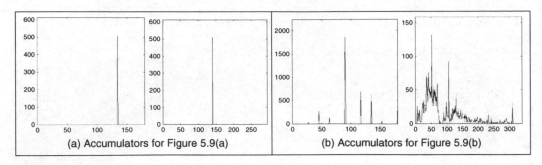

(a) Accumulators for Figure 5.9(a)   (b) Accumulators for Figure 5.9(b)

Figure 5.18 Parameter space reduction for the Hough transform for lines

5.4.5.2 *Parameter space reduction for circles*

In the case of lines, the relationship between local information computed from an image and the inclusion of a group of points (pairs) is in an alternative analytical description which can readily be established. For more complex primitives, it is possible to include several geometric relationships. These relationships are not defined for an arbitrary set of points, but include angular constraints that define relative positions between them. In general, we can consider different geometric properties of the circle to decompose the parameter space. This has motivated the development of many methods of parameter space decomposition (Aguado et al., 1996). An important geometric relationship is given by the geometry of the second directional derivatives. This relationship can be obtained by considering that Equation 5.31 defines a position vector function. That is,

$$\omega(\theta) = x(\theta) \begin{bmatrix} 1 \\ 0 \end{bmatrix} + y(\theta) \begin{bmatrix} 0 \\ 1 \end{bmatrix} \qquad (5.41)$$

where

$$x(\theta) = x_0 + r\cos(\theta) \quad y(\theta) = y_0 + r\sin(\theta) \qquad (5.42)$$

In this definition, we have included the parameter of the curve as an argument to highlight the fact that the function defines a vector for each value of θ. The endpoints of all the vectors trace a circle. The derivatives of Equation 5.41 with respect to θ define the first and second directional derivatives. That is,

$$v'(\theta) = x'(\theta) \begin{bmatrix} 1 \\ 0 \end{bmatrix} + y'(\theta) \begin{bmatrix} 0 \\ 1 \end{bmatrix}$$

$$v''(\theta) = x''(\theta) \begin{bmatrix} 1 \\ 0 \end{bmatrix} + y''(\theta) \begin{bmatrix} 0 \\ 1 \end{bmatrix} \qquad (5.43)$$

where

$$x'(\theta) = -r\sin(\theta) \quad y'(\theta) = r\cos(\theta)$$

$$x''(\theta) = -r\cos(\theta) \quad y''(\theta) = -r\sin(\theta) \qquad (5.44)$$

Figure 5.19 illustrates the definition of the first and second directional derivatives. The first derivative defines a tangential vector while the second one is similar to the vector function, but it has reverse direction. The fact that the edge direction measured for circles can be arranged so as to point towards the centre was the basis of one of the early approaches to reducing the computational load of the HT for circles (Kimme et al., 1975).

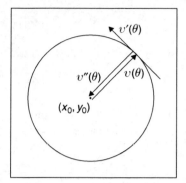

Figure 5.19 Definition of the first and second directional derivatives for a circle

According to Equations 5.42 and 5.44, we observe that the tangent of the angle of the first directional derivative denoted as $\phi'(\theta)$ is given by

$$\phi'(\theta) = \frac{y'(\theta)}{x'(\theta)} = -\frac{1}{\tan(\theta)} \tag{5.45}$$

Angles will be denoted by using the symbol $\wedge$. That is,

$$\hat{\phi}'(\theta) = \tan^{-1}(\phi'(\theta)) \tag{5.46}$$

Similarly, for the tangent of the second directional derivative we have:

$$\phi''(\theta) = \frac{y''(\theta)}{x''(\theta)} = \tan(\theta) \quad \text{and} \quad \hat{\phi}''(\theta) = \tan^{-1}(\phi''(\theta)) \tag{5.47}$$

By observing the definition of $\phi''(\theta)$, we have:

$$\phi''(\theta) = \frac{y''(\theta)}{x''(\theta)} = \frac{y(\theta) - y_0}{x(\theta) - x_0} \tag{5.48}$$

This equation defines a straight line passing through the points $(x(\theta), y(\theta))$ and (x_0, y_0) and is perhaps the most important relation in parameter space decomposition. The definition of the line is more evident by rearranging terms. That is,

$$y(\theta) = \phi''(\theta)(x(\theta) - x_0) + y_0 \tag{5.49}$$

This equation is independent of the radius parameter. Thus, it can be used to gather evidence of the location of the shape in a 2D accumulator. The HT mapping is defined by the dual form given by

$$y_0 = \phi''(\theta)(x_0 - x(\theta)) + y(\theta) \tag{5.50}$$

That is, given an image point $(x(\theta), y(\theta))$ and the value of $\phi''(\theta)$ we can generate a line of votes in the 2D accumulator (x_0, y_0). Once the centre of the circle is known, a 1D accumulator can be used to locate the radius. The key aspect of the parameter space decomposition is the method used to obtain the value of $\phi''(\theta)$ from image data. We will consider two alternative ways. First, we will show that $\phi''(\theta)$ can be obtained by edge direction information. Secondly, we will show how it can be obtained from the information of a pair of points.

To obtain $\phi''(\theta)$, we can use the definition in Equations 5.46 and 5.47. According to these equations, the tangents $\phi''(\theta)$ and $\phi'(\theta)$ are perpendicular. Thus,

$$\phi''(\theta) = -\frac{1}{\phi'(\theta)} \tag{5.51}$$

Thus, the HT mapping in Equation 5.50 can be written in terms of gradient direction $\phi'(\theta)$ as

$$y_0 = y(\theta) + \frac{x(\theta) - x_0}{\phi'(\theta)} \tag{5.52}$$

This equation has a simple geometric interpretation, illustrated in Figure 5.20(a). We can see that the line of votes passes through the points $(x(\theta), y(\theta))$ and (x_0, y_0). The slope of the line is perpendicular to the direction of gradient direction.

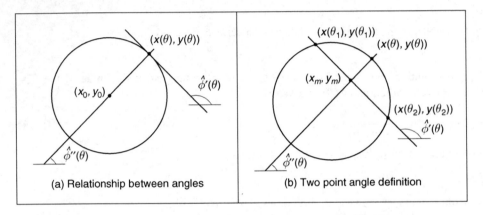

(a) Relationship between angles   (b) Two point angle definition

Figure 5.20   Geometry of the angle of the first and second directional derivatives

An alternative decomposition can be obtained by considering the geometry shown in Figure 5.20(b). In the figure we can see that if we take a pair of points (x_1, y_1) and (x_2, y_2), where $x_i = x(\theta_i)$, then the line that passes through the points has the same slope as the line at a point $(x(\theta), y(\theta))$. Accordingly,

$$\phi'(\theta) = \frac{y_2 - y_1}{x_2 - x_1} \tag{5.53}$$

where

$$\theta = \frac{1}{2}(\theta_1 + \theta_2) \tag{5.54}$$

Based on Equation 5.53 we have that

$$\phi''(\theta) = -\frac{x_2 - x_1}{y_2 - y_1} \tag{5.55}$$

The problem with using a pair of points is that by Equation 5.54 we cannot know the location of the point $(x(\theta), y(\theta))$. Fortunately, the voting line also passes through the midpoint of the line between the two selected points. Let us define this point as

$$x_m = \frac{1}{2}(x_1 + x_2) \quad y_m = \frac{1}{2}(y_1 + y_2) \tag{5.56}$$

Thus, by substitution of Equation 5.53 in Equation 5.52 and by replacing the point $(x(\theta), y(\theta))$ by (x_m, y_m), the HT mapping can be expressed as

$$y_0 = y_m + \frac{(x_m - x_0)(x_2 - x_1)}{(y_2 - y_1)} \tag{5.57}$$

This equation does not use gradient direction information, but it is based on pairs of points. This is analogous to the parameter space decomposition of the line presented in Equation 5.40. In that case, the slope can be computed by using gradient direction or, alternatively, by taking a pair of points. In the case of the circle, the tangent (and therefore the angle of the second directional derivative) can be computed by the gradient direction (i.e. Equation 5.51) or by a pair of points (i.e. Equation 5.55). However, it is important to notice that there are some other combinations of parameter space decomposition (Aguado, 1996).

Code 5.8 shows the implementation of the parameter space decomposition for the HT for circles. The implementation only detects the position of the circle and it gathers evidence by using the mapping in Equation 5.57. Pairs of points are restricted to a neighbourhood between 10×10 pixels and 12×12 pixels. We avoid using pixels that are close to each other since they do not produce accurate votes. We also avoid using pixels that are far away from each other, since by distance it is probable that they do not belong to the same circle and would only increase the noise in the accumulator. To trace the line, we use two equations that are selected according to the slope.

Figure 5.21 shows the accumulators obtained by the implementation of Code 5.8 for the images in Figure 5.14(a) and (b). Both accumulators show a clear peak that represents the location

```
%Parameter Decomposition for the Hough Transform for Circles

function HTDCircle(inputimage)

%image size
[rows,columns]=size(inputimage);

%accumulator
acc=zeros(rows,columns);

%gather evidence
for x1=1:columns
    for y1=1:rows
        if(inputimage(y1,x1)==0)
            for x2=x1-12:x1+12
                for y2=y1-12:y1+12
                    if(abs(x2-x1)>10 | abs(y2-y1)>10)
                        if(x2>0 & y2>0 & x2<columns & y2<rows)
                            if(inputimage(y2,x2)==0)
                                xm=(x1+x2)/2;    ym=(y1+y2)/2;
                                if(y2-y1~=0)    m=((x2-x1)/(y2-y1));
                                else    m=99999999;
                            end
```

```
                         if(m>-1 & m<1)
                            for x0=1:columns
                                y0=round(ym+m*(xm-x0));
                                if(y0>0 & y0<rows)
                                    acc(y0,x0)=acc(y0,x0)+1;
                                end
                            end
                         else
                            for y0=1:rows
                                x0= round(xm+(ym-y0)/m);
                                if(x0>0 & x0<columns)
                                    acc(y0,x0)=acc(y0,x0)+1;
                                end
                            end
                         end
                      end
                   end
                end
             end
          end
       end
    end
 end
```

Code 5.8   Parameter space reduction for the Hough transform for circles

of the circle. Small peaks in the background of the accumulator in Figure 5.21(b) correspond to circles with only a few points. In general, there is a compromise between the *width* of the peak and the *noise* in the accumulator. The peak can be made narrower by considering pairs of points that are more widely spaced. However, this can also increase the level of background noise. Background noise can be reduced by taking points that are closer together, but this makes the peak wider.

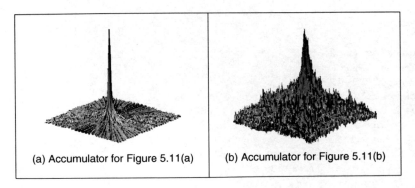

| (a) Accumulator for Figure 5.11(a) | (b) Accumulator for Figure 5.11(b) |

Figure 5.21   Parameter space reduction for the Hough transform for circles

5.4.5.3 Parameter space reduction for ellipses

Part of the simplicity in the parameter decomposition for circles comes from the fact that circles are isotropic. Ellipses have more free parameters and are geometrically more complex. Thus, geometrical properties involve more complex relationships between points, tangents and angles. However, they maintain the geometric relationship defined by the angle of the second derivative. According to Equations 5.41 and 5.43, the vector position and directional derivatives of an ellipse in Equation 5.35 have the components

$$x'(\theta) = -a_x \sin(\theta) + b_x \cos(\theta) \quad y'(\theta) = -a_y \sin(\theta) + b_y \cos(\theta)$$
$$x''(\theta) = -a_x \cos(\theta) - b_x \sin(\theta) \quad y''(\theta) = -a_y \cos(\theta) - b_y \sin(\theta)$$

(5.58)

The tangent of angle of the first and second directional derivatives are given by

$$\phi'(\theta) = \frac{y'(\theta)}{x'(\theta)} = \frac{-a_y \cos(\theta) + b_y \sin(\theta)}{-a_x \cos(\theta) + b_x \sin(\theta)}$$

$$\phi''(\theta) = \frac{y''(\theta)}{x''(\theta)} = \frac{-a_y \cos(\theta) - b_y \sin(\theta)}{-a_x \cos(\theta) - b_x \sin(\theta)}$$

(5.59)

By considering Equation 5.58, Equation 5.48 is also valid for an ellipse. That is,

$$\frac{y(\theta) - y_0}{x(\theta) - x_0} = \phi''(\theta)$$

(5.60)

The geometry of the definition in this equation is illustrated in Figure 5.22(a). As in the case of circles, this equation defines a line that passes through the points $(x(\theta), y(\theta))$ and (x_0, y_0). However, in the case of the ellipse the angles $\hat{\phi}'(\theta)$ and $\hat{\phi}''(\theta)$ are not orthogonal. This makes the computation of $\phi''(\theta)$ more complex. To obtain $\phi''(\theta)$ we can extend the geometry presented in Figure 5.20(b). That is, we take a pair of points to define a line whose slope defines the value of $\phi''(\theta)$ at another point. This is illustrated in Figure 5.22(b). The line in Equation 5.60 passes through the middle point (x_m, y_m). However, it is not orthogonal to the tangent line. To obtain an expression of the HT mapping, we will first show that the relationship in Equation 5.54 is also valid for ellipses. Then we will use this equation to obtain $\phi''(\theta)$.

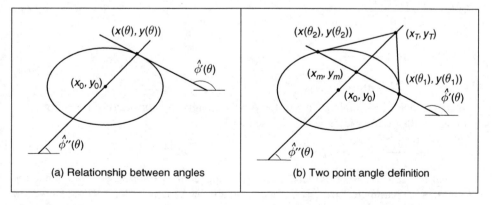

(a) Relationship between angles   (b) Two point angle definition

Figure 5.22 Geometry of the angle of the first and second directional derivatives

The relationships in Figure 5.22(b) do not depend on the orientation or position of the ellipse. Thus, three points can be defined by

$$x_1 = a_x \cos(\theta_1) \quad x_2 = a_x \cos(\theta_2) \quad x(\theta) = a_x \cos(\theta)$$
$$y_1 = b_x \sin(\theta_1) \quad y_2 = b_x \sin(\theta_2) \quad y(\theta) = b_x \sin(\theta) \tag{5.61}$$

The point $(x(\theta), y(\theta))$ is given by the intersection of the line in Equation 5.60 with the ellipse. That is,

$$\frac{y(\theta) - y_0}{x(\theta) - x_0} = \frac{a_x}{b_y} \frac{y_m}{x_m} \tag{5.62}$$

By substitution of the values of (x_m, y_m) defined as the average of the coordinates of the points (x_1, y_1) and (x_2, y_2) in Equation 5.56, we have:

$$\tan(\theta) = \frac{a_x}{b_y} \frac{b_y \sin(\theta_1) + b_y \sin(\theta_2)}{a_x \cos(\theta_1) + a_x \cos(\theta_2)} \tag{5.63}$$

Thus,

$$\tan(\theta) = \tan\left(\frac{1}{2}(\theta_1 + \theta_2)\right) \tag{5.64}$$

From this equation is evident that the relationship in Equation 5.54 is also valid for ellipses. Based on this result, the tangent angle of the second directional derivative can be defined as

$$\phi''(\theta) = \frac{b_y}{a_x} \tan(\theta) \tag{5.65}$$

By substitution in Equation 5.62, we have:

$$\phi''(\theta) = \frac{y_m}{x_m} \tag{5.66}$$

This equation is valid when the ellipse is not translated. If the ellipse is translated then the tangent of the angle can be written in terms of the points (x_m, y_m) and (x_T, y_T) as

$$\phi''(\theta) = \frac{y_T - y_m}{x_T - x_m} \tag{5.67}$$

By considering that the point (x_T, y_T) is the intersection point of the tangent lines at (x_1, y_1) and (x_2, y_2), we obtain

$$\phi''(\theta) = \frac{AC + 2BD}{2A + BC} \tag{5.68}$$

where

$$A = y_1 - y_2 \quad B = x_1 - x_2$$
$$C = \phi_1 + \phi_2 \quad D = \phi_1 \cdot \phi_2 \tag{5.69}$$

and ϕ_1, ϕ_2 are the slope of the tangent line to the points. Finally, by considering Equation 5.60, the HT mapping for the centre parameter is defined as

$$y_0 = y_m + \frac{AC + 2BD}{2A + BC}(x_0 - x_m) \tag{5.70}$$

This equation can be used to gather evidence that is independent of rotation or scale. Once the location is known, a 3D parameter space is needed to obtain the remaining parameters. However, these parameters can also be computed independently using two 2D parameter spaces (Aguado et al., 1996). You can avoid using the gradient direction in Equation 5.68 by including more points. In fact, the tangent $\phi''(\theta)$ can be computed by taking four points (Aguado, 1996). However, the inclusion of more points generally leads to more background noise in the accumulator.

Code 5.9 shows the implementation of the ellipse location mapping in Equation 5.57. As in the case of the circle, pairs of points need to be restricted to a neighbourhood. In the implementation, we consider pairs at a fixed distance given by the variable i. Since we are including gradient direction information, the resulting peak is generally quite wide. Again, the selection of the distance between points is a compromise between the level of background noise and the width of the peak.

```
%Parameter Decomposition for Ellipses
function HTDEllipse(inputimage)

%image size
[rows,columns]=size(inputimage);

%edges
[M,Ang]=Edges(inputimage);
M=MaxSupr(M,Ang);

%accumulator
acc=zeros(rows,columns);

%gather evidence
for x1=1:columns
  for y1=1:1:rows
    if(M(y1,x1)~=0)
       for i=60:60
         x2=x1-i; y2=y1-I;
         incx=1; incy=0;
         for k=0: 8*i-1
            if(x2>0 & y2>0 & x2<columns & y2<rows)
            if M(y2,x2)~=0

              m1=Ang(y1,x1); m2=Ang(y2,x2);

              if(abs(m1-m2)>.2)

                xm=(x1+x2)/2; ym=(y1+y2)/2;
                m1=tan(m1); m2=tan(m2);

                A=y1-y2; B=x2-x1;
                C=m1+m2; D=m1*m2;
                N=(2*A+B*C);
                if N~=0
                  m=(A*C+2*B*D)/N;
                else
                  m=99999999;
                end;
```

```
        if(m>-1 & m<1)
            for x0=1:columns
                y0=round(ym+m*(x0-xm));
                if(y0>0 & y0<rows)
                    acc(y0,x0)=acc(y0,x0)+1;
                end
            end
        else
            for y0=1:rows
                x0= round(xm+(y0-ym)/m);
                if(x0>0 & x0<columns)
                    acc(y0,x0)=acc(y0,x0)+1;
                end
            end
        end % if abs
    end   % if M
    end

    x2=x2+incx; y2=y2+incy;

    if x2>x1+I
        x2=x1+i;
        incx=0;        incy=1;
        y2=y2+incy;
    end

    if y2>y1+i
        y2=y1+i;
        incx=-1;       incy=0;
        x2=x2+incx;
    end

    if x2<x1-i
        x2=x1-i;
        incx=0;        incy=-1;
        y2=y2+incy;
    end
    end % for k
    end % for I
    end % if (x1,y1)
    end % y1
end %x1
```

Code 5.9 Implementation of the parameter space reduction for the Hough transform for ellipses

Figure 5.23 shows the accumulators obtained by the implementation of Code 5.9 for the images in Figure 5.17(a) and (b). The peak represents the location of the ellipses. In general, there is noise and the accumulator is wide, for two main reasons. First, when the gradient direction is not accurate, then the line of votes does not pass exactly over the centre of the ellipse. This forces the peak to become wider with less height. Secondly, to avoid numerical instabilities we need to select points that are well separated. However, this increases the probability that the points do not belong to the same ellipse, thus generating background noise in the accumulator.

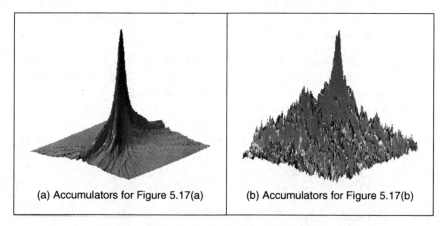

(a) Accumulators for Figure 5.17(a)   (b) Accumulators for Figure 5.17(b)

Figure 5.23   Parameter space reduction for the Hough transform for ellipses

5.5   Generalized Hough transform

Many shapes are far more complex than lines, circles or ellipses. It is often possible to partition a complex shape into several geometric primitives, but this can lead to a highly complex data structure. In general, is more convenient to extract the whole shape. This has motivated the development of techniques that can find *arbitrary* shapes using the evidence-gathering procedure of the HT. These techniques give results equivalent to those delivered by matched template filtering, but with the computational advantage of the evidence-gathering approach. An early approach offered only limited capability only for arbitrary shapes (Merlin and Faber 1975). The full mapping is called the *generalized Hough transform* (GHT) (Ballard, 1981) and can be used to locate arbitrary shapes with unknown *position*, *size* and *orientation*. The GHT can be formally defined by considering the duality of a curve. One possible implementation can be based on the discrete representation given by tabular functions. These two aspects are explained in the following two sections.

5.5.1   Formal definition of the GHT

The formal analysis of the HT provides the route for generalizing it to arbitrary shapes. We can start by generalizing the definitions in Equation 5.41. In this way a model shape can be defined by a curve

$$v(\theta) = x(\theta) \begin{bmatrix} 1 \\ 0 \end{bmatrix} + y(\theta) \begin{bmatrix} 0 \\ 1 \end{bmatrix} \tag{5.71}$$

For a circle, for example, we have $x(\theta) = r\cos(\theta)$ and $y(\theta) = r\sin(\theta)$. Any shape can be represented by following a more complex definition of $x(\theta)$ and $y(\theta)$.

In general, we are interested in matching the model shape against a shape in an image. However, the shape in the image has a different location, orientation and scale. Originally, the

GHT defined a scale parameter in the x and y directions, but owing to computational complexity and practical relevance the use of a single scale has become much more popular. Analogous to Equation (5.33), we can define the image shape by considering translation, rotation and change of scale. Thus, the shape in the image can be defined as

$$\omega(\theta, b, \lambda, \rho) = b + \lambda \mathbf{R}(\rho) v(\theta) \tag{5.72}$$

where $b = (x_0, y_0)$ is the translation vector, λ is a scale factor and $\mathbf{R}(\rho)$ is a rotation matrix (as in Equation 5.31). Here, we have included explicitly the parameters of the transformation as arguments, but to simplify the notation they will be omitted later. The shape of $\omega(\theta, b, \lambda, \rho)$ depends on four parameters. Two parameters define the location b, plus the rotation and scale. It is important to notice that θ does not define a free parameter, it only traces the curve.

To define a mapping for the HT we can follow the approach used to obtain Equation 5.35. Thus, the location of the shape is given by

$$b = \omega(\theta) - \lambda \mathbf{R}(\rho) v(\theta) \tag{5.73}$$

Given a shape $\omega(\theta)$ and a set of parameters b, λ and ρ, this equation defines the location of the shape. However, we do not know the shape $\omega(\theta)$ (since it depends on the parameters that we are looking for), but we only have a point in the curve. If we call $\omega_i = (\omega_{xi}, \omega_{yi})$ the point in the image, then

$$b = \omega_i - \lambda \mathbf{R}(\rho) v(\theta) \tag{5.74}$$

defines a system with four unknowns and with as many equations as points in the image. To find the solution we can gather evidence by using a four-dimensional (4D) accumulator space. For each potential value of b, λ and ρ, we trace a point spread function by considering all the values of θ. That is, all the points in the curve $v(\theta)$.

In the GHT the gathering process is performed by adding an extra constraint to the system that allows us to match points in the image with points in the model shape. This constraint is based on gradient direction information and can be explained as follows. We said that ideally, we would like to use Equation 5.73 to gather evidence. For that we need to know the shape $\omega(\theta)$ and the model $v(\theta)$, but we only know the discrete points ω_i and we have supposed that these are the same as the shape, i.e. that $\omega(\theta) = \omega_i$. Based on this assumption, we then consider all the potential points in the model shape, $v(\theta)$. However, this is not necessary since we only need the point in the model, $v(\theta)$, that corresponds to the point in the shape, $\omega(\theta)$. We cannot know the point in the shape, $v(\theta)$, but we can compute some properties from the model and from the image. Then, we can check whether these properties are similar at the point in the model and at a point in the image. If they are indeed *similar*, the points might *correspond*: if they do we can gather evidence on the parameters of the shape. The GHT considers as feature the gradient direction at the point. We can generalize Equations 5.45 and 5.46 to define the gradient direction at a point in the arbitrary model. Thus,

$$\phi'(\theta) = \frac{y'(\theta)}{x'(\theta)} \quad \text{and} \quad \hat{\phi}'(\theta) = \tan^{-1}(\phi'(\theta)) \tag{5.75}$$

Thus, Equation 5.73 is true only if the gradient direction at a point in the image matches the rotated gradient direction at a point in the (rotated) model, that is

$$\phi'_i = \hat{\phi}'(\theta) - \rho \tag{5.76}$$

where $\hat{\varphi}'_i$ is the angle at the point ω_i. Note that according to this equation, gradient direction is independent of scale (in theory at least) and it changes in the same ratio as rotation. We can constrain Equation 5.74 to consider only the points $v(\theta)$ for which

$$\phi'_i - \hat{\phi}'(\theta) + \rho = 0 \qquad (5.77)$$

That is, a point spread function for a given edge point ω_i is obtained by selecting a subset of points in $v(\theta)$ such that the edge direction at the image point rotated by ρ equals the gradient direction at the model point. For each point ω_i and selected point in $v(\theta)$ the point spread function is defined by the HT mapping in Equation 5.74.

5.5.2 Polar definition

Equation 5.74 defines the mapping of the HT in Cartesian form. That is, it defines the votes in the parameter space as a pair of coordinates (x, y). There is an alternative definition in polar form. The polar implementation is more common than the Cartesian form (Hecker and Bolle, 1994; Sonka et al., 1994). The advantage of the polar form is that it is easy to implement since changes in rotation and scale correspond to addition in the angle-magnitude representation. However, ensuring that the polar vector has the correct direction incurs more complexity.

Equation 5.74 can be written in a form that combines rotation and scale as

$$b = \omega(\theta) - \gamma(\lambda, \rho) \qquad (5.78)$$

where $\gamma^T(\lambda, \rho) = [\gamma_x(\lambda, \rho)\gamma_y(\lambda, \rho)]$ and where the combined rotation and scale is

$$\gamma_x(\lambda, \rho) = \lambda(x(\theta)\cos(\rho) - y(\theta)\sin(\rho))$$
$$\gamma_y(\lambda, \rho) = \lambda(x(\theta)\sin(\rho) + y(\theta)\cos(\rho)) \qquad (5.79)$$

This combination of rotation and scale defines a vector, $\gamma(\lambda, \rho)$, whose tangent angle and magnitude are given by

$$\tan(\alpha) = \frac{\gamma_y(\lambda, \rho)}{\gamma_x(\lambda, \rho)} \qquad r = \sqrt{\gamma_x^2(\lambda, \rho) + \gamma_y^2(\lambda, \rho)} \qquad (5.80)$$

The main idea here is that if we know the values for α and r, then we can gather evidence by considering Equation 5.78 in polar form. That is,

$$b = \omega(\theta) - re^{\alpha} \qquad (5.81)$$

Thus, we should focus on computing values for α and r. After some algebraic manipulation, we have:

$$\alpha = \phi(\theta) + \rho \qquad r = \lambda\Gamma(\theta) \qquad (5.82)$$

where

$$\phi(\theta) = \tan^{-1}\left(\frac{y(\theta)}{x(\theta)}\right) \qquad \Gamma(\theta) = \sqrt{x^2(\theta) + y^2(\theta)} \qquad (5.83)$$

In this definition, we must include the constraint defined in Equation 5.77. That is, we gather evidence only when the gradient direction is the *same*. Notice that the square root in the definition of the magnitude in Equation 5.83 can have positive and negative values. The sign must be selected in a way that the vector has the correct direction.

5.5.3 The GHT technique

Equations 5.74 and 5.81 define an HT mapping function for arbitrary shapes. The geometry of these equations is shown in Figure 5.24. Given an image point ω_i we have to find a displacement vector $\gamma(\lambda, \rho)$. When the vector is placed at ω_i, its end is at the point b. In the GHT jargon, this point called the reference point. The vector $\gamma(\lambda, \rho)$ can be easily obtained as $\lambda R(\rho) \upsilon(\theta)$ or alternatively as re^α. However, to evaluate these equations, we need to know the point $\upsilon(\theta)$. This is the crucial step in the evidence gathering process. Notice the remarkable similarity between Figures 5.20(a), 5.22(a) and 5.24(a). This is not a coincidence, but Equation 5.60 is a particular case of Equation 5.73.

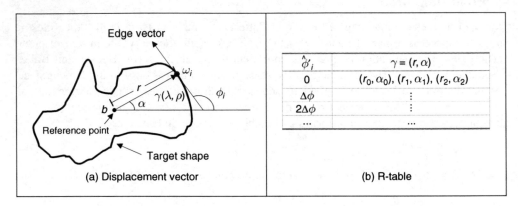

Figure 5.24 Geometry of the GHT

The process of determining $\upsilon(\theta)$ centres on solving Equation 5.76. According to this equation, since we know $\hat{\phi}'_i$, then we need to find the point $\upsilon(\theta)$ whose gradient direction is $\hat{\phi}'_i + \rho = 0$. Then we must use $\upsilon(\theta)$ to obtain the displacement vector $\gamma(\lambda, \rho)$. The GHT precomputes the solution of this problem and stores it an array called the *R-table*. The R-table stores for each value of $\hat{\phi}'_i$ the vector $\gamma(\lambda, \rho)$ for $\rho = 1$ and $\lambda = 1$. In polar form, the vectors are stored as a magnitude direction pair and in Cartesian form as a coordinate pair.

The possible range for $\hat{\phi}'_i$ is between $-\pi/2$ and $\pi/2$ radians. This range is split into N equispaced slots, or bins. These slots become rows of data in the R-table. The edge direction at each border point determines the appropriate row in the R-table. The length, r, and direction, α, from the reference point are entered into a new column element, at that row, for each border point in the shape. In this manner, the N rows of the R-table have elements related to the border information; elements for which there is no information contain null vectors. The length of each row is given by the number of edge points that have the edge direction corresponding to that row; the total number of elements in the R-table equals the number of edge points above a chosen threshold. The *structure* of the R-table for N edge direction bins and m template border points is illustrated in Figure 5.23 (b).

The process of *building* the R-table is illustrated in Code 5.10. In this code, we implement the Cartesian definition given in Equation 5.74. According to this equation the displacement vector is given by

$$\gamma(1, 0) = \omega(\theta) - b \tag{5.84}$$

```
%R-Table
function T=RTable(entries,inputimage)

%image size
[rows,columns]=size(inputimage);

%edges
[M,Ang]=Edges(inputimage);
M=MaxSupr(M,Ang);

%compute reference point
xr=0; yr=0; p=0;
for x=1:columns
   for y=1:rows
      if(M(y,x)~=0)
         xr=xr+x;
         yr=yr+y;
         p=p+1;
      end
   end
end
xr=round(xr/p);
yr=round(yr/p);

%accumulator
D=pi/entries;

s=0;             % number of entries in the table
t=[];
F=zeros(entries,1); % number of entries in the row

% for each edge point
for x=1:columns
   for y=1:rows
      if(M(y,x)~=0)

         phi=Ang(y,x);
         i=round((phi+(pi/2))/D);
         if(i==0) i=1; end;

         V=F(i)+1;

         if(V>s)
            s=s+1;
            T(:,:,s)=zeros(entries,2);
         end;

         T(i,1,V)=x-xr;
         T(i,2,V)=y-yr;
         F(i)=F(I)+1;

      end %if
   end % y
end% x
```

Code 5.10   Implementation of the construction of the R-table

The matrix **T** stores the coordinates of γ (1, 0). This matrix is expanded to accommodate all of the computed entries.

Code 5.11 shows the implementation of the gathering process of the GHT. In this case we use the Cartesian definition in Equation 5.74. The coordinates of points given by evaluation of all R-table points for the particular row indexed by the gradient magnitude are used to increment cells in the accumulator array. The maximum number of votes occurs at the location of the original reference point. After all edge points have been inspected, the location of the shape is given by the maximum of an accumulator array.

```
%Generalized Hough Transform

function GHT(inputimage,RTable)

%image size
[rows,columns]=size(inputimage);

%table size
[rowsT,h,columnsT]=size(RTable);
D=pi/rowsT;

%edges
[M,Ang]=Edges(inputimage);
M=MaxSupr(M,Ang);

%accumulator
acc=zeros(rows,columns);

%for each edge point
for x=1:columns
    for y=1:rows
        if(M(y,x)~=0)

            phi=Ang(y,x);
            i=round(((phi+(pi/2))/D);
            if(i==0) i=1; end;

            for j=1:columnsT
                if(RTable(i,1,j)==0 & RTable(i,2,j)==0)
                    j=columnsT; %no more entries
                else
                    a0=x-RTable(i,1,j); b0=y-RTable(i,2,j);
                    if(a0>0 & a0<columns & b0>0 & b0<rows)
                        acc(b0,a0)=acc(b0,a0)+1;
                    end
                end
            end
        end %if
    end % y
end% x
```

Code 5.11 Implementation of the GHT

Note that if we want to try other values for rotation and scale, then it is necessary to compute a table $\gamma(\lambda, \rho)$ for all potential values. However, this can be avoided by considering that $\gamma(\lambda, \rho)$ can be computed from $\gamma(1, 0)$. That is, if we want to accumulate evidence for $\gamma(\lambda, \rho)$, then we use the entry indexed by $\hat{\phi}'_i + \rho$ and we rotate and scale the vector $\gamma(1, 0)$. That is,

$$\gamma_x(\lambda, \rho) = \lambda(\gamma_x(1, 0)\cos(\rho) - \gamma_y(1, 0)\sin(\rho))$$

$$\gamma_y(\lambda, \rho) = \lambda(\gamma_x(1, 0)\sin(\rho) + \gamma_y(1, 0)\cos(\rho))$$

(5.85)

In the case of the polar form, the angle and magnitude need to be defined according to Equation 5.82.

The application of the GHT to detect an arbitrary shape with unknown translation is illustrated in Figure 5.25. We constructed an R-table from the template shown in Figure 5.2(a). The table contains 30 rows. The accumulator in Figure 5.25(c) was obtained by applying the GHT to the image in Figure 5.25(b). Since the table was obtained from a shape with the same scale and rotation as the primitive in the image, the GHT produces an accumulator with a clear peak at the centre of mass of the shape.

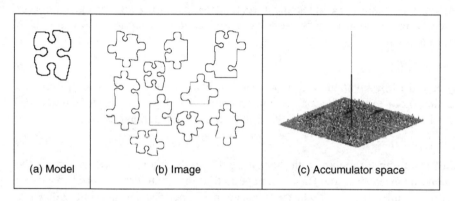

| (a) Model | (b) Image | (c) Accumulator space |

Figure 5.25 Example of the GHT

Although the example in Figure 5.25 shows that the GHT is an effective method for shape extraction, there are several inherent difficulties in its formulation (Grimson and Huttenglocher, 1990; Aguado et al., 2000b). The most evident problem is that the table does not provide an accurate representation when objects are scaled and translated. This is because the table implicitly assumes that the curve is represented in discrete form. Thus, the GHT maps a discrete form into a discrete parameter space. In addition, the transformation of scale and rotation can induce other discretization errors. This is because when discrete images are mapped to be larger, or when they are rotated, loci which are unbroken sets of points rarely map to unbroken sets in the new image. Another important problem is the excessive computations required by the 4D parameter space. This makes the technique impractical. In addition, the GHT is clearly dependent on the accuracy of directional information. By these factors, the results provided by the GHT can become less reliable. A solution is to use of an analytic form instead of a table (Aguado et al., 1998). This avoids discretization errors and makes the technique more reliable. This also allows the extension to affine or other transformations. However, this technique requires solving for the point $v(\theta)$ in an analytic way, increasing the computational load. A solution is to reduce the number of points by considering characteristics points defined as points of high curvature. However, this still requires the use of a 4D accumulator. An alternative to reduce this computational load is to include the concept of invariance in the GHT mapping.

5.5.4 Invariant GHT

The problem with the GHT (and other extensions of the HT) is that they are very general. That is, the HT gathers evidence for a single point in the image. However, a point on its own provides little information. Thus, it is necessary to consider a large parameter space to cover all the potential shapes defined by a given image point. The GHT improves evidence gathering by considering a point and its gradient direction. However, since gradient direction changes with rotation, the evidence gathering is improved in terms of noise handling, but little is done about computational complexity.

To reduce the computational complexity of the GHT, we can consider replacing the gradient direction by another feature; that is, by a feature that is not affected by *rotation*. Let us explain this idea in more detail. The main aim of the constraint in Equation 5.77 is to include gradient direction to reduce the number of votes in the accumulator by identifying a point $v(\theta)$. Once this point is known, we obtain the displacement vector $\gamma(\lambda, \rho)$. However, for each value of rotation, we have a different point in $v(\theta)$. Now let us replace that constraint in Equation 5.76 by a constraint of the form

$$Q(\omega_i) = Q(v(\theta)) \tag{5.86}$$

The function Q is said to be invariant and it computes a feature at the point. This feature can be, for example, the colour of the point, or any other property that does not change in the model and in the image. By considering Equation 5.86, Equation 5.77 is redefined as

$$Q(\omega_i) = Q(v(\theta)) = 0 \tag{5.87}$$

That is, instead of searching for a point with the same gradient direction, we will search for the point with the same invariant feature. The advantage is that this feature will not change with rotation or scale, so we only require a 2D space to locate the shape. The definition of Q depends on the application and the type of transformation. The most general invariant properties can be obtained by considering geometric definitions. In the case of rotation and scale changes (i.e. similarity transformations) the fundamental invariant property is given by the concept of angle.

An angle is defined by three points and its value remains unchanged when it is rotated and scaled. Thus, if we associate to each edge point ω_i a set of other two points $\{\omega_j, \omega_T\}$, we can compute a geometric feature that is invariant to similarity transformations. That is,

$$Q(\omega_i) = \frac{X_j Y_i - X_i Y_j}{X_i X_j + Y_i Y_j}, \tag{5.88}$$

where $X_k = \omega_k - \omega_T$, $Y_k = \omega_k - \omega_T$. Equation 5.88 defines the tangent of the angle at the point ω_T. In general, we can define the points $\{\omega_j, \omega_T\}$ in different ways. An alternative geometric arrangement is shown in Figure 5.26(a). Given the points ω_i and a fixed angle ϑ, then we determine the point ω_j such that the angle between the tangent line at ω_i and the line that joins the points is ϑ. The third point is defined by the intersection of the tangent lines at ω_i and ω_j. The tangent of the angle β is defined by Equation 5.88. This can be expressed in terms of the points and its gradient directions as

$$Q(\omega_i) = \frac{\phi'_i - \phi'_j}{1 + \phi'_i \phi'_j} \tag{5.89}$$

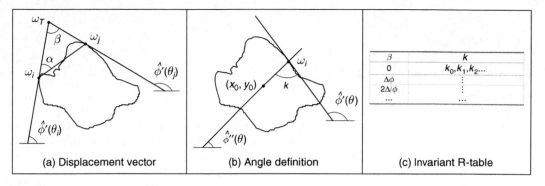

	β		k
	0		$k_0, k_1, k_2 \ldots$
	$\Delta\phi$		$\vdots$
	$2\Delta\phi$		$\vdots$
	$\ldots$		$\ldots$

(a) Displacement vector	(b) Angle definition	(c) Invariant R-table

Figure 5.26   Geometry of the invariant GHT

We can replace the gradient angle in the R-table, by the angle β. The form of the new invariant table is shown in Figure 5.26(c). Since the angle β does not change with rotation or change of scale, then we do not need to change the index for each potential rotation and scale. However, the displacement vectors changes according to rotation and scale (i.e. Equation 5.85) that. Thus, if we want an invariant formulation, we must also change the definition of the position vector.

To locate the point b we can generalize the ideas presented in Figures 5.20(a) and 5.22(a). Figure 5.26(b) shows this generalization. As in the case of the circle and ellipse, we can locate the shape by considering a line of votes that passes through the point b. This line is determined by the value of ϕ_i''. We will do two things. First, we will find an invariant definition of this value. Secondly, we will include it on the GHT table.

We can develop Equation 5.73 as

$$\begin{bmatrix} x_0 \\ y_0 \end{bmatrix} = \begin{bmatrix} \omega_{xi} \\ \omega_{yi} \end{bmatrix} + \lambda \begin{bmatrix} \cos(\rho) & \sin(\rho) \\ -\sin(\rho) & \cos(\rho) \end{bmatrix} \begin{bmatrix} x(\theta) \\ y(\theta) \end{bmatrix} \tag{5.90}$$

Thus, Equation 5.60 generalizes to

$$\phi_i'' = \frac{\omega_{yi} - y_0}{\omega_{xi} - x_0} = \frac{[-\sin(\rho) \ \cos(\rho)] y(\theta)}{[\cos(\rho) \ \sin(\rho)] x(\theta)} \tag{5.91}$$

By some algebraic manipulation, we have:

$$\phi_i'' = \tan(\xi - \rho) \tag{5.92}$$

where

$$\xi = \frac{y(\theta)}{x(\theta)} \tag{5.93}$$

To define ϕ_i'', we can consider the tangent angle at the point ω_i. By considering the derivative of Equation 5.72, we have:

$$\phi_i' = \frac{[-\sin(\rho) \ \cos(\rho)] y'(\theta)}{[\cos(\rho) \ \sin(\rho)] x'(\theta)} \tag{5.94}$$

Thus,

$$\phi_i' = \tan(\phi - \rho) \tag{5.95}$$

where

$$\phi = \frac{y'(\theta)}{x'(\theta)} \tag{5.96}$$

By considering Equations 5.92 and 5.95, we define

$$\hat{\phi}''_i = k + \hat{\phi}'_i \tag{5.97}$$

The important point in this definition is that the value of k is invariant to rotation. Thus, if we use this value in combination with the tangent at a point we can have an invariant characterization. To see that k is invariant, we solve it for Equation 5.97. That is,

$$k = \hat{\phi}'_i - \hat{\phi}''_i \tag{5.98}$$

Thus,

$$k = \xi - \rho - (\phi - \rho) \tag{5.99}$$

That is,

$$k = \xi - \phi \tag{5.100}$$

This is independent of rotation. The definition of k has a simple geometric interpretation illustrated in Figure 5.26(b).

To obtain an invariant GHT, it is necessary to know for each point ω_i, the corresponding point $v(\theta$ and then compute the value of ϕ''_i. Then evidence can be gathered by the line in Equation 5.91. That is,

$$y_0 = \phi''_i (x_0 - \omega_{xi}) + \omega_{yi} \tag{5.101}$$

To compute ϕ''_i we can obtain k and then use Equation 5.100. In the standard tabular form the value of k can be precomputed and stored as function of the angle β.

Code 5.12 illustrates the implementation to obtain the invariant R-table. This code is based on Code 5.10. The value of α is set to $\pi/4$ and each element of the table stores a single value computed according to Equation 5.98. The more cumbersome part of the code is to search for the point ω_j. We search in two directions from ω_i and we stop once an edge point has been located. This search is performed by tracing a line. The trace is dependent on the slope. When the slope is between -1 and $+1$ we determine a value of y for each value of x, otherwise we determine a value of x for each value of y.

Code 5.13 illustrates the evidence-gathering process according to Equation 5.101. This code is based in the implementation presented in Code 5.11. We use the value of β defined in Equation 5.89 to index the table passed as parameter to the function GHTInv. The data k recovered from the table is used to compute the slope of the angle defined in Equation 5.97. This is the slope of the line of votes traced in the accumulators.

Figure 5.27 shows the accumulator obtained by the implementation of Code 5.13. Figure 5.27(a) shows the template used in this example. This template was used to construct the R-Table in Code 5.12. The R-table was used to accumulate evidence when searching for the piece of the puzzle in the image in Figure 5.27(b). Figure 5.27(c) shows the result of the evidence-gathering process. We can observe a peak in the location of the object. However, this accumulator contains significant noise. The noise is produced since rotation and scale change the value of the computed gradient. Thus, the line of votes is only approximated. Another problem

```
%Invariant R-Table

function T=RTableInv(entries,inputimage)

%image size
[rows,columns]=size(inputimage);

%edges
[M,Ang]=Edges(inputimage);
M=MaxSupr(M,Ang);

alfa=pi/4;
D=pi/entries;
s=0;                %number of entries in the table
t=0;
F=zeros(entries,1);  %number of entries in the row

%compute reference point
xr=0; yr=0; p=0;
for x=1:columns
    for y=1:rows
      if(M(y,x)~=0)
         xr=xr+x;
         yr=yr+y;
         p=p+1;
      end
    end
end
xr=round(xr/p);
yr=round(yr/p);

%for each edge point
for x=1:columns
   for y=1:rows
     if(M(y,x)~=0)
        %search for the second point
        x1=-1; y1=-1;
        phi=Ang(y,x);
        m=tan(phi-alfa);

        if(m>-1 & m<1)
            for i=3:columns
               c=x+i;
               j=round(m*(c-x)+y);
               if(j>0 & j<rows & c>0 & c<columns & M(j,c)~=0)
                 x1=c ; y1=j;
                 i= columns;
               end

               c=x-i;
               j=round(m*(c-x)+y);
               if(j>0 & j<rows & c>0 & c<columns & M(j,c)~=0)
                 x1=i ; y1=j;
                 i=columns;
               end
            end
```

```
        else
            for j=3:rows
                c=y+j;
                i=round(x+(c-y)/m);
                if(c>0 & c<rows & i>0 & i< columns & M(c,i)~=0)
                    x1=i ; y1=c;
                    i=rows;
                end
                c=y-j;
                i=round(x+(c-y)/m);
                if(c>0 & c<rows & i>0 & i< columns & M(c,i)~=0)
                    x1=i ; y1=c;
                    i= rows;
                end
            end
        end

        if( x1~=-1)
            %compute beta
                phi=tan(Ang(y,x));
                phj= tan(Ang(y1,x1));
                if((1+phi*phj)~=0)
                    beta=atan((phi-phj)/(1+phi*phj));
                else
                    beta=1.57;
                end

                %compute k
                if((x-xr)~=0)
                    ph=atan((y-yr)/(x-xr));
                else
                    ph=1.57;
                end
                k=ph-Ang(y,x);

                %insert in the table
                i=round((beta+(pi/2))/D);
                if(i==0) i=1; end;

                V=F(i)+1;

        if(V>s)
            s=s+1;
            T(:,s)=zeros(entries,1);
                end;

                T(i,V)=k;
                F(i)=F(i)+1;
            end

        end %if
    end % y
end % x
```

Code 5.12 Construction of the invariant R-table

```
%Invariant Generalized Hough Transform

function GHTInv(inputimage,RTable)

%image size
[rows,columns]=size(inputimage);

%table size
[rowsT,h,columnsT]=size(RTable);
D=pi/rowsT;

%edges
[M,Ang]=Edges(inputimage);
M=MaxSupr(M,Ang);

alfa=pi/4;

%accumulator
acc=zeros(rows,columns);

% for each edge point
for x=1:columns
  for y=1:rows
    if(M(y,x)~=0)
       % search for the second point
       x1=-1; y1=-1;
       phi=Ang(y,x);
       m=tan(phi-alfa);

       if(m>-1 & m<1)
         for i=3:columns
           c=x+i;
           j=round(m*(c-x)+y);
           if(j>0 & j<rows & c>0 & c<columns & M(j,c)~=0)
             x1=c ;y1=j;
             i= columns;
           end
           c=x-i;
           j=round(m*(c-x)+y);
           if(j>0 & j<rows & c>0 & c<columns & M(j,c)~=0)
             x1=c ;y1=j;
             i=columns;
           end
         end
       else
         for j=3:rows
           c=y+j;
           i=round(x+(c-y)/m);
           if(c>0 & c<rows & i>0 & i< columns & M(c,i)~=0)
             x1=i ;y1=c;
             i=rows;
           end
```

```
            c=y-j;
            i=round(x+(c-y)/m);
            if(c>0 & c<rows & i>0 & i< columns & M(c,i)~=0)
                x1=i ;y1=c;
                i=rows;
            end
        end
    end

  if( x1~=-1)
     %compute beta
            phi=tan(Ang(y,x));
            phj=tan(Ang(y1,x1));
            if((1+phi*phj)~=0)
                beta=atan((phi-phj)/(1+phi*phj));
            else
                beta=1.57;
            end

            i=round((beta+(pi/2))/D);
            if(i==0) i=1; end;

            %search for k
            for j=1:columnsT
                if(RTable(i,j)==0)
                    j=columnsT; % no more entries
                else
                    k=RTable(i,j);
                    %lines of votes
                    m=tan(k+Ang(y,x));
                if(m>-1 & m<1)
                    for x0=1:columns
                        y0=round(y+m*(x0-x));
                        if(y0>0 & y0<rows)
                            acc(y0,x0)=acc(y0,x0)+1;
                        end
                    end
                else
                    for y0=1:rows
                        x0= round(x+(y0-y)/m);
                        if(x0>0 & x0<columns)
                            acc(y0,x0)=acc(y0,x0)+1;
                        end
                    end
                end
              end
            end
        end
     end %if
  end % y
end % x
```

Code 5.13   Implementation of the invariant GHT

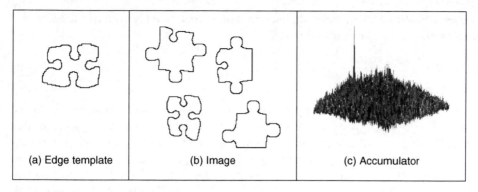

| (a) Edge template | (b) Image | (c) Accumulator |

Figure 5.27 Applying the invariant GHT

is that pairs of points ω_i and ω_j might not be found in an image, thus the technique is more sensitive to occlusion and noise than the GHT.

5.6 Other extensions to the Hough transform

The motivation for extending the HT is clear: *keep* the *performance*, but *improve* the *speed*. There are other approaches to reduce the computational load of the HT. These approaches aim to improve speed and reduce memory, focusing on smaller regions of the accumulator space. These approaches have included the *fast HT* (Li and Lavin, 1986), which uses successively splits the accumulator space into quadrants and continues to study the quadrant with most evidence; the *adaptive HT* (Illingworth and Kittler, 1987), which uses a fixed accumulator size to focus iteratively on potential maxima in the accumulator space; the *randomized HT* (Xu et al., 1990) and the *probabilistic HT* (Kälviäinen et al., 1995), which use a random search of the accumulator space; and other pyramidal techniques. One main problem with techniques that do not search the full accumulator space, but a reduced version to save speed, is that the wrong shape can be extracted (Princen et al., 1992a), a problem known as *phantom shape location*. These approaches can also be used (with some variation) to improve speed of performance in template matching. There have been many approaches aimed to improve performance of the HT and the GHT.

There has been a comparative study on the GHT (including efficiency) (Kassim et al., 1999) and alternative approaches to the GHT include two *fuzzy* HTs: Philip (1991) and Sonka et al. (1994) include uncertainty of the perimeter points within a GHT structure, and Han et al. (1994) approximately fits a shape but requires application-specific specification of a fuzzy membership function. There have been two major reviews of the state of research in the HT (Illingworth and Kittler, 1988; Leavers, 1993), but they are rather dated now, and a textbook (Leavers, 1992) which cover many of these topics. The analytic approaches to improving the HT's performance use mathematical analysis to reduce size, and more importantly dimensionality, of the accumulator space. This concurrently improves speed. A review of HT-based techniques for circle extraction (Yuen et al., 1990) covered some of the most popular techniques available at the time.

As techniques move to analysing moving objects, there is the *velocity Hough transform* for detecting moving shapes (Nash et al., 1997). As in any HT, in the velocity HT a moving shape needs a parameterization which includes the motion. For a circle moving with (linear) velocity we have points which are a function of time t as

$$x(t) = c_x + v_x t + r \cos \theta$$

$$y(t) = c_y + v_y t + r \sin \theta$$

(5.102)

where c_x, c_y are the coordinates of the circle's centre, v_x, v_y describe the velocity along the x- and the y-axes, respectively, r is the circle's radius, and θ allows us to draw the locus of the circle at time t. We then construct a 5D accumulator array in terms of the unknown parameters c_x, c_y, v_x, v_y, r and vote for each image of the sequence (after edge detection and thresholding) in this accumulator array. By grouping the information across a sequence the technique was shown to be more reliable in occlusion than extracting a single circle for each frame and determining the track as the locus of centres of the extracted circles. This was extended to a technique for finding *moving lines*, as in the motion of the thigh in a model-based approach to recognizing people by the way they walk (gait biometrics) (Cunado et al., 2003). This is illustrated in Figure 5.28, which shows a walking subject on whom is superimposed a line showing the extracted position and orientation of the (moving) human thigh. It was also used in a generalized HT for *moving shapes* (Grant et al., 2002), which imposed a motion trajectory on the GHT extraction, and in a GHT which includes *deforming* moving shapes (Mowbray and Nixon, 2004).

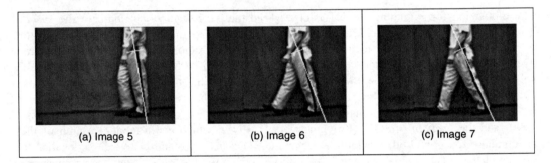

| (a) Image 5 | (b) Image 6 | (c) Image 7 |

Figure 5.28   Detecting moving lines

5.7   Further reading

The majority of further reading in finding shapes concerns papers, many of which have already been referenced. An excellent survey of the techniques used for feature extraction (including template matching, deformable templates, etc.) can be found in Trier et al. (1996). Few of the textbooks devote much space to shape extraction, sometimes dismissing it in a couple of pages. This contrasts with the volume of research there has been in this area, and the HT finds increasing application as computational power continues to increase (and storage cost reduces). One text alone is dedicated to shape analysis (van Otterloo, 1991) and contains many discussions on symmetry. For implementation, Parker (1994) includes C code for template matching and for the HT for lines, but no more (the more recent version, Parker, 1996, omits it entirely). Other techniques use a similar evidence-gathering process to the HT. These techniques are referred to as geometric hashing and clustering techniques (Stockman, 1987; Lamdan et al., 1988).

In contrast to the HT, these techniques do not define an analytic mapping, but they gather evidence by grouping a set of features computed from the image and from the model.

5.8 References

Aguado, A. S., *Primitive Extraction via Gathering Evidence of Global Parameterised Models*, PhD Thesis, University of Southampton, 1996

Aguado, A. S., Montiel, E. and Nixon, M. S., On Using Directional Information for Parameter Space Decomposition in Ellipse Detection, *Pattern Recog.* **28**(3), pp. 369–381, 1996

Aguado, A. S., Nixon, M. S. and Montiel, M. E., Parameterizing Arbitrary Shapes via Fourier Descriptors for Evidence-Gathering Extraction, *Comput. Vision Image Understand.*, **69**(2), pp. 202–221, 1998

Aguado, A. S., Montiel, E. and Nixon, M. S., On the Intimate Relationship Between the Principle of Duality and the Hough Transform, *Proc. R. Soc. A*, **456**, pp. 503–526, 2000

Aguado, A. S., Montiel, E. and Nixon, M. S., Bias Error Analysis of the Generalized Hough Transform, *J. Math. Imaging Vision*, **12**, pp. 25–42, 2000

Altman, J. and Reitbock, H. J. P., A Fast Correlation Method for Scale- and Translation-Invariant Pattern Recognition, *IEEE Trans. PAMI*, **6**(1), pp. 46–57

Ballard, D. H., Generalizing the Hough Transform to Find Arbitrary Shapes, *CVGIP*, **13**, pp. 111–122, 1981

Bracewell, R. N., *The Fourier Transform and its Applications*, 2nd edn, McGraw-Hill Book Co., Singapore, 1986

Bresenham, J. E., Algorithm for Computer Control of a Digital Plotter, *IBM Syst. J.*, **4**(1), pp. 25–30, 1965

Bresenham, J. E., A Linear Algorithm for Incremental Digital Display of Circular Arcs, *Communs ACM*, **20**(2), pp. 750–752, 1977

Brown, C. M., Inherent Bias and Noise in the Hough Transform, *IEEE Trans. PAMI*, **5**, pp. 493–505, 1983

Casasent, D. and Psaltis, D., New Optical Transforms for Pattern Recognition, *Proc. IEEE*, **65**(1), pp. 77–83, 1977

Cunado, D., Nixon, M. S. and Carter, J. N., Automatic Extraction and Description of Human Gait Models for Recognition Purposes, *Comput. Vision Image Understand.*, **90**(1), pp. 1–41, 2003

Deans, S. R., Hough Transform from the Radon Transform, *IEEE Trans. PAMI*, **13**, pp. 185–188, 1981

Duda, R. O. and Hart, P. E., Use of the Hough Transform to Detect Lines and Curves in Pictures, *Communs. ACM*, **15**, pp. 11–15, 1972

Gerig, G. and Klein, F., Fast Contour Identification Through Efficient Hough Transform and Simplified Interpretation Strategy, *Proc. 8th Int. Conf. Pattern Recog*, pp. 498–500, 1986

Grant, M. G., Nixon., M. S. and Lewis, P. H., Extracting Moving Shapes by Evidence Gathering, *Pattern Recog.*, **35**, pp. 1099–1114, 2002

Grimson, W. E. L. and Huttenglocher, D. P., On the Sensitivity of the Hough Transform for Object Recognition, *IEEE Trans. PAMI*, **12**, pp. 255–275, 1990

Han, J. H., Koczy, L. T. and Poston, T., Fuzzy Hough Transform, *Pattern Recog. Lett.*, **15**, pp. 649–659, 1994

Hecker, Y. C. and Bolle, R. M., On Geometric Hashing and the Generalized Hough Transform, *IEEE Trans. SMC*, **24**, pp. 1328–1338, 1994

Hough, P. V. C., Method and Means for Recognizing Complex Patterns, *US Patent 3969654*, 1962

Illingworth, J. and Kittler, J., The Adaptive Hough Transform, *IEEE Trans. PAMI*, **9**(5), pp. 690–697, 1987

Illingworth, J. and Kittler, J., A Survey of the Hough Transform, *CVGIP*, **48**, pp. 87–116, 1988

Kälviäinen, H., Hirvonen, P., Xu, L. and Oja, E., Probabilistic and Non-Probabilistic Hough Transforms: Overview and Comparisons, *Image Vision Comput.*, **13**(4), May, 239–252, 1995

Kassim, A. A., Tan, T. and Tan K. H., A Comparative Study of Efficient Generalized Hough Transform Techniques, *Image Vision Comput.*, **17**(10), pp. 737–748, 1999

Kimme, C., Ballard, D. and Sklansky, J., Finding Circles by an Array of Accumulators, *Communs ACM*, **18**(2), pp. 120–1222, 1975

Kiryati, N. and Bruckstein, A. M., Antialiasing the Hough Transform, *CVGIP: Graphical Models Image Process.*, **53**, pp. 213–222, 1991

Lamdan, Y., Schawatz, J. and Wolfon, H., Object Recognition by Affine Invariant Matching, *Proc. IEEE Conf. Comput. Vision Pattern Recog.*, pp. 335–344, 1988

Leavers, V., *Shape Detection in Computer Vision using the Hough Transform*, Springer, London, 1992

Leavers, V., Which Hough Transform? *CVGIP: Image Understand.*, **58**, pp. 250–264, 1993

Li, H. and Lavin, M. A., Fast Hough Transform: A Hierarchical Approach, *CVGIP*, **36**, pp. 139–161, 1986

Merlin, P. M. and Farber, D. J., A Parallel Mechanism for Detecting Curves in Pictures, *IEEE Trans. Comput.*, **24**, pp. 96–98, 1975

Mowbray, S. D. and Nixon, M. S., Extraction and Recognition of Periodically Deforming Objects by Continuous, Spatio-temporal Shape Description, *Proc. CVPR 2004*, **2**, pp. 895–901, 2004

Nash, J. M., Carter, J. N. and Nixon, M. S., Dynamic Feature Extraction via the Velocity Hough Transform, *Pattern Recog. Lett.*, **18**(10), pp. 1035–1047, 1997

O'Gorman, F. and Clowes, M. B., Finding Picture Edges Through Collinearity of Feature Points, *IEEE Trans. Comput.*, **25**(4), pp. 449–456, 1976

Parker, J. R., *Practical Computer Vision Using C*, Wiley & Sons, New York, 1994

Parker, J. R., *Algorithms for Image Processing and Computer Vision*, Wiley & Sons, New York, 1996

Philip, K. P., *Automatic Detection of Myocardial Contours in Cine Computed Tomographic Images*, PhD Thesis, University of Iowa, 1991

Princen, J., Yuen, H. K., Illingworth, J. and Kittler, J., Properties of the Adaptive Hough Transform, *Proc. 6th Scandinavian Conf. Image Analysis*, Oulu, Finland, June 1992a

Princen, J., Illingworth, J. and Kittler, J., A Formal Definition of the Hough Transform: Properties and Relationships, *J. Math. Imaging Vision*, **1**, pp. 153–168, 1992b

Rosenfeld, A., *Picture Processing by Computer*, Academic Press, London, 1969

Sklansky, J., On the Hough Technique for Curve Detection, *IEEE Trans. Comput.*, **27**, pp. 923–926, 1978

Sonka, M., Hllavac, V. and Boyle, R, *Image Processing, Analysis and Computer Vision*, Chapman Hall, London, 1994

Stockman, G. C. and Agrawala, A. K., Equivalence of Hough Curve Detection to Template Matching, *Communs ACM*, **20**, pp. 820–822, 1977

Stockman, G., Object Recognition and Localization via Pose Clustering, *CVGIP*, **40**, pp. 361–387, 1987

Traver V. J. and Pla, F., The Log-Polar Image Representation in Pattern Recognition Tasks, *Lecture Notes Comput. Sci.*, **2652**, pp. 1032–1040, 2003

Trier, O. D., Jain, A. K. and Taxt, T., Feature Extraction Methods for Character Recognition – A Survey, *Pattern Recog.*, **29**(4), pp. 641–662, 1996

van Otterloo, P. J., *A Contour-Oriented Approach to Shape Analysis*, Prentice Hall International (UK), Hemel Hempstead, 1991

Yuen, H. K., Princen, J., Illingworth, J. and Kittler, J., Comparative Study of Hough Transform Methods for Circle Finding, *Image Vision Comput.*, **8**(1), pp 71–77, 1990

Xu, L., Oja, E. and Kultanen, P., A New Curve Detection Method: Randomized Hough Transform, *Pattern Recog. Lett.*, **11**, pp. 331–338, 1990

Zokai, S. and Wolberg, G., Image Registration using Log-Polar Mappings for Recovery of Large-Scale Similarity and Projective Transformations, *IEEE Trans. Image Process.*, **14**, pp. 1422–1434, 2005

6

Flexible shape extraction (snakes and other techniques)

6.1 Overview

The previous chapter covered finding shapes by matching. This implies knowledge of a model (mathematical or template) of the target shape (feature). The shape is *fixed* in that it is flexible only in terms of the parameters that define the shape, or the parameters that define a template's appearance. Sometimes, however, it is not possible to model a shape with sufficient accuracy, or to provide a template of the target as needed for the generalized Hough transform (GHT). It might be that the exact shape is unknown or that the perturbation of that shape is *impossible* to parameterize. In this case, we seek techniques that can *evolve* to the target solution, or adapt their result to the data. This implies the use of flexible shape formulations. This chapter presents four techniques that can be used to find flexible shapes in images. These are summarized in Table 6.1 and can be distinguished by the matching functional used to indicate the extent of

Table 6.1 Overview of Chapter 6

Main topic	Sub topics	Main points
Deformable templates	*Template matching* for *deformable* shapes. Defining a way to *analyse* the best match.	Energy maximization, computational considerations, optimization.
Active contours and snakes	Finding shapes by *evolving contours*. *Discrete* and *continuous* formulations. *Operational* considerations and new active contour approaches.	Energy minimization for curve evolution. Greedy algorithm. Kass snake. Parameterization; initialization and performance. Gradient vector field and level set approaches.
Shape skeletonization	Notions of *distance, skeletons* and *symmetry* and its measurement. Application of symmetry detection by *evidence gathering*. *Performance* factors.	Distance transform and shape skeleton. Discrete symmetry operator. Accumulating evidence of symmetrical point arrangements. Performance: speed and noise.
Active shape models	Expressing shape variation by *statistics*. Capturing shape *variation* within feature extraction.	Active shape model. Active appearance model. Principal components analysis.

match between image data and a shape. If the shape is flexible or *deformable*, so as to match the image data, we have a *deformable template*. This is where we shall start. Later, we shall move to techniques that are called *snakes*, because of their movement. We shall explain two different implementations of the snake model. The first one is based on discrete minimization and the second one on finite element analysis. We shall also look at determining a shape's skeleton, by *distance* analysis and by the *symmetry* of their appearance. This technique finds any symmetric shape by gathering evidence by considering features between pairs of points. Finally, we shall consider approaches that use of the *statistics* of a shape's possible appearance to control selection of the final shape, called *active shape models*.

6.2 Deformable templates

One of the earlier approaches to deformable template analysis (Yuille, 1991) aimed to find facial features for purposes of recognition. The approach considered an eye to be comprised of an iris that sits within the sclera and which can be modelled as a combination of a circle that lies within a parabola. Clearly, the circle and a version of the parabola can be extracted by using Hough transform techniques, but this cannot be achieved in combination. When we combine the two shapes and allow them to change in size and orientation, while retaining their spatial relationship (that the iris or circle should reside within the sclera or parabola), then we have a deformable template.

The parabola is a shape described by a set of points (x, y) related by

$$y = a - \frac{a}{b^2}x^2 \tag{6.1}$$

where, as illustrated in Figure 6.1(a), a is the height of the parabola and b is its radius. As such, the maximum height is a and the minimum height is zero. A similar equation describes the lower parabola, it terms of b and c. The 'centre' of both parabolas is $\mathbf{c}_p$. The circle is as defined earlier, with centre coordinates $\mathbf{c}_c$ and radius r. We then seek values of the parameters that give a best match of this template to the image data. Clearly, one match we would like to make concerns matching the edge data to that of the template, as in the Hough transform. The set of values for the parameters that give a template which matches the most edge points

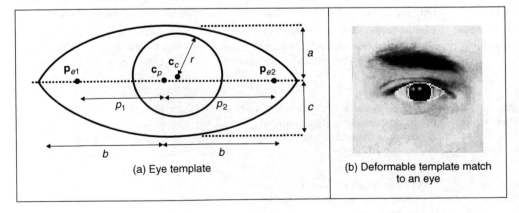

(a) Eye template (b) Deformable template match to an eye

Figure 6.1 Finding an eye with a deformable template

(since edge points are found at the boundaries of features) could then be deemed to be the best set of parameters describing the eye in an image. We then seek values of parameters that maximize

$$\{\mathbf{c}_p, a, b, c, \mathbf{c}_c, r\} = \max \left(\sum_{x,y \in \text{circle.perimeter, parabolas.perimeter}} \mathbf{E}_{x,y} \right) \tag{6.2}$$

This would prefer the larger shape to the smaller ones, so we could divide the contribution of the circle and the parabolas by their perimeter to give an edge energy contribution E_e

$$E_e = \frac{\displaystyle\sum_{x,y \in \text{circle.perimeter}} \mathbf{E}_{x,y}}{\text{circle.perimeter}} + \frac{\displaystyle\sum_{x,y \in \text{parabolas.perimeter}} \mathbf{E}_{x,y}}{\text{parabolas.perimeter}} \tag{6.3}$$

and we seek a combination of values for the parameters $\{\mathbf{c}_p, a, b, c, \mathbf{c}_c, r\}$ which maximize this energy. This, however, implies little knowledge of the *structure* of the eye. Since we know that the sclera is white (usually) and the iris is darker than it, we could build this information into the process. We can form an energy E_v functional for the circular region which averages the brightness over the circle area as

$$E_v = -\frac{\displaystyle\sum_{x,y \in \text{circle}} \mathbf{P}_{x,y}}{\text{circle.area}} \tag{6.4}$$

This is formed in the negative, since maximizing its value gives the best set of parameters. Similarly, we can form an energy functional for the light regions where the eye is white as E_p

$$E_p = \frac{\displaystyle\sum_{x,y \in \text{parabolae}-\text{circle}} \mathbf{P}_{x,y}}{\text{parabolas} - \text{circle.area}} \tag{6.5}$$

where parabolas–circle implies points within the parabolas, but not within the circle. We can then choose a set of parameters which maximize the combined energy functional formed by adding each energy when weighted by some chosen factors as

$$E = c_e \cdot E_e + c_v \cdot E_v + c_p \cdot E_p \tag{6.6}$$

where c_e, c_v and c_p are the weighting factors. In this way, we are choosing values for the parameters which simultaneously maximize the chance that the edges of the circle and the perimeter coincide with the image edges, that the inside of the circle is dark and that the insides of the parabolas are light. The value chosen for each of the weighting factors controls the influence of that factor on the eventual result.

The energy fields are shown in Figure 6.2 when computed over the entire image. The valley image shows up regions with low image intensity and the peak image shows regions of high image intensity, like the whites of the eyes. In its original formulation, this approach had five energy terms and the extra two are associated with the points $\mathbf{p}_{e1}$ and $\mathbf{p}_{e2}$, either side of the iris in Figure 6.1(a).

This is where the problem starts, as we now have 11 parameters (eight for the shapes and three for the weighting coefficients). We could simply cycle through every possible value. Given, say, 100 possible values for each parameter, we then have to search 10^{22} combinations of parameters, which would be no problem given multithread computers with terrahertz

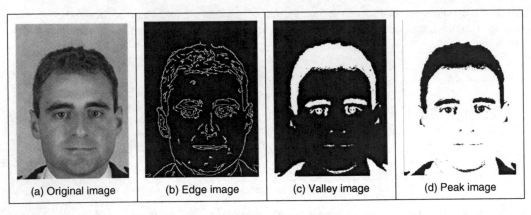

| (a) Original image | (b) Edge image | (c) Valley image | (d) Peak image |

Figure 6.2 Energy fields over whole face image (Benn, 1999)

processing speed achieved via optical interconnect, but computers like that are not ready yet (on our budgets at least). We can reduce the number of combinations by introducing constraints on the relative size and position of the shapes, e.g. the circle should lie wholly within the parabolas, but this will not reduce the number of combinations much. We can seek two alternatives: one is to use *optimization* techniques. The original approach (Yuille, 1991) favoured the use of gradient descent techniques; currently, the *genetic algorithm* approach (Goldberg, 1988) seems to be most favoured and this has been shown to good effect for deformable template eye extraction on a database of 1000 faces (Benn, 1999) (this is the source of the images shown here). The alternative is to seek a different technique that uses fewer parameters. This is where we move to *snakes*, which are a much more popular approach. These snakes evolve a set of *points* (a contour) to match the image data, rather than evolving a shape.

6.3 Active contours (snakes)

6.3.1 Basics

Active contours or *snakes* (Kass et al., 1988) are a completely different approach to feature extraction. An active contour is a set of points that aims to enclose a target feature, the feature to be extracted. It is a bit like using a balloon to 'find' a shape: the balloon is placed outside the shape, enclosing it. Then by taking air out of the balloon, making it smaller, the shape is found when the balloon stops shrinking, when it fits the target shape. By this manner, active contours arrange a set of points so as to describe a target feature, by enclosing it. Snakes are quite recent compared with many computer vision techniques and their original formulation was as an interactive extraction process, although they are now usually deployed for automatic feature extraction.

An initial contour is placed outside the target feature, and is then evolved so as to enclose it. The process is illustrated in Figure 6.3, where the target feature is the perimeter of the iris. First, an initial contour is placed outside the iris (Figure 6.3a). The contour is then minimized to find a new contour which shrinks so as to be closer to the iris (Figure 6.3b). After seven iterations, the contour points can be seen to match the iris perimeter well (Figure 6.3d).

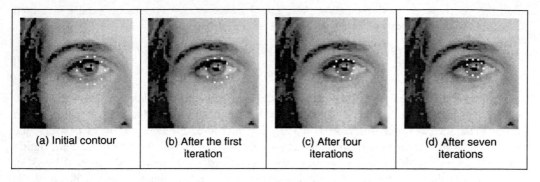

| (a) Initial contour | (b) After the first iteration | (c) After four iterations | (d) After seven iterations |

Figure 6.3 Using a snake to find an eye's iris

Active contours are expressed as an *energy minimization* process. The target feature is a minimum of a suitably formulated energy functional. This energy functional includes more than just edge information: it includes properties that control the way in which the contour can stretch and curve. In this way, a snake represents a *compromise* between its *own* properties (such as its ability to bend and stretch) and *image* properties (such as the edge magnitude). Accordingly, the energy functional is the addition of a function of the contour's internal energy, its constraint energy and the image energy; these are denoted E_{int}, E_{con} and E_{image}, respectively. These are functions of the set of points that make up a snake, $\mathbf{v}(s)$, which is the set of x and y coordinates of the points in the snake. The energy functional is the integral of these functions of the snake, given $s \in [0, 1)$ is the normalized length around the snake. The energy functional E_{snake} is then:

$$E_{snake} = \int_{s=0}^{1} E_{int}(\mathbf{v}(s)) + E_{image}(\mathbf{v}(s)) + E_{con}(\mathbf{v}(s)) ds \qquad (6.7)$$

In this equation, the internal energy, E_{int}, controls the natural behaviour of the snake and hence the arrangement of the snake points; the image energy, E_{image}, attracts the snake to chosen low-level features (such as edge points); and the constraint energy, E_{con}, allows higher level information to control the snake's evolution. The aim of the snake is to evolve by *minimizing* Equation 6.7. New snake contours are those with lower energy and are a better match to the target feature (according to the values of E_{int}, E_{image} and E_{con}) than the original set of points from which the active contour has evolved. In this manner, we seek to choose a set of points $\mathbf{v}(s)$ such that

$$\frac{dE_{snake}}{d\mathbf{v}(s)} = 0 \qquad (6.8)$$

This can select a maximum rather than a minimum, and a second order derivative can be used to discriminate between a maximum and a minimum. However, this is not usually necessary as a minimum is usually the only stable solution (on reaching a maximum, it would then be likely to pass over the top to then minimize the energy). Before investigating how we can minimize Equation 6.7, let us first consider the parameters that can control a snake's behaviour.

The energy functionals are expressed in terms of functions of the snake, and of the image. These functions contribute to the snake energy according to values chosen for respective weighting coefficients. In this manner, the internal image energy is defined to be a weighted summation of first and second order derivatives around the contour.

$$E_{int} = \alpha(s) \left| \frac{d\mathbf{v}(s)}{ds} \right|^2 + \beta(s) \left| \frac{d^2\mathbf{v}(s)}{ds^2} \right|^2 \tag{6.9}$$

The first order differential, $d\mathbf{v}(s)/ds$, measures the energy due to *stretching*, which is the *elastic* energy since high values of this differential imply a high rate of change in that region of the contour. The second order differential, $d^2\mathbf{v}(s)/ds^2$, measures the energy due to *bending*, the *curvature* energy. The first order differential is weighted by $\alpha(s)$, which controls the contribution of the elastic energy due to point spacing; the second order differential is weighted by $\beta(s)$, which controls the contribution of the curvature energy due to point variation. Choice of the values of α and β controls the shape the snake aims to attain. Low values for α imply that the points can change in spacing greatly, whereas higher values imply that the snake aims to attain evenly spaced contour points. Low values for β imply that curvature is not minimized and the contour can form corners in its perimeter, whereas high values predispose the snake to smooth contours. These are the properties of the contour itself, which is just part of a snake's compromise between its own properties and measured features in an image.

The image energy attracts the snake to low-level features, such as brightness or edge data, aiming to select those with least contribution. The original formulation suggested that lines, edges and terminations could contribute to the energy function. Their energies are denoted E_{line}, E_{edge} and E_{term}, respectively, and are controlled by weighting coefficients w_{line}, w_{edge} and w_{term}, respectively. The image energy is then:

$$E_{image} = w_{line}E_{line} + w_{edge}E_{edge} + w_{term}E_{term} \tag{6.10}$$

The *line* energy can be set to the image intensity at a particular point. If black has a lower value than white, then the snake will be extracted to dark features. Altering the sign of w_{line} will attract the snake to brighter features. The *edge* energy can be that computed by application of an edge detection operator, the magnitude, say, of the output of the Sobel edge detection operator. The *termination* energy, E_{term} as measured by Equation 4.57, can include the curvature of level image contours [as opposed to the curvature of the snake, controlled by $\beta(s)$], but this is rarely used. It is most common to use the edge energy, although the line energy can find application.

6.3.2 The greedy algorithm for snakes

The implementation of a snake, to evolve a set of points to minimize Equation 6.7, can use finite elements, or finite differences, which is complicated and follows later. It is easier to start with the *greedy algorithm* (Williams and Shah, 1992) which implements the energy minimization process as a purely discrete algorithm, illustrated in Figure 6.4. The process starts by specifying an initial contour. Earlier, Figure 6.3(a) used a circle of 16 points along the perimeter of a circle. Alternatively, these can be specified manually. The greedy algorithm then evolves the snake in an iterative manner by local neighbourhood search around contour points to select new ones which have lower snake energy. The process is called greedy by virtue of the way the search propagates around the contour. At each iteration, all contour points are evolved and the process is repeated for the first contour point. The index to snake points is computed modulo S (the number of snake points).

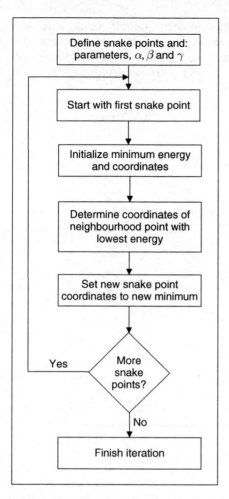

Figure 6.4 Operation of the greedy algorithm

For a set of snake points $\mathbf{v}_s$, $\forall s \in 0, S-1$, the energy functional minimized for each snake point is:

$$E_{\text{snake}}(s) = E_{\text{int}}(\mathbf{v}_s) + E_{\text{image}}(\mathbf{v}_s) \tag{6.11}$$

This is expressed as

$$E_{\text{snake}}(s) = \alpha(s) \left| \frac{d\mathbf{v}_s}{ds} \right|^2 + \beta(s) \left| \frac{d^2\mathbf{v}_s}{ds^2} \right|^2 + \gamma(s) E_{\text{edge}} \tag{6.12}$$

where the first order and second order differentials are approximated for each point searched in the local neighbourhood of the currently selected contour point. The weighting parameters, α, β and γ, are all functions of the contour. Accordingly, each contour point has associated values for α, β and γ. An implementation of the specification of an initial contour by a function point is given in Code 6.1. In this implementation, the contour is stored as a matrix of vectors. Each vector has five elements: two are the x and y coordinates of the contour point, the remaining three parameters are the values of α, β and γ for that contour point, set here to be 0.5, 0.5

and 1.0, respectively. The `no` contour points are arranged to be in a circle, radius `rad` and centre `(xc, yc)`. As such, a vector is returned for each snake point, point$_s$, where (point$_s$)$_0$, (point$_s$)$_1$, (point$_s$)$_2$, (point$_s$)$_3$, (point$_s$)$_4$ are the x coordinate, the y coordinate and α, β and γ for the particular snake point s: x_s, y_s, α_s, β_s and γ_s, respectively.

```
points(rad,no,xc,yc):=  for  s∈0..no-1

                            xₛ←xc+floor( rad·cos ( s·2·π / no )+0.5 )

                            yₛ←yc+floor( rad·sin ( s·2·π / no )+0.5 )

                            αₛ←0.5
                            βₛ←0.5
                            γₛ←1

                                       ⎡ xₛ ⎤
                                       ⎢ yₛ ⎥
                            pointₛ← ⎢ αₛ ⎥
                                       ⎢ βₛ ⎥
                                       ⎣ γₛ ⎦

                        point
```

Code 6.1   Specifying an initial contour

The first order differential is approximated as the modulus of the difference between the average spacing of contour points (evaluated as the Euclidean distance between them), and the Euclidean distance between the currently selected image point $\mathbf{v}_s$ and the next contour point. By selection of an appropriate value of $\alpha(s)$ for each contour point $\mathbf{v}_s$, this can control the spacing between the contour points:

$$\left| \frac{d\mathbf{v}_s}{ds} \right|^2 = \left| \sum_{i=0}^{S-1} \|\mathbf{v}_i - \mathbf{v}_{i+1}\| \Big/ S - \|\mathbf{v}_s - \mathbf{v}_{s+1}\| \right|$$

$$= \left| \sum_{i=0}^{S-1} \sqrt{(\mathbf{x}_i - \mathbf{x}_{i+1})^2 + (\mathbf{y}_i - \mathbf{y}_{i+1})^2} \Big/ S - \sqrt{(\mathbf{x}_s - \mathbf{x}_{s+1})^2 + (\mathbf{y}_s - \mathbf{y}_{s+1})^2} \right| \qquad (6.13)$$

as evaluated from the x and the y coordinates of the adjacent snake point $(\mathbf{x}_{s+1}, \mathbf{y}_{s+1})$ and the coordinates of the point currently inspected $(\mathbf{x}_s, \mathbf{y}_s)$. Clearly, the first order differential, as evaluated from Equation 6.13, drops to zero when the contour is evenly spaced, as required. This is implemented by the function `Econt` in Code 6.2, which uses a function `dist` to evaluate the average spacing and a function `dist2` to evaluate the Euclidean distance between the currently searched point $(\mathbf{v}_s)$ and the next contour point $(\mathbf{v}_{s+1})$. The arguments to E_{cont} are the x and

```
dist(s,contour):=    | s1←mod(s,rows(contour))
                      | s2←mod(s+1,rows(contour))
                      |  _____
                      | √[(contour_{s1})_0 - (contour_{s2})_0]^2 + [(contour_{s1})_1 - (contour_{s2})_1]^2

dist2(x,y,s,contour):= | s2←mod(s+1,rows(contour))
                       |  _____
                       | √[(contour_{s2})_0 - x]^2 + [(contour_{s2})_1 - y]^2

                       |            1          rows(cont)-1
Econt(x,y,s,cont):=    | D← ─────────────  ·      Σ       dist(s1, cont)
                       |      rows(cont)         s1=0
                       |
                       | |D - dist2(x,y,s,cont)|
```

Code 6.2 Evaluating the contour energy

y coordinates of the point currently being inspected, x and y, the index of the contour point currently under consideration, s, and the contour itself, con.

The second order differential can be implemented as an estimate of the curvature between the next and previous contour points, v_{s+1} and v_{s-1}, respectively, and the point in the local neighbourhood of the currently inspected snake point v_s:

$$\left|\frac{d^2 v_s}{ds^2}\right|^2 = |(v_{s+1} - 2v_s + v_{s-1})|^2$$

$$= (x_{s+1} - 2x_s + x_{s-1})^2 + (y_{s+1} - 2y_s + y_{s-1})^2 \tag{6.14}$$

This is implemented by a function Ecur in Code 6.3, whose arguments again are the x and y coordinates of the point currently being inspected, x and y, the index of the contour point currently under consideration, s, and the contour itself, cont.

```
Ecur(x,y,s,con) := | s1←mod(s-1+rows(con),rows(con))
                   | s3←mod(s+1,rows(con))
                   | [(con_{s1})_0 - 2·x+(con_{s3})_0]^2 +[(con_{s1})_1 - 2·y+(con_{s3})_1]^2
```

Code 6.3 Evaluating the contour curvature

E_{edge} can be implemented as the magnitude of the Sobel edge operator at point x, y. This is normalized to ensure that its value lies between zero and unity. This is also performed for the elastic and curvature energies in the current region of interest. This is achieved by normalization using Equation 3.2 arranged to provide an output ranging between 0 and 1. The edge image could also be normalized within the current window of interest, but this makes it more possible that the result is influenced by noise. Since the snake is arranged to be a minimization process, the edge image is inverted so that the points with highest edge strength are given the lowest edge value (0), whereas the areas where the image is constant are given a high value (1). Accordingly, the snake will be attracted to the edge points with greatest magnitude. The normalization process ensures that the contour energy and curvature and the edge strength are balanced forces and eases appropriate selection of values for α, β and γ. This is achieved by a balancing function (balance) that normalizes the contour and curvature energy within the window of interest.

The greedy algorithm then uses these energy functionals to minimize the composite energy functional (Equation 6.12), given in the function grdy in Code 6.4. This gives a single iteration in the evolution of a contour wherein all snake points are searched. The energy for each snake point is first determined and is stored as the point with minimum energy. This ensures that if any other point is found to have equally small energy, then the contour point will remain in the same position. Then, the local 3×3 neighbourhood is searched to determine whether any other point has a lower energy than the current contour point. If it does, that point is returned as the new contour point.

```
grdy(edg,con) :=   for  s1∈0..rows(con)
                      s←mod(s1,rows(con))
                      xmin←(con_s)_0
                      ymin←(con_s)_1
                      forces←balance[(con_s)_0,(con_s)_1,edg,s,con]
                      Emin←(con_s)_2.Econt(xmin,ymin,s,con)
                      Emin←Emin+(con_s)_3·Ecur(xmin,ymin,s,con)
                      Emin←Emin+(con_s)_4·(edg_0)_{(con_s)_1,(con_s)_0}

                      for  x∈(con_s)_0-1..(con_s)_0+1
                        for  y∈(con_s)_1-1..(con_s)_1+1
                          if  check(x,y,edg_0)
                                  xx←x-(con_s)_0+1
                                  yy←y-(con_s)_1+1
                                  Ej←(con_s)_2·(forces_{0,0})_{yy,xx}
                                  Ej←Ej+(con_s)_3·(forces_{0,1})_{yy,xx}
                                  Ej←Ej+(con_s)_4·(edg_0)_{y,x}
                                  if  Ej<Emin
                                          Emin←Ej
                                          xmin←x
                                          ymin←y

                                    ⎡  xmin     ⎤
                                    ⎢  ymin     ⎥
                           con_s ←  ⎢  (con_s)_2 ⎥
                                    ⎢  (con_s)_3 ⎥
                                    ⎣  (con_s)_4 ⎦

                   con
```

Code 6.4   The greedy algorithm

A verbatim implementation of the greedy algorithm would include three thresholds. One is a threshold on tangential direction and another on edge magnitude. If an edge point were adjudged to be of direction above the chosen threshold, and with magnitude above its corresponding threshold, then β can be set to zero for that point to allow corners to form. This has not been included in Code 6.4, in part because there is mutual dependence between α and β. Also, the original presentation of the greedy algorithm proposed to continue evolving the snake until it becomes static, when the number of contour points moved in a single iteration is below the third threshold value. This can lead to instability since it can lead to a situation where contour points merely oscillate between two solutions and the process would appear not to converge. Again, this has not been implemented here.

The effect of varying α and β is shown in Figures 6.5 and 6.6. Setting α to zero removes influence of *spacing* on the contour points' arrangement. In this manner, the points will become to be unevenly spaced (Figure 6.5b), and eventually can be placed on top of each other. Reducing the control by spacing can be desirable for features that have high localized curvature. Low values of α can allow for bunching of points in such regions, giving a better feature description.

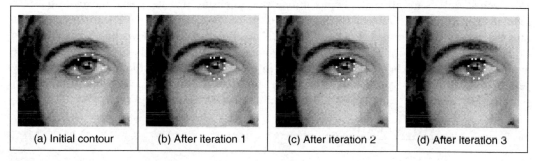

| (a) Initial contour | (b) After iteration 1 | (c) After iteration 2 | (d) After iteration 3 |

Figure 6.5   Effect of removing control by spacing

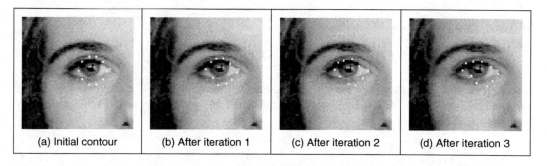

| (a) Initial contour | (b) After iteration 1 | (c) After iteration 2 | (d) After iteration 3 |

Figure 6.6   Effect of removing low curvature control

Setting β to zero removes influence of *curvature* on the contour points' arrangement, allowing corners to form in the contour, as illustrated in Figure 6.6. This is manifest in the first iteration (Figure 6.6b), and since with β set to zero for the whole contour, each contour point can become a corner with high curvature (Figure 6.6c), leading to the rather ridiculous result in Figure 6.6(d).

Reducing the control by curvature can clearly be desirable for features that have high localized curvature. This illustrates the mutual dependence between α and β, since low values of α can accompany low values of β in regions of high localized curvature. Setting γ to zero would force the snake to ignore image data and evolve under its own forces. This would be rather farcical. The influence of γ is reduced in applications where the image data used is known to be noisy. Note that one fundamental problem with a discrete version is that the final solution can oscillate when it swaps between two sets of points which both have equally low energy. This can be prevented by detecting the occurrence of oscillation. A further difficulty is that as the contour becomes smaller, the number of contour points constrains the result as they cannot be compressed into too small a space. The only solution to this is to resample the contour.

6.3.3 Complete (Kass) snake implementation

The greedy method iterates around the snake to find local minimum energy at snake points. This is an *approximation*, since it does not necessarily determine the 'best' local minimum in the region of the snake points, by virtue of iteration. A *complete snake implementation*, or *Kass snake*, solves for all snake points in one step to ensure that the snake moves to the best local energy minimum. We seek to choose snake points $(\mathbf{v}(s) = (\mathbf{x}(s), \mathbf{y}(s)))$ in such a manner that the energy is minimized (Equation 6.8). Calculus of variations shows how the solution to Equation 6.7 reduces to a pair of differential equations that can be solved by finite difference analysis (Waite and Welsh, 1990). This results in a set of equations that iteratively provide new sets of contour points. By calculus of variation, we shall consider an admissible solution $\hat{\mathbf{v}}(s)$ perturbed by a small amount, $\varepsilon\delta\mathbf{v}(s)$, which achieves minimum energy, as:

$$\frac{\mathrm{d}E_{\mathrm{snake}}\left(\hat{\mathbf{v}}(s) + \varepsilon\delta\mathbf{v}(s)\right)}{\mathrm{d}\varepsilon} = 0 \tag{6.15}$$

where the perturbation is spatial, affecting the x and y coordinates of a snake point:

$$\delta\mathbf{v}(s) = (\delta_x(s), \delta_y(s)) \tag{6.16}$$

This gives the perturbed snake solution as

$$\hat{\mathbf{v}}(s) + \varepsilon\delta\mathbf{v}(s) = \left(\hat{x}(s) + \varepsilon\delta_x(s), \hat{y}(s) + \varepsilon\delta_y(s)\right) \tag{6.17}$$

where $\hat{x}(s)$ and $\hat{y}(s)$ are the x and y coordinates, respectively, of the snake points at the solution $(\hat{\mathbf{v}}(s) = (\hat{x}(s), \hat{y}(s)))$. By setting the constraint energy E_{con} to zero, the snake energy (Equation 6.7) becomes:

$$E_{\mathrm{snake}}(\mathbf{v}(s)) = \int_{s=0}^{1} \left\{E_{\mathrm{int}}(\mathbf{v}(s)) + E_{\mathrm{image}}(\mathbf{v}(s))\right\} \mathrm{d}s \tag{6.18}$$

Edge magnitude information is often used (so that snakes are attracted to edges found by an edge detection operator), so we shall replace E_{image} by E_{edge}. By substitution for the perturbed snake points, we obtain

$$E_{\mathrm{snake}}(\hat{\mathbf{v}}(s) + \varepsilon\delta\mathbf{v}(s)) = \int_{s=0}^{1} \left\{E_{\mathrm{int}}\left(\hat{\mathbf{v}}(s) + \varepsilon\delta\mathbf{v}(s)\right) + E_{\mathrm{edge}}\left(\hat{\mathbf{v}}(s) + \varepsilon\delta\mathbf{v}(s)\right)\right\} \mathrm{d}s \tag{6.19}$$

By substitution from Equation 6.9, we obtain

$$E_{\text{snake}}(\hat{\mathbf{v}}(s) + \varepsilon \delta \mathbf{v}(s)) = \int_{s=0}^{s=1} \left\{ \alpha(s) \left| \frac{\mathrm{d}\,(\hat{\mathbf{v}}(s) + \varepsilon \delta \mathbf{v}(s))}{\mathrm{d}s} \right|^2 \right.$$

$$\left. + \beta(s) \left| \frac{\mathrm{d}^2\,(\hat{\mathbf{v}}(s) + \varepsilon \delta \mathbf{v}(s))}{\mathrm{d}s^2} \right|^2 + E_{\text{edge}}\,(\hat{\mathbf{v}}(s) + \varepsilon \delta \mathbf{v}(s)) \right\} \mathrm{d}s \tag{6.20}$$

By substitution from Equation 6.17,

$$E_{\text{snake}}\,(\hat{\mathbf{v}}\,(s) + \varepsilon \delta \mathbf{v}\,(s))$$

$$= \int_{s=0}^{s=1} \left[\begin{array}{l} \alpha(s) \left\{ \begin{array}{l} \left(\dfrac{\mathrm{d}\hat{x}(s)}{\mathrm{d}s} \right)^2 + 2\varepsilon \dfrac{\mathrm{d}\hat{x}(s)}{\mathrm{d}s} \dfrac{\mathrm{d}\delta_x(s)}{\mathrm{d}s} + \left(\varepsilon \dfrac{\mathrm{d}\delta_x(s)}{\mathrm{d}s} \right)^2 \\[3mm] + \left(\dfrac{\mathrm{d}\hat{y}(s)}{\mathrm{d}s} \right)^2 + 2\varepsilon \dfrac{\mathrm{d}\hat{y}(s)}{\mathrm{d}s} \dfrac{\mathrm{d}\delta_y(s)}{\mathrm{d}s} + \left(\varepsilon \dfrac{\mathrm{d}\delta_y(s)}{\mathrm{d}s} \right)^2 \end{array} \right\} \\[10mm] + \beta(s) \left\{ \begin{array}{l} \left(\dfrac{\mathrm{d}^2\hat{x}(s)}{\mathrm{d}s^2} \right)^2 + 2\varepsilon \dfrac{\mathrm{d}^2\hat{x}(s)}{\mathrm{d}s^2} \dfrac{\mathrm{d}^2\delta_x(s)}{\mathrm{d}s^2} + \left(\varepsilon \dfrac{\mathrm{d}^2\delta_x(s)}{\mathrm{d}s^2} \right)^2 \\[3mm] + \left(\dfrac{\mathrm{d}^2\hat{y}(s)}{\mathrm{d}s^2} \right)^2 + 2\varepsilon \dfrac{\mathrm{d}^2\hat{y}(s)}{\mathrm{d}s^2} \dfrac{\mathrm{d}^2\delta_y(s)}{\mathrm{d}s^2} + \left(\varepsilon \dfrac{\mathrm{d}^2\delta_y(s)}{\mathrm{d}s^2} \right)^2 \end{array} \right\} \\[8mm] \qquad\qquad + E_{\text{edge}}\,(\hat{\mathbf{v}}\,(s) + \varepsilon \delta \mathbf{v}\,(s)) \end{array} \right] \mathrm{d}s \tag{6.21}$$

By expanding E_{edge} at the perturbed solution by Taylor series, we obtain

$$E_{\text{edge}}\,(\hat{\mathbf{v}}(s) + \varepsilon \delta \mathbf{v}(s)) = E_{\text{edge}}\,(\hat{x}(s) + \varepsilon \delta_x(s),\ \hat{y}(s) + \varepsilon \delta_y(s))$$

$$= E_{\text{edge}}\,(\hat{x}(s),\ \hat{y}(s)) + \varepsilon \delta_x(s) \left. \frac{\partial E_{\text{edge}}}{\partial x} \right|_{\hat{x},\hat{y}} + \varepsilon \delta_y(s) \left. \frac{\partial E_{\text{edge}}}{\partial y} \right|_{\hat{x},\hat{y}}$$

$$+ O(\varepsilon^2) \tag{6.22}$$

This implies that the image information must be twice differentiable, which holds for edge information, but not for some other forms of image energy. Ignoring higher order terms in ϵ (since ϵ is small), by reformulation Equation 6.21 becomes

$$E_{\text{snake}}\,(\hat{\mathbf{v}}(s) + \varepsilon \delta \mathbf{v}(s)) = E_{\text{snake}}\,(\hat{\mathbf{v}}(s))$$

$$+ 2\varepsilon \int_{s=0}^{s=1} \alpha(s) \frac{\mathrm{d}\hat{x}(s)}{\mathrm{d}s} \frac{\mathrm{d}\delta_x(s)}{\mathrm{d}s} + \beta(s) \frac{\mathrm{d}^2\hat{x}(s)}{\mathrm{d}s^2} \frac{\mathrm{d}^2\delta_x(s)}{\mathrm{d}s^2} + \frac{\delta_x(s)}{2} \left. \frac{\partial E_{\text{edge}}}{\partial x} \right|_{\hat{x},\hat{y}} \mathrm{d}s$$

$$+ 2\varepsilon \int_{s=0}^{s=1} \alpha(s) \frac{\mathrm{d}\hat{y}(s)}{\mathrm{d}s} \frac{\mathrm{d}\delta_y(s)}{\mathrm{d}s} + \beta(s) \frac{\mathrm{d}^2\hat{y}(s)}{\mathrm{d}s^2} \frac{\mathrm{d}^2\delta_y(s)}{\mathrm{d}s^2} + \frac{\delta_y(s)}{2} \left. \frac{\partial E_{\text{edge}}}{\partial y} \right|_{\hat{x},\hat{y}} \mathrm{d}s \tag{6.23}$$

Since the perturbed solution is at a minimum, the integration terms in Equation 6.23 must be identically zero:

$$\int_{s=0}^{s=1} \alpha(s)\frac{\mathrm{d}\hat{x}(s)}{\mathrm{d}s}\frac{\mathrm{d}\delta_x(s)}{\mathrm{d}s} + \beta(s)\frac{\mathrm{d}^2\hat{x}(s)}{\mathrm{d}s^2}\frac{\mathrm{d}^2\delta_x(s)}{\mathrm{d}s^2} + \frac{\delta_x(s)}{2}\frac{\partial E_{edge}}{\partial x}\bigg|_{\hat{x},\hat{y}} \mathrm{d}s = 0 \tag{6.24}$$

$$\int_{s=0}^{s=1} \alpha(s)\frac{\mathrm{d}\hat{y}(s)}{\mathrm{d}s}\frac{\mathrm{d}\delta_y(s)}{\mathrm{d}s} + \beta(s)\frac{\mathrm{d}^2\hat{y}(s)}{\mathrm{d}s^2}\frac{\mathrm{d}^2\delta_y(s)}{\mathrm{d}s^2} + \frac{\delta_y(s)}{2}\frac{\partial E_{edge}}{\partial y}\bigg|_{\hat{x},\hat{y}} \mathrm{d}s = 0 \tag{6.25}$$

By integration we obtain

$$\left[\alpha(s)\frac{\mathrm{d}\hat{x}(s)}{\mathrm{d}s}\delta_x(s)\right]_{s=0}^{1} - \int_{s=0}^{s=1}\frac{\mathrm{d}}{\mathrm{d}s}\left\{\alpha(s)\frac{\mathrm{d}\hat{x}(s)}{\mathrm{d}s}\right\}\delta_x(s)\mathrm{d}s$$

$$\left[\beta(s)\frac{\mathrm{d}^2\hat{x}(s)}{\mathrm{d}s^2}\frac{\mathrm{d}\delta_x(s)}{\mathrm{d}s}\right]_{s=0}^{1} - \left[\frac{\mathrm{d}}{\mathrm{d}s}\left\{\beta(s)\frac{\mathrm{d}^2\hat{x}(s)}{\mathrm{d}s^2}\right\}\delta_x(s)\right]_{s=0}^{1}$$

$$+ \int_{s=0}^{s=1}\frac{\mathrm{d}^2}{\mathrm{d}s^2}\left\{\beta(s)\frac{\mathrm{d}^2\hat{x}(s)}{\mathrm{d}s^2}\right\}\delta_x(s)\mathrm{d}s + \frac{1}{2}\int_{s=0}^{1}\frac{\partial E_{edge}}{\partial x}\bigg|_{\hat{x},\hat{y}}\delta_x(s)\mathrm{d}s = 0 \tag{6.26}$$

As the first, third and fourth terms are zero (since for a closed contour, $\delta_x(1) - \delta_x(0) = 0$ and $\delta_y(1) - \delta_y(0) = 0$), this reduces to

$$\int_{s=0}^{s=1}\left\{-\frac{\mathrm{d}}{\mathrm{d}s}\left\{\alpha(s)\frac{\mathrm{d}\hat{x}(s)}{\mathrm{d}s}\right\} + \frac{\mathrm{d}^2}{\mathrm{d}s^2}\left\{\beta(s)\frac{\mathrm{d}^2\hat{x}(s)}{\mathrm{d}s^2}\right\} + \frac{1}{2}\frac{\partial E_{edge}}{\partial x}\bigg|_{\hat{x},\hat{y}}\right\}\delta_x(s)\mathrm{d}s = 0 \tag{6.27}$$

Since this equation holds for all $\delta_x(s)$,

$$-\frac{\mathrm{d}}{\mathrm{d}s}\left\{\alpha(s)\frac{\mathrm{d}\hat{x}(s)}{\mathrm{d}s}\right\} + \frac{\mathrm{d}^2}{\mathrm{d}s^2}\left\{\beta(s)\frac{\mathrm{d}^2\hat{x}(s)}{\mathrm{d}s^2}\right\} + \frac{1}{2}\frac{\partial E_{edge}}{\partial x}\bigg|_{\hat{x},\hat{y}} = 0 \tag{6.28}$$

By a similar development of Equation 6.25 we obtain

$$-\frac{\mathrm{d}}{\mathrm{d}s}\left\{\alpha(s)\frac{\mathrm{d}\hat{y}(s)}{\mathrm{d}s}\right\} + \frac{\mathrm{d}^2}{\mathrm{d}s^2}\left\{\beta(s)\frac{\mathrm{d}^2\hat{y}(s)}{\mathrm{d}s^2}\right\} + \frac{1}{2}\frac{\partial E_{edge}}{\partial y}\bigg|_{\hat{x},\hat{y}} = 0 \tag{6.29}$$

This has reformulated the original energy minimization framework (Equation 6.7) into a pair of differential equations. To implement a complete snake, we seek the solution to Equations 6.28 and 6.29. By the method of finite differences, we substitute for $\mathrm{d}x(s)/\mathrm{d}s \cong \mathbf{x}_{s+1} - \mathbf{x}_s$, the first order difference, and the second order difference is $\mathrm{d}^2x(s)/\mathrm{d}s^2 \cong \mathbf{x}_{s+1} - 2\mathbf{x}_s + \mathbf{x}_{s-1}$ (as in Equation 6.12), which by substitution into Equation 6.28, for a contour discretized into S points equally spaced by an arc length h, (remembering that the indices $s \in [1, S)$ to snake points are computed modulo S) gives

$$
- \frac{1}{h} \left\{ \alpha_{s+1} \frac{(\mathbf{x}_{s+1} - \mathbf{x}_s)}{h} - \alpha_s \frac{(\mathbf{x}_s - \mathbf{x}_{s-1})}{h} \right\}
$$

$$
+ \frac{1}{h^2} \left\{ \beta_{s+1} \frac{(\mathbf{x}_{s+2} - 2\mathbf{x}_{s+1} + \mathbf{x}_s)}{h^2} - 2\beta_s \frac{(\mathbf{x}_{s+1} - 2\mathbf{x}_s + \mathbf{x}_{s-1})}{h^2} + \beta_{s-1} \frac{(\mathbf{x}_s - 2\mathbf{x}_{s-1} + \mathbf{x}_{s-2})}{h^2} \right\}
$$

$$
+ \frac{1}{2} \left. \frac{\partial E_{edge}}{\partial x} \right|_{x_s, y_s} = 0 \tag{6.30}
$$

By collecting the coefficients of different points, Equation 6.30 can be expressed as

$$
f_s = a_s \mathbf{x}_{s-2} + b_s \mathbf{x}_{s-1} + c_s \mathbf{x}_s + d_s \mathbf{x}_{s+1} + e_s \mathbf{x}_{s+2} \tag{6.31}
$$

where

$$
f_s = -\frac{1}{2} \left. \frac{\partial E_{edge}}{\partial x} \right|_{x_s, y_s} \quad a_s = \frac{\beta_{s-1}}{h^4} \quad b_s = -\frac{2(\beta_s + \beta_{s-1})}{h^4} - \frac{\alpha_s}{h^2}
$$

$$
c_s = \frac{\beta_{s+1} + 4\beta_s + \beta_{s-1}}{h^4} + \frac{\alpha_{s+1} + \alpha_s}{h^2} \quad d_s = -\frac{2(\beta_{s+1} + \beta_s)}{h^4} - \frac{\alpha_{s+1}}{h^2} \quad e_s = \frac{\beta_{s+1}}{h^4}
$$

This is now in the form of a linear (matrix) equation:

$$
\mathbf{Ax} = fx(\mathbf{x}, \mathbf{y}) \tag{6.32}
$$

where $fx(\mathbf{x}, \mathbf{y})$ is the first order differential of the edge magnitude along the x-axis and where

$$
\mathbf{A} = \begin{bmatrix}
c_1 & d_1 & e_1 & 0 & .. & a_1 & b_1 \\
b_2 & c_2 & d_2 & e_2 & 0 & .. & a_2 \\
a_3 & b_3 & c_3 & d_3 & e_3 & 0 & \\
\vdots & \vdots & \vdots & \vdots & \vdots & & \\
e_{S-1} & 0 & .. & a_{S-1} & b_{S-1} & c_{S-1} & d_{S-1} \\
d_S & e_S & 0 & .. & a_S & b_S & c_S
\end{bmatrix}
$$

Similarly, by analysis of Equation 6.29 we obtain:

$$
\mathbf{Ay} = fy(\mathbf{x}, \mathbf{y}) \tag{6.33}
$$

where $fy(\mathbf{x}, \mathbf{y})$ is the first order difference of the edge magnitude along the y-axis. These equations can be solved iteratively to provide a new vector $\mathbf{v}^{<i+1>}$ from an initial vector $\mathbf{v}^{<i>}$, where i is an evolution index. The iterative solution is

$$
\frac{(\mathbf{x}^{<i+1>} - \mathbf{x}^{<i>})}{\Delta} + \mathbf{Ax}^{<i+1>} = fx(\mathbf{x}^{<i>}, \mathbf{y}^{<i>}) \tag{6.34}
$$

where the control factor Δ is a scalar chosen to control convergence. The control factor, Δ, controls the rate of evolution of the snake: large values make the snake move quickly, small values make for slow movement. As usual, fast movement implies that the snake can pass over features of interest without noticing them, whereas slow movement can be rather tedious. So, the appropriate choice for Δ is again a compromise, this time between selectivity and time. The formulation for the vector of y coordinates is:

$$\frac{(\mathbf{y}^{<i+1>} - \mathbf{y}^{<i>})}{\Delta} + \mathbf{A}y^{<i+1>} = fy(\mathbf{x}^{<i>}, \mathbf{y}^{<i>}) \tag{6.35}$$

By rearrangement, this gives the final pair of equations that can be used to evolve a contour iteratively; the complete snake solution is then:

$$\mathbf{x}^{<i+1>} = \left(\mathbf{A} + \frac{1}{\Delta}\mathbf{I}\right)^{-1} \left(\frac{1}{\Delta}\mathbf{x}^{<i>} + fx(\mathbf{x}^{<i>}, \mathbf{y}^{<i>})\right) \tag{6.36}$$

where $\mathbf{I}$ is the identity matrix. This implies that the new set of x coordinates is a weighted sum of the initial set of contour points and the image information. The fraction is calculated according to specified snake properties, the values chosen for α and β. For the y coordinates we have

$$\mathbf{y}^{<i+1>} = \left(\mathbf{A} + \frac{1}{\Delta}\mathbf{I}\right)^{-1} \left(\frac{1}{\Delta}\mathbf{y}^{<i>} + fy(\mathbf{x}^{<i>}, \mathbf{y}^{<i>})\right) \tag{6.37}$$

The new set of contour points then become the starting set for the *next* iteration. Note that this is a continuous formulation, as opposed to the discrete (greedy) implementation. One *penalty* is the need for matrix inversion, affecting speed. Clearly, the benefits are that coordinates are calculated as *real* functions and the *complete* set of new contour points is provided at each iteration. The result of implementing the complete solution is illustrated in Figure 6.7. The initialization (Figure 6.7a) is the same as for the greedy algorithm, but with 32 contour points. At the first iteration (Figure 6.7b) the contour begins to shrink, and move towards the eye's iris. By the sixth iteration (Figure 6.7c) some of the contour points have snagged on strong edge data, particularly in the upper part of the contour. At this point, however, the excessive curvature becomes inadmissible, and the contour releases these points to achieve a smooth contour again, one that is better matched to the edge data and the chosen snake features. Finally, Figure 6.7(e) is where the contour ceases to move. Part of the contour has been snagged on strong edge data in the eyebrow, whereas the remainder of the contour matches the chosen feature well.

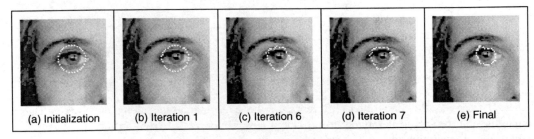

| (a) Initialization | (b) Iteration 1 | (c) Iteration 6 | (d) Iteration 7 | (e) Final |

Figure 6.7   Illustrating the evolution of a complete snake

Clearly, a different solution could be obtained by using different values for the snake parameters; in application the choice of values for α, β and Δ must be made very carefully. This is part of the difficulty in using snakes for practical feature extraction; a further difficulty is that the result depends on where the initial contour is placed. These difficulties are called *parameterization* and *initialization*, respectively. These problems have motivated much research and development.

6.3.4 Other snake approaches

There are many further considerations to implementing snakes and there is a great wealth of material. One consideration is that we have only considered *closed* contours. There are also *open contours*. These require slight difference in formulation for the Kass snake (Waite and Welsh, 1990) and only minor modification for implementation in the greedy algorithm. One difficulty with the greedy algorithm is its sensitivity to noise owing to its local neighbourhood action. In addition, the greedy algorithm can end up in an oscillatory position where the final contour simply jumps between two equally attractive energy minima. One solution (Lai and Chin, 1994) resolved this difficulty by increasing the size of the snake neighbourhood, but this incurs much greater complexity. To allow snakes to *expand*, as opposed to contracting, a *normal force* can be included which inflates a snake and pushes it over unattractive features (Cohen, 1991; Cohen and Cohen, 1993). The force is implemented by addition of

$$F_{\text{normal}} = \rho\mathbf{n}(s) \tag{6.38}$$

to the evolution equation, where $\mathbf{n}(s)$ is the normal force and ρ weights its effect. This is inherently sensitive to the magnitude of the normal force that, if too large, can force the contour to pass over features of interest. Another way to allow expansion is to modify the elasticity constraint (Berger, 1991) so that the internal energy becomes

$$E_{\text{int}} = \alpha(s)\left(\left|\frac{d\mathbf{v}(s)}{ds}\right|^2 - (L+\varepsilon)\right)^2 + \beta(s)\left|\frac{d^2\mathbf{v}(s)}{ds^2}\right|^2 \tag{6.39}$$

where the length adjustment ϵ when positive, $\epsilon > 0$, and added to the contour length L causes the contour to expand. When negative, $\epsilon < 0$, this causes the length to reduce and so the contour contracts. To avoid imbalance due to the contraction force, the technique can be modified to remove it (by changing the continuity and curvature constraints) without losing the controlling properties of the internal forces (Xu et al., 1994) (and which, incidentally, allowed corners to form in the snake). This gives a contour no prejudice to expansion or contraction as required. The technique allowed for integration of prior shape knowledge; methods have also been developed to allow *local shape* to influence contour evolution (Berger, 1991; Williams and Shah, 1992).

Some snake approaches have included factors that attract contours to regions using statistical models (Ronfard, 1994) or texture (Ivins and Porrill, 1995), to complement operators that combine edge detection with region growing. The snake model can also be generalized to higher dimensions and there are three-dimensional (3D) snake *surfaces* (Wang and Wang, 1992; Cohen et al., 1992). Finally, a new approach has introduced shapes for moving objects, by including velocity (Peterfreund, 1999).

6.3.5 Further snake developments

Snakes have not only been formulated to include local shape, but also phrased in terms of *regularization* (Lai and Chin, 1995), where a single parameter controls snake evolution, emphasizing a snake's natural compromise between its own forces and the image forces. Regularization involves using a single parameter to control the balance between the external and the internal forces. Given a regularization parameter λ, the snake energy of Equation 6.36 can be given as

$$E_{\text{snake}}(\mathbf{v}(s)) = \int\limits_{s=0}^{1} \left\{ \lambda E_{\text{int}}(\mathbf{v}(s)) + (1-\lambda) E_{\text{image}}(\mathbf{v}(s)) \right\} ds \qquad (6.40)$$

Clearly, if $\lambda = 1$, the snake will use the internal energy only, whereas if $\lambda = 0$, the snake will be attracted to the selected image function only. Usually, regularization concerns selecting a value in between zero and one guided, say, by knowledge of the likely confidence in the edge information. In fact, Lai's approach calculates the regularization parameter at contour points as

$$\lambda_i = \frac{\sigma_\eta^2}{\sigma_i^2 + \sigma_\eta^2} \qquad (6.41)$$

where σ_i^2 appears to be the variance of the point i and σ_η^2 is the variance of the noise at the point (even digging into Lai's PhD thesis provided no explicit clues here, save that 'these parameters may be learned from training samples'; if this is impossible a procedure can be invoked). As before, λ_i lies between zero and one, and where the variances are bounded as

$$\frac{1}{\sigma_i^2} + \frac{1}{\sigma_\eta^2} = 1 \qquad (6.42)$$

This does actually link these *generalized* active contour models to an approach we shall meet later, where the target shape is extracted conditional upon its expected variation. Lai's approach also addressed *initialization*, and showed how a GHT could be used to initialize an active contour and built into the extraction process. A major development of new external force model, which is called the *gradient vector flow* (GVF) (Xu and Prince, 1998). The GVF is computed as a diffusion of the gradient vectors of an edge map. There is, however, a limitation on using a single contour for extraction, since it is never known precisely where to stop.

Many of the problems with initialization with active contours can be resolved by using a *dual contour* approach (Gunn and Nixon, 1997), which also includes local shape and regularization. This approach aims to enclose the target shape within an inner and an outer contour. The outer contour *contracts* while the inner contour *expands*. A balance is struck between the two contours to allow them to let the target shape to be extracted. Gunn showed how shapes could be extracted successfully, even when the target contour was far from the two initial contours. Further, the technique was shown to provide better immunity to initialization, in comparison with the results of a Kass snake, and Xu's approach.

Later, the dual approach was extended to a discrete space (Gunn and Nixon, 1998), using an established search algorithm. The search algorithm used dynamic programming which has already been used within active contours to find a global solution (Lai and Chin, 1995) and in matching and tracking contours (Geiger et al., 1995). This new approach has already been used within an enormous study (using a database of over 20 000 images, no less) on automated cell segmentation for cervical cancer screening (Bamford and Lovell, 1998), achieving more than 99% accurate segmentation. The approach is formulated as a discrete search using a dual contour approach, illustrated in Figure 6.8. The inner and the outer contour aim to be inside and outside the target shape, respectively. The space between the inner and the outer contour is divided into lines (like the spokes on the wheel of a bicycle) and M points are taken along each of the N lines. We then have a grid of $M \times N$ points, in which the target contour (shape) is expected to lie. The full lattice of points is shown in Figure 6.9(a). Should we need higher resolution, we can choose large values of M and of N, but this in turn implies more computational effort. One can envisage strategies that allow for linearization of the coverage of the space between the two contours, but these can make implementation much more complex.

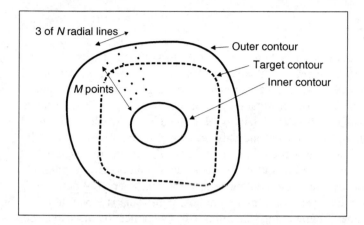

Figure 6.8   Discrete dual contour point space

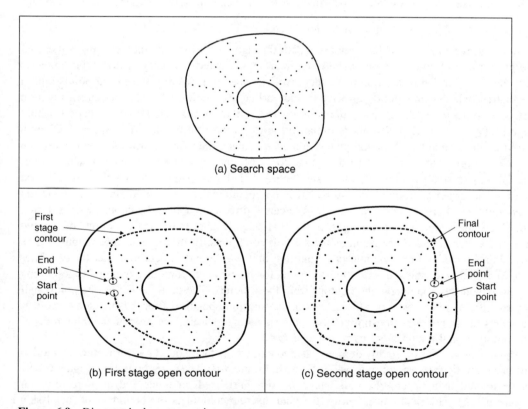

Figure 6.9   Discrete dual contour point space

The approach again uses *regularization*, where the snake energy is a discrete form to Equation 6.40, so the energy at a snake point (unlike earlier formulations, e.g. Equation 6.11) is

$$E(\mathbf{v}_i) = \lambda E_{\text{int}}(\mathbf{v}_i) + (1 - \lambda)E_{\text{ext}}(\mathbf{v}_i) \qquad (6.43)$$

where the internal energy is formulated as

$$E_{\text{int}}(\mathbf{v}_i) = \left(\frac{|\mathbf{v}_{i+1} - 2\mathbf{v}_i + \mathbf{v}_{i-1}|}{|\mathbf{v}_{i+1} - \mathbf{v}_{i-1}|} \right)^2 \qquad (6.44)$$

The numerator expresses the curvature, seen earlier in the greedy formulation. It is scaled by a factor that ensures the contour is *scale invariant* with no *prejudice* as to the *size* of the contour. If there is no prejudice, the contour will be attracted to smooth contours, given appropriate choice of the regularization parameter. As such, the formulation is simply a more sophisticated version of the greedy algorithm, dispensing with several factors of limited value (such as the need to choose values for three weighting parameters: one only now need be chosen; the elasticity constraint has also been removed, and that is perhaps more debatable). The interest here is that the search for the optimal contour is constrained to be between two contours, as in Figure 6.8. By way of a snake's formulation, we seek the contour with minimum energy. When this is applied to a contour which is bounded, we seek a minimum cost path. This is a natural target for the well-known Viterbi (dynamic programming) algorithm (for its application in vision, see, for example, Geiger et al., 1995). This is designed precisely to do this: to find a minimum cost path within specified bounds. To formulate it by dynamic programming we seek a cost function to be minimized. We formulate a cost function C between one snake element and the next as

$$C_i(\mathbf{v}_{i+1}, \mathbf{v}_i) = \min[C_{i-1}(\mathbf{v}_i, \mathbf{v}_{i-1}) + \lambda E_{\text{int}}(\mathbf{v}_i) + (1 - \lambda)E_{\text{ext}}(\mathbf{v}_i)] \qquad (6.45)$$

In this way, we should be able to choose a path through a set of snakes that minimizes the total energy, formed by the compromise between internal and external energy at that point, together with the path that led to the point. As such, we will need to store the energies at points within the matrix, which corresponds directly to the earlier tessellation. We also require a position matrix to store for each stage (i) the position ($\mathbf{v}_{i-1}$) that minimizes the cost function at that stage ($C_i(\mathbf{v}_{i+1}, \mathbf{v}_i)$). This also needs initialization to set the first point, $C_1(\mathbf{v}_1, \mathbf{v}_0) = 0$. Given a *closed* contour (one which is completely joined together), then for an arbitrary start point, we a separate optimization routine to determine the best starting and end points for the contour. The full search space is illustrated in Figure 6.9(a). Ideally, this should be searched for a closed contour, the target contour of Figure 6.8. It is computationally less demanding to consider an *open* contour, where the ends do not join. We can approximate a closed contour by considering it to be an open contour in *two* stages. In the first stage (Figure 6.9b), the midpoints of the two lines at the start and end are taken as the starting conditions. In the second stage (Figure 6.9c), the points determined by dynamic programming half way round the contour (i.e. for two lines at $N/2$) are taken as the start and the end points for a new open-contour dynamic programming search, which then optimizes the contour from these points. The premise is that the points half way round the contour will be at, or close to, their optimal position after the first stage and it is the points at, or near, the starting points in the first stage that require refinement. This reduces the computational requirement by a factor of M^2.

The technique was originally demonstrated to extract the face boundary, for feature extraction within automatic face recognition, as illustrated in Figure 6.10. The outer boundary (Figure 6.10a) was extracted using a convex hull which in turn initialized an inner and an outer contour (Figure 6.10b). The final extraction by the dual discrete contour is the boundary of facial skin (Figure 6.10c). The number of points in the mesh limits the accuracy with which the final contour is extracted, but application could be followed by use of a continuous Kass snake to improve final resolution. It has been shown that human faces could be discriminated by the contour extracted by this technique, although the study highlighted potential difficult with facial organs and illumination. As already mentioned, it was later deployed in cell analysis, where the inner and the outer contour were derived by analysis of the stained cell image.

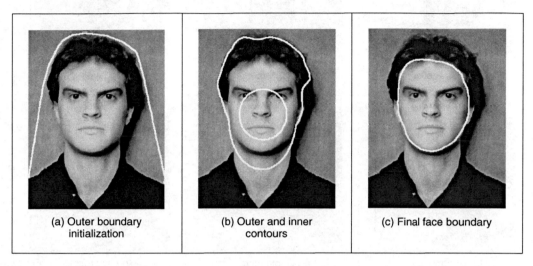

| (a) Outer boundary initialization | (b) Outer and inner contours | (c) Final face boundary |

Figure 6.10 Extracting the face outline by a discrete dual contour

6.3.6 Geometric active contours

Problems discussed so far with active contours include initialization and poor convergence to concave regions. In addition, parametric active contours (the snakes discussed previously) can have difficulty in segmenting multiple objects simultaneously, because of the explicit representation of curve. *Geometric active contour* (GAC) models have been introduced to solve this problem, where the curve is represented implicitly in a *level set function*. Essentially, the main argument is that by changing the representation, we can improve the result, and there have indeed been some very impressive results presented. Consider, for example, the result in Figure 6.11, where we are extracting the boundary of the hand by using the initialization shown in Figure 6.11(a). This would be hard to achieve by the active contour models discussed so far: there are concavities, sharp corners and background contamination which it is difficult for parametric techniques to handle. It is not perfect, but it is clearly much better (there are techniques to improve on this result, but this is far enough for the moment). However, there are no panaceas in engineering, and we should not expect them to exist. The new techniques can be found to be complex to implement, even to understand, although by virtue of their impressive results there are new approaches aimed to speed application and to ease implementation. As yet, the techniques do not find routine deployment (certainly not in real-time applications), but this is part of the evolution of any technique. The complexity and scope of this book mandate a short description of these new approaches here, but as usual we shall provide pointers to more in-depth source material.

Level set methods (Osher and Sethian, 1988) essentially find the shape without parameterizing it, so the curve description is *implicit* rather than explicit, by finding it as the zero level set

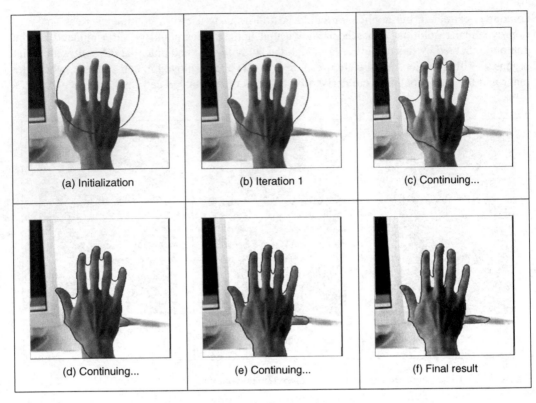

(a) Initialization | (b) Iteration 1 | (c) Continuing...

(d) Continuing... | (e) Continuing... | (f) Final result

Figure 6.11 Extraction by curve evolution (a diffusion snake) (Cremers et al., 2002)

of a function (Sethian, 1999; Osher and Paragios, 2003). The zero level set is the interface between two regions in an image. This can be visualized as taking slices through a surface shown in Figure 6.12(a). As we take slices at different levels (as the surface evolves) then the shape can split (Figure 6.12b). This would be difficult to parameterize (we would have to detect when it splits), but it can be handled within a level set approach by considering the underlying

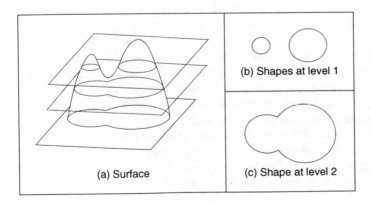

(a) Surface

(b) Shapes at level 1

(c) Shape at level 2

Figure 6.12 Surfaces and level sets

surface. At a lower level (Figure 6.12c), we have a single composite shape. As such, we have an extraction which evolves with time (to change the level). The initialization is a closed curve and we shall formulate how we want the curve to move in a way analogous to minimizing its energy.

The level set function is the signed distance to the contour. This distance is arranged to be negative inside the contour and positive outside it. The contour itself, the target shape, is where the distance is zero, at the interface between the two regions. Accordingly, we store values for each pixel representing this distance. We then determine new values for this surface, say by expansion. As we evolve the surface, the level sets evolve accordingly, equivalent to moving the surface where the slices are taken, in Figure 6.12. Since the distance map needs renormalization after each iteration, it can make the technique slow in operation (or needs a fast computer).

Let us assume that the interface C is controlled to change in a constant manner and evolves with time t by propagating along its normal direction with speed F, where F is a function of, say, curvature (Equation 4.61) and speed, according to

$$\frac{\partial C}{\partial t} = F \cdot \frac{\nabla \phi}{|\nabla \phi|} \tag{6.46}$$

Here, the term $\dfrac{\nabla \phi}{|\nabla \phi|}$ is a vector pointing in the direction normal to the surface, as previously discussed in Section 4.6, Equation 4.48. (The curvature at a point is measured perpendicular to the level set function at that point.) The curve is then evolving in a normal direction, controlled by the curvature. At all times, the interface C is the zero level set

$$\phi(C(t), t) = 0 \tag{6.47}$$

The level set function ϕ is positive outside the region and negative when it is inside, and it is zero on the boundary of the shape. As such, by differentiation we get

$$\frac{\partial \phi(C(t), t)}{\partial t} = 0 \tag{6.48}$$

and by the chain rule we obtain

$$\frac{\partial \phi}{\partial C} \frac{\partial C}{\partial t} + \frac{\partial \phi}{\partial t} = 0 \tag{6.49}$$

By rearrangement, and substitution from Equation 6.46, we obtain

$$\frac{\partial \phi}{\partial t} = -F \frac{\partial \phi}{\partial C} \cdot \frac{\nabla \phi}{|\nabla \phi|} = -F|\nabla \phi| \tag{6.50}$$

which suggests that the propagation of a curve depends on its gradient. In fact, we can include a (multiplicative) stopping function of the form

$$S = \frac{1}{1 + |\nabla \mathbf{P}|^n} \tag{6.51}$$

where $\nabla \mathbf{P}$ is the magnitude of the image gradient giving a stopping function (like the one in anisotropic diffusion in Equation 3.42) which is zero at edge points (hence stopping evolution) and near unity when there is no edge data (allowing movement). This is a form of the Hamilton–Jacobi equation, which is a partial differential equation that needs to be solved so as to obtain the solution. One way to achieve this is by finite differences (as earlier approximating the differential operation) and a spatial grid (the image itself). We then obtain a solution that differences the contour at iterations $< n + 1 >$ and $< n >$ (separated by an interval Δt) as

$$\frac{\phi(i, j, \Delta t)^{<n+1>} - \phi(i, j, \Delta t)^{<n>}}{\Delta t} = -F \left| \nabla_{ij} \phi(i, j)^{<n>} \right| \tag{6.52}$$

where $\nabla_{ij} \phi$ represents a spatial derivative, leading to the solution

$$\phi(i, j, \Delta t)^{<n+1>} = \phi(i, j, \Delta t)^{<n>} - \Delta t \left(F \left| \nabla_{ij} \phi(i, j)^{<n>} \right| \right) \tag{6.53}$$

This is only an introductory view, rather simplifying a complex scenario, and much greater detail is to be found in the two major texts in this area (Sethian, 1999; Osher and Paragios, 2003). The real poser is how to solve it all. We shall concentrate on some of the major techniques, but not go into their details. Caselles et al. (1993) and Malladi et al. (1995) were the first to propose geometric active contour models, which use gradient-based information for segmentation. The gradient-based geometric active contour can detect multiple objects simultaneously, but it has other important problems, which are boundary leakage, noise sensitivity, computational inefficiency and difficulty of implementation. There have been formulations (Caselles et al., 1997; Siddiqi et al., 1998; Xie and Mirmehdi, 2004) introduced to solve these problems; however, they can just increase the tolerance rather than achieve an exact solution. Several numerical schemes have also been proposed to improve computational efficiency of the level set method, including *narrow band* (Adalsteinsson and Sethian, 1995) (to find the solution within a constrained distance, i.e. to compute the level set only near the contour), *fast marching methods* (Sethian, 1999) (to constrain movement) and *additive operator splitting* (Weickert et al., 1998). Despite substantial improvements in efficiency, they can be difficult to implement. These approaches show excellent results, but they are not for the less than brave, although there are numerous tutorials and implementations available on the web. Clearly there is a need for unified presentation, and some claim this (e.g. Caselles et al., 1997), and linkage to parametric active contour models.

The technique with which many people compare the result of their own new approach is a GAC called the *active contour without edges*, introduced by Chan and Vese (2001), which is based on the Mumford–Shah functional (Mumford and Shah, 1989). Their model uses *regional statistics* for segmentation, and as such is a region-based level set model. The overall premise is to *avoid* using gradient (edge) information since this can lead to boundary leakage and cause the contour to collapse. A further advantage is that it can find objects when boundary data is weak or diffuse. The main strategy is to minimize energy, as in an active contour. To illustrate the model, let us presume that we have a bimodal image $\mathbf{P}$ which contains an object and a background. The object has pixels of intensity $\mathbf{P}^i$ within its boundary and the intensity of the background is $\mathbf{P}^0$, outside the boundary. We can then measure a fit of a contour, or curve, C to the image as

$$F^i(C) + F^o(C) = \int_{inside(C)} |\mathbf{P}(x, y) - c^i|^2 + \int_{outside(C)} |\mathbf{P}(x, y) - c^o|^2 \tag{6.54}$$

where the constant c^i is the average brightness inside the curve, depending on the curve, and c^o is the brightness outside it. The boundary of the object C_O is the curve that minimizes the fit derived by expressing the regions inside and outside the curve as

$$C_O = \min_C \left(F^i(C) + F^o(C) \right) \tag{6.55}$$

[Note that the original description is excellent, although Chan and Vese are from a maths department, which makes the presentation a bit terse. Also, the strict version of minimization is the infimum or greatest lower bound; inf(X) is the biggest real number that is smaller than or equal to every number in X.] The minimum is when

$$F^i(C_O) + F^o(C_O) \approx 0 \tag{6.56}$$

when the curve is at the boundary of the object. When the curve C is inside the object $F^i(C) \approx 0$ and $F^o(C) > 0$; conversely when the curve is outside the object $F^i(C) > 0$ and $F^o(C) \approx 0$. When the curves straddles the two and is both inside and outside the object then $F^i(C) > 0$ and $F^o(C) > 0$; the function is zero when C is placed on the boundary of the object. By using regions, we are avoiding using edges and the process depends finding the best separation between the *regions* (and by the *averaging* operation in the region, we have better *noise immunity*). If we constrain this process by introducing terms that depend on the length of the contour and the area of the contour, we extend the energy functional from Equation 6.54 as

$$F(c^i, c^o, C) = \mu \cdot length(C) + v \cdot area(C) + \lambda_1 \cdot \int_{inside(C)} |\mathbf{P}(x, y) - c^i|^2 + \lambda_2 \cdot \int_{outside(C)} |\mathbf{P}(x, y) - c^o|^2 \tag{6.57}$$

where μ, v, λ_1 and λ_2 are parameters controlling selectivity. The contour is then, for a fixed set of parameters, chosen by minimization of the energy functional as

$$C_O = \min_{c^i, c^o, C} \left(F\left(c^i c^o, C\right) \right) \tag{6.58}$$

A level set formulation is then used wherein an approximation to the unit step function (the Heaviside function) is defined to control the influence of points within and without the contour, which by differentiation gives an approximation to an impulse (the Dirac function), and with a solution to a form of Equation 6.50 (in discrete form) is used to update the level set.

The active contour without edges model can address problems with initialization, noise and boundary leakage (since it uses regions, not gradients), but still suffers from computational inefficiency and difficulty in implementation, because of the level set method. An example result is shown in Figure 6.13, where the target aim is to extract the hippo: the active contour without edges aims to split the image into the extracted object (the hippo) and its background (the grass). To do this, we need to specify an initialization which we shall choose to be within a small circle inside the hippo, as shown in Figure 6.13(a). The result of extraction is shown in Figure 6.13(b) and we can see that the technique has detected much of the hippo, but the result is not perfect. The values used for the parameters here were: $\lambda_1 = \lambda_2 = 1.0$ (i.e. area was not used to control evolution); $\mu = 0.1 * 255^2$ (the length parameter was controlled according to the image resolution) and some internal parameters were $h = 1$ (a 1 pixel step space); $\Delta t = 1$ (a small time spacing) and $\epsilon = 1$ (a parameter within the step, and hence the impulse functions). Alternative choices are possible and can affect the result achieved. The result here has been selected to show performance attributes; the earlier result (Figure 6.11) was selected to demonstrate finesse.

The regions with intensity and appearance that are most similar to the selected initialization have been identified in the result: this is much of the hippo, including the left ear and the region around the left eye, but omitting some of the upper body. There are some small potential problems too: there are some birds extracted on the top of the hippo and a small region underneath it (was this hippo's breakfast, we wonder?). Note that by virtue of the regional level set formulation the image is treated in its entirety and multiple shapes are detected, some well away from the target shape. By and large, the result looks encouraging as much of the hippo is extracted in the result and the largest shape contains much of the target; if we were to seek to obtain an

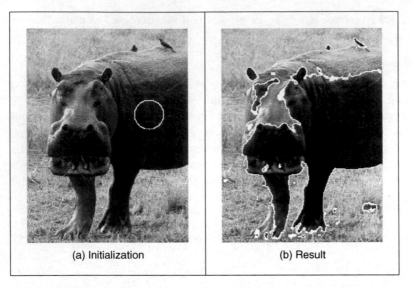

| (a) Initialization | (b) Result |

Figure 6.13 Extraction by a level set-based approach

exact match then we would need to use an exact model such as the GHT, or impose a model on the extraction, such as a statistical shape prior. That the technique can operate best when the image is bimodal is reflected in the fact that extraction is most successful when there is a clear difference between the target and the background, such as in the lower body. An alternative interpretation is that the technique clearly can handle situations where the edge data is weak and diffuse, such as in the upper body.

The technique have moved on and can now include statistical priors to guide shape extraction (Cremers et al., 2007). One study shows the relationship between parametric and geometric active contours (Xu et al., 2000). As such snakes and evolutionary approaches to shape extraction remain an attractive and stimulating area of research, so as ever it is well worth studying the literature to find new, accurate, techniques with high performance and low computational cost. We shall now move to determining *skeletons* which, though more a form of low-level operation, can use evidence gathering in implementation thus motivating its inclusion rather late in this book.

6.4 Shape skeletonization

6.4.1 Distance transforms

It is possible to describe a shape not just by its perimeter, or its area, but also by it *skeleton*. Here we do not mean an anatomical skeleton, more a central *axis* to a shape. This is then the axis which is equidistant from the borders of a shape, and can be determined by a *distance transform*. In this way we have a representation that has the same topology, the same size and orientation, but contains just the essence of the shape. As such, we are again in morphology and there has been interest for some time in binary shape analysis (Borgefors, 1986).

Essentially, the distance transform shows the distance from each point in an image shape to its central axis. (We are measuring distance here by difference in coordinate values; other measures of distance such as Euclidean are considered in Chapter 8.) Intuitively, the distance transform can be achieved by successive erosion and each pixel is labelled with the number of erosions before it disappeared. Accordingly, the pixels at the border of a shape will have a distance transform of unity, those adjacent inside will have a value of two, and so on. This is illustrated in Figure 6.14, where Figure 6.14(a) shows the analysed shape (a rectangle derived by, say, thresholding an image; the superimposed pixel values are arbitrary here as it is simply a binary image) and Figure 6.14(b) shows the distance transform, where the pixel values are the distance. Here, the central axis has a value of 3, as it takes that number of erosions to reach it from either side.

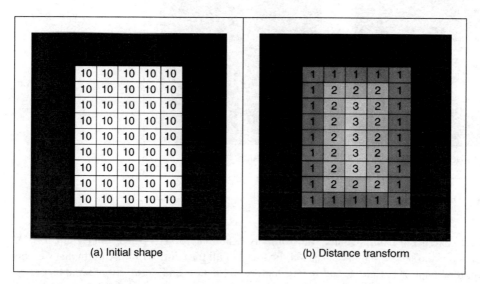

Figure 6.14 Illustrating distance transformation

The application to a rectangle at higher resolution is shown in Figure 6.15(a) and (b). Here, we can see that the central axis is quite clear and includes parts that reach towards the corners; and the central axis can be detected (Niblack et al., 1992) from the transform data. The application to a more irregular shape is shown applied to that of a card suit in Figure 6.15(c) and (d).

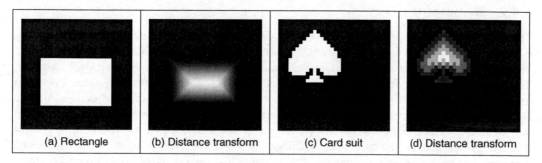

Figure 6.15 Applying the distance transformation

Flexible shape extraction (snakes and other techniques) 267

The natural difficulty is the effect of noise. This can change the result, as shown in Figure 6.16. This can certainly be ameliorated by using the earlier morphological operators (Section 3.6) to clean the image, but this can obscure the shape when the noise is severe. The major point is that this noise shows that the effect of a small change in the object can be quite severe on the resulting distance transform. As such, it has little tolerance of occlusion or change to its perimeter.

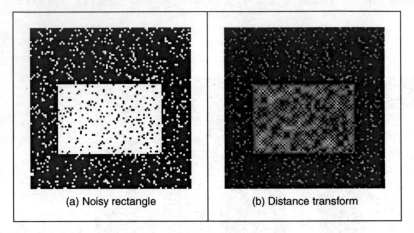

(a) Noisy rectangle   (b) Distance transform

Figure 6.16 Distance transformation on noisy images

The natural extension from distance transforms is to the *medial axis transform* (Blum, 1967), which determines the skeleton that consists of the locus of all the centres of maximum disks in the analysed region/shape. This has found use in feature extraction and description, so approaches have considered improvement in *speed* (Lee, 1982). One more recent study (Katz and Pizer, 2003) noted the practically difficulty experienced in *noisy* imagery: 'It is well documented how a tiny change to an object's boundary can cause a large change in its Medial Axis Transform'. To handle this, and hierarchical shape decomposition, the new approach 'provides a natural parts-hierarchy while eliminating instabilities due to small boundary changes'. An alternative is to seek an approach that is designed explicitly to handle noise, say by averaging, and we shall consider this type of approach next.

6.4.2 Symmetry

The *discrete symmetry operator* (Reisfeld et al., 1995) uses a totally different basis to find shapes, is intuitively very appealing and has links with human perception. Rather than rely on finding the border of a shape, or its shape, it locates features according to their *symmetrical properties*. The operator essentially forms an *accumulator* of points that are measures of symmetry between image points. Pairs of image points are attributed symmetry values that are derived from a *distance* weighting function, a *phase* weighting function and the *edge* magnitude at each of the pair of points. The distance weighting function controls the scope of the function, to control whether points that are more distant contribute in a similar manner to those that are close together. The phase weighting function shows when edge vectors at the pair of points point to each other. The symmetry accumulation is at the centre of each pair of points. In this way, the accumulator measures the degree of symmetry between image points, controlled by the edge strength. The distance weighting function D is

$$D(i, j, \sigma) = \frac{1}{\sqrt{2\pi}\sigma} e^{-\frac{|\mathbf{P}_i - \mathbf{P}_j|}{2\sigma}} \tag{6.59}$$

where i and j are the indices to two image points $\mathbf{P}_i$ and $\mathbf{P}_j$ and the deviation σ controls the scope of the function, by scaling the contribution of the distance between the points in the exponential function. A small value for the deviation σ implies local operation and detection of local symmetry. Larger values of σ imply that points that are further apart contribute to the accumulation process, as well as ones that are close together. In, say, application to the image of a face, large and small values of σ will aim for the whole face or the eyes, respectively.

The effect of the value of σ on the scalar distance weighting function expressed as Equation 6.60 is illustrated in Figure 6.17.

$$Di(j, \sigma) = \frac{1}{\sqrt{2\pi}\sigma} e^{\frac{j}{\sqrt{2}\sigma}} \tag{6.60}$$

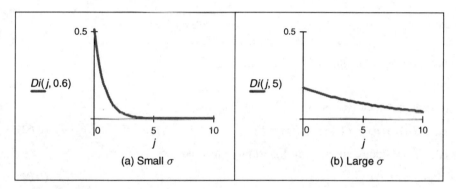

Figure 6.17 Effect of σ on distance weighting

Figure 6.17(a)shows the effect of a small value for the deviation, $\sigma = 0.6$, and shows that the weighting is greatest for closely spaced points and drops rapidly for points with larger spacing. Larger values of σ imply that the distance weight drops less rapidly for points that are more widely spaced, as in Figure 6.17(b) where $\sigma = 5$, allowing points that are spaced further apart to contribute to the measured symmetry. The phase weighting function P is

$$P(i, j) = (1 - \cos(\theta_i + \theta_j - 2\alpha_{ij})) \times (1 - \cos(\theta_i - \theta_j)) \tag{6.61}$$

where θ is the edge direction at the two points and α_{ij} measures the direction of a line joining the two points:

$$\alpha_{ij} = \tan^{-1}\left(\frac{y(\mathbf{P}_j) - y(\mathbf{P}_i)}{x(\mathbf{P}_j) - x(\mathbf{P}_i)}\right) \tag{6.62}$$

where $x(\mathbf{P}_i)$ and $y(\mathbf{P}_i)$ are the x and y coordinates of the point $\mathbf{P}_i$, respectively. This function is minimum when the edge direction at two points is in the same direction $(\theta_i = \theta_i)$, and is a maximum when the edge direction is away from each other $(\theta_i = \theta_j + \pi)$, along the line joining the two points, $(\theta_j = \alpha_{ij})$.

The effect of relative edge direction on phase weighting is illustrated in Figure 6.18, where Figure 6.18(a) concerns two edge points that point towards each other and describes the effect on the phase weighting function by varying α_{ij}. This shows how the phase weight is maximum when the edge direction at the two points is along the line joining them, in this case when $\alpha_{ij} = 0$ and $\theta_i = 0$. Figure 6.18(b) concerns one point with edge direction along the line joining two points, where the edge direction at the second point is varied. The phase weighting function is maximum when the edge direction at each point is towards that of the other, in this case when $|\theta_j| = \pi$.

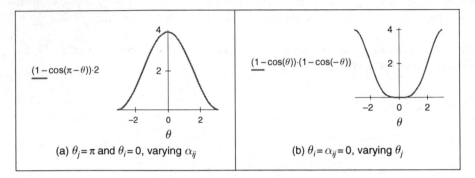

(a) $\theta_j = \pi$ and $\theta_i = 0$, varying α_{ij} (b) $\theta_i = \alpha_{ij} = 0$, varying θ_j

Figure 6.18   Effect of relative edge direction on phase weighting

The symmetry relation between two points is then defined as

$$C(i, j, \sigma) = D(i, j, \sigma) \times P(i, j) \times E(i) \times E(j) \tag{6.63}$$

where E is the edge magnitude expressed in logarithmic form as

$$E(i) = \log(1 + M(i)) \tag{6.64}$$

where M is the edge magnitude derived by application of an edge detection operator. The symmetry contribution of two points is accumulated at the midpoint of the line joining the two points. The total symmetry $S_{\mathbf{P}_m}$ at point $\mathbf{P}_m$ is the sum of the measured symmetry for all pairs of points which have their midpoint at $\mathbf{P}_m$, i.e. those points $\Gamma(\mathbf{P}_m)$ given by

$$\Gamma(\mathbf{P}_m) = \left[(i, j) \left| \frac{\mathbf{P}_i + \mathbf{P}_j}{2} = \mathbf{P}_m \wedge i \neq j \right. \right] \tag{6.65}$$

and the accumulated symmetry is then

$$S_{\mathbf{P}_m}(\sigma) = \sum_{i,j \in \Gamma(\mathbf{P}_m)} C(i, j, \sigma) \tag{6.66}$$

The result of applying the symmetry operator to two images is shown in Figure 6.19, for small and large values of σ. Figure 6.19(a) and (d) show the image of a rectangle and the image of the club, respectively, to which the symmetry operator was applied, and Figure 6.19(b) and (e) for the symmetry operator with a *low* value for the deviation parameter, showing detection of areas with high localized symmetry. Figure 6.19(c) and (f) are for a *large* value of the deviation parameter which detects overall symmetry and places a peak near the centre of the target shape. In Figure 6.19(b) and (e) the symmetry operator acts as a corner detector where the edge direction is discontinuous. (Note that this rectangle is one of the synthetic images we can use

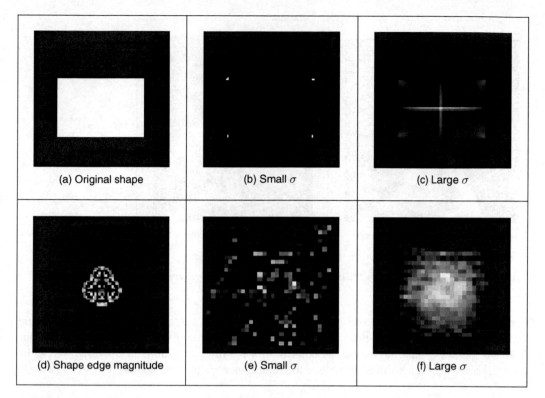

(a) Original shape	(b) Small σ	(c) Large σ
(d) Shape edge magnitude	(e) Small σ	(f) Large σ

Figure 6.19 Applying the symmetry operator for feature extraction

to test techniques, since we can understand its output easily. We also tested the operator on the image of a circle; since the circle is completely symmetrical, its symmetry plot is a single point, at the centre of the circle.) In Figure 6.19(e), the discrete symmetry operator provides a peak close to the position of the accumulator space peak in the GHT. Note that if the reference point specified in the GHT is the centre of symmetry, the results of the discrete symmetry operator and the GHT would be the same for large values of deviation.

This is a discrete operator; a *continuous symmetry operator* has been developed (Zabrodsky et al., 1995), and a later clarification (Kanatani, 1997) aimed to address potential practical difficulty associated with *hierarchy* of symmetry (namely that symmetrical shapes have subsets of regions, also with symmetry). There has also been a number of sophisticated approaches to detection of *skewed symmetry* (Gross and Boult, 1994; Cham and Cipolla, 1995), with later extension to detection in *orthographic projection* (Vangool et al., 1995). Another generalization addresses the problem of *scale* (Reisfeld, 1996) and extracts points of symmetry, together with scale. A *focusing* ability has been added to the discrete symmetry operator by reformulating the distance weighting function (Parsons and Nixon, 1999) and we able were to deploy this when using symmetry to in an approach which recognizes people by their gait (the way they walk) (Hayfron-Acquah et al., 2003). Why symmetry was chosen for this task is illustrated in Figure 6.20: this shows the main axes of symmetry of the walking subject (Figure 6.20b), that exist within the body, largely defining the skeleton. There is another axis of symmetry, between the legs. When the symmetry operator is applied to a sequence of images, this axis grows and retracts. By agglomerating the sequence and describing it by a (low-pass filtered)

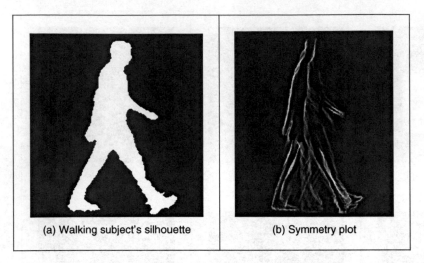

| (a) Walking subject's silhouette | (b) Symmetry plot |

Figure 6.20   Applying the symmetry operator for recognition by gait (Hayfron-Acquah et al., 2003)

Fourier transform, we can determine a set of numbers which are the same for the same person and different from those for other people, thus achieving recognition. No approach as yet has alleviated the computational burden associated with the discrete symmetry operator, and some of the process used can be used to reduce the requirement (e.g. judicious use of thresholding).

6.5   Flexible shape models: active shape and active appearance

So far, our approaches to analysing shape have concerned a match to image data. This has usually concerned a match between a model (either a template that can deform or a shape that can evolve) and a single image. An active contour is flexible, but its evolution is essentially controlled by local properties, such as the local curvature or edge strength. The chosen value for, or the likely range of, the parameters to weight these functionals may have been learnt by extensive testing on a database of images of similar type to the one used in application, or selected by experience. A completely different approach is to consider that if the database contains all possible *variations* of a shape, such as its appearance or pose, the database can form a model of the likely variation of that shape. As such, if we can incorporate this as a global constraint, while also guiding the match to the most likely version of a shape, then we have a deformable approach that is guided by the *statistics* of the likely variation in a shape. These approaches are termed *flexible templates* and use *global* shape constraints formulated from exemplars in training data.

This major new approach is called *active shape modelling*. The essence of this approach concerns a model of a shape made up of points; the variation in these points is called the *point distribution model*. The chosen *landmark* points are labelled on the training images. The set of training images aims to capture all possible variations of the shape. Each point describes a particular point on the boundary, so order is important in the labelling process. Example choices for these points include where the curvature is high (e.g. the corner of an eye) or at the apex of an arch where the contrast is high (e.g. the top of an eyebrow). The statistics of the variations in position of these points describe the ways in which a shape can appear. Example applications include finding the human face in images (e.g. for purposes of automatic face recognition). The only part of the face for which a distinct model is available is the round circle in the iris, and this can be small except at very high resolution. The rest of the face is made of unknown shapes and these can change with changes in facial expression. As such, they are well suited to a technique that combines shape with distributions, since we have a known set of shapes and a fixed interrelationship, but some of the detail can change. The variation in detail is what is captured in an active shape model.

Naturally, there is a lot of data. If we choose lots of points and we have lots of training images, we shall end up with an enormous number of points. That is where *principal components analysis* comes in, as it can compress data into the most significant items. Principal components analysis is an established mathematical tool; help is available in Appendix 4, on the web and in the literature (Press et al., *Numerical Recipes*, 1992). Essentially, it rotates a coordinate system so as to achieve maximal discriminatory capability: we might not be able to see something if we view it from two distinct points, but if we view it from some point in between then it is quite clear. That is what is done here: the coordinate system is rotated so as to work out the most significant variations in the morass of data. Given a set of N training examples where each example is a set of n points, for the ith training example $\mathbf{x}_i$ we have

$$\mathbf{x}_i = (x_{1i}, x_{2i}, \ldots x_{ni}) \quad i \in 1, N \tag{6.67}$$

where x_{ki} is the kth variable in the ith training example. When this is applied to shapes, each element is the two coordinates of each point. The average is then computed over the whole set of training examples as

$$\bar{\mathbf{x}} = \frac{1}{N} \sum_{i=1}^{N} \mathbf{x}_i \tag{6.68}$$

The deviation of each example from the mean $\delta\mathbf{x}_i$ is then

$$\delta\mathbf{x}_i = \mathbf{x}_i - \bar{\mathbf{x}} \tag{6.69}$$

This difference reflects how far each example is from the mean at a point. The $2n \times 2n$ covariance matrix $\mathbf{S}$ shows how far all the differences are from the mean as

$$\mathbf{S} = \frac{1}{N} \sum_{i=1}^{N} \delta\mathbf{x}_i \delta\mathbf{x}_i^T \tag{6.70}$$

Principal components analysis of this covariance matrix shows by how much these examples, and hence a shape, can change. In fact, any of the exemplars of the shape can be approximated as

$$\mathbf{x}_i = \bar{\mathbf{x}} + \mathbf{P}\mathbf{w} \tag{6.71}$$

where $\mathbf{P} = (\mathbf{p}_1, \mathbf{p}_2 \ldots \mathbf{p}_t)$ is a matrix of the first t eigenvectors, and $\mathbf{w} = (w_1, w_2 \ldots w_t)^T$ is a corresponding vector of weights where each weight value controls the contribution of a particular eigenvector. Different values in $\mathbf{w}$ give different occurrences of the model, or shape. Given that these changes are within specified limits, the new model or shape will be similar to the basic (mean) shape. This is because the modes of variation are described by the (unit) eigenvectors of $\mathbf{S}$, as

$$\mathbf{S}\mathbf{p}_k = \lambda_k \mathbf{p}_k \tag{6.72}$$

where λ_k denotes the eigenvalues and the eigenvectors obey orthogonality such that

$$\mathbf{p}_k \mathbf{p}_k^T = 1 \tag{6.73}$$

and where the eigenvalues are rank ordered such that $\lambda_k \geq \lambda_{k+1}$. Here, the largest eigenvalues correspond to the most significant modes of variation in the data. The proportion of the variance in the training data, corresponding to each eigenvector, is proportional to the corresponding eigenvalue. As such, a limited number of eigenvalues (and eigenvectors) can be used to encompass the majority of the data. The remaining eigenvalues (and eigenvectors) correspond to modes of variation that are hardly present in the data (like the proportion of very high-frequency contribution of an image; we can reconstruct an image mainly from the low-frequency components, as used in image coding). Note that in order to examine the statistics of the labelled landmark points over the training set applied to a new shape, the points need to be aligned, and established procedures are available (Cootes et al., 1995).

The process of application (to find instances of the modelled shape) involves an iterative approach to bring about increasing match between the points in the model and the image. This is achieved by examining regions around model points to determine the best nearby match. This provides estimates of the appropriate translation, scale rotation and eigenvectors to best fit the model to the data. This is repeated until the model converges to the data, when there is little change to the parameters. Since the models only change to fit the data better, and are controlled by the expected appearance of the shape, they were called active shape models. The application of an active shape model to find the face features of one of the technique's inventors (yes, that's Tim behind the target shapes) is shown in Figure 6.21, where the initial position is shown in Figure 6.21(a), the result after five iterations in Figure 6.21(b) and the final result in Figure 6.21(c). The technique can operate in a coarse-to-fine manner, working at low resolution initially (and making relatively fast moves) while slowing to work at finer resolution before the techniques result improves no further, at convergence. Clearly, the technique has not been misled by the spectacles, or by the presence of other features in the background. This can be used either for enrolment (finding the face, automatically) or for automatic face recognition (finding and describing the features). The technique cannot handle initialization which is too poor, although clearly by Figure 6.21(a) the initialization does not need to be too close either.

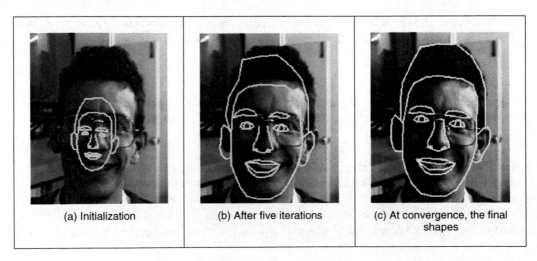

| (a) Initialization | (b) After five iterations | (c) At convergence, the final shapes |

Figure 6.21 Finding face features using an active shape model

Active shape models (ASMs) have been applied in face recognition (Lanitis et al., 1997), medical image analysis (Cootes et al., 1994), including 3D analysis (Hill et al., 1994), and industrial inspection (Cootes et al., 1995). A similar theory has been used to develop a new approach that incorporates texture, called *active appearance models* (AAMs) (Cootes et al., 1998a,b). This approach again represents a shape as a set of landmark points and uses a set of training data to establish the potential range of variation in the shape. One major difference is that AAMs explicitly include texture and update model parameters to move landmark points closer to image points by matching texture in an iterative search process. The essential differences between ASMs and AAMs include:

- ASMs use texture information local to a point, whereas AAMs use texture information in a whole region.
- ASMs seek to minimize the distance between model points and the corresponding image points, whereas AAMs seek to minimize distance between a synthesized model and a target image.
- AAMs search around the current position, typically along profiles normal to the boundary, whereas AAMs consider the image only at the current position.

One comparison (Cootes et al., 1999) has shown that although ASMs can be faster in implementation than AAMs, the AAMs can require fewer landmark points and can converge to a better result, especially in terms of texture (wherein the AAM was formulated). We await with interest further developments in these approaches to flexible shape modelling. An example result by an AAM for face feature finding is shown in Figure 6.22. Although this cannot demonstrate computational advantage, we can see that the inclusion of hair in the eyebrows has improved segmentation there. Inevitably, interest has concerned improving computational requirements, in one case by an efficient fitting algorithm based on the inverse compositional image alignment algorithm (Matthews and Baker, 2004). Recent interest has concerned the ability to handle occlusion (Gross et al., 2006), as occurring either by changing (3D) orientation or by gesture.

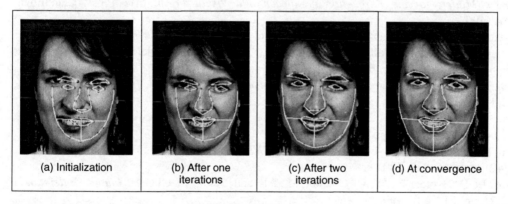

| (a) Initialization | (b) After one iterations | (c) After two iterations | (d) At convergence |

Figure 6.22 Finding face features using an active appearance model

6.6 Further reading

The majority of further reading in finding shapes concerns papers, many of which have already been referenced. An excellent survey of the techniques used for *feature extraction* (including template matching, deformable templates, etc.) can be found in Trier et al. (1996), while a broader view was taken later (Jain et al., 1998). A comprehensive survey of flexible extractions from *medical* imagery (McInerney and Terzopolous, 1996) reinforces the dominance of snakes in medical image analysis, to which they are particularly suited given a target of smooth shapes. (An excellent survey of history and progress of medical image analysis is available (Duncan and Ayache, 2000).) Few of the textbooks devote much space to shape extraction, and snakes, especially level set methods, are too recent a development to be included in many textbooks. One text alone is dedicated to shape analysis (van Otterloo, 1991) and contains many discussions on symmetry. A visit to Professor Cootes' website (http://www.isbe.man.ac.uk/~bim/) reveals a lengthy report on flexible shape modelling and a lot of support material (including Windows and Linux code) in active shape modelling. For work on level set methods for image segmentation, see Cremers et al. (2007).

6.7 References

Adalsteinsson, D. and Sethian, J., A Fast Level Set Method for Propagating Interfaces, *J. Computational Physics*, **118**(2), pp. 269–277, 1995

Bamford, P. and Lovell, B., Unsupervised Cell Nucleus Segmentation with Active Contours, *Signal Process.*, **71**, pp. 203–213, 1998

Benn, D. E., Nixon, M. S. and Carter, J. N., Extending Concentricity Analysis by Deformable Templates for Improved Eye Extraction. *Proc. of the 2nd Int. Conf. on Audio- and Video-Based Biometric Person Authentication AVBPA99*, pp. 1–6, 1999

Berger, M. O., Towards Dynamic Adaption of Snake Contours, *Proc. 6th Int. Conf. Image Analysis and Processing*, Como, Italy, pp. 47–54, 1991

Blum, H., A Transformation for Extracting New Descriptors of Shape, In: W. Wathen-Dunn (Ed.), *Models for the Perception of Speech and Visual Form*, MIT Press, Cambridge, MA, 1967

Borgefors, G., Distance Transformations in Digital Images, *CVGIP*, **34**(3), pp. 344–371, 1986

Caselles, V., Catte, F., Coll, T. and Dibos, F., A Geometric Model for Active Contours. *Numerische Mathematic*, **66**, pp. 1–31, 1993

Caselles, V., Kimmel, R. and Sapiro, G., Geodesic Active Contours, *Int. J. Comput. Vision*, **22**(1), pp. 61–79, 1997

Cham, T. J. and Cipolla, R., Symmetry Detection through Local Skewed Symmetries, *Image Vision Comput.*, **13**(5), pp. 439–450, 1995

Chan, T. F. and Vese, L. A., Active Contours Without Edges, *IEEE Trans. Image Process.*, **10**(2), pp. 266–277, 2001

Cohen, L. D., Note: On Active Contour Models and Balloons, *CVGIP: Image Understand.*, **53**(2), pp. 211–218, 1991

Cohen, I., Cohen, L. D. and Ayache, N., Using Deformable Surfaces to Segment 3D Images and Inter Differential Structures, *CVGIP: Image Understand.*, **56**(2), pp. 242–263, 1992

Cohen, L. D. and Cohen, I., Finite-Element Methods for Active Contour Models and Balloons for 2D and 3D Images, *IEEE Trans. PAMI*, **15**(11), pp. 1131–1147, 1993

Cootes, T. F., Hill, A., Taylor, C. J. and Haslam, J., The Use of Active Shape Models for Locating Structures in Medical Images, *Image Vision Comput.*, **12**(6), pp. 355–366, 1994

Cootes, T. F., Taylor, C. J., Cooper, D. H. and Graham, J., Active Shape Models – Their Training and Application, *CVIU*, **61**(1), pp. 38–59, 1995

Cootes, T. F., Edwards, G. J. and Taylor, C. J., A Comparative Evaluation of Active Appearance Model Algorithms, In: P. H. Lewis and M. S. Nixon (Eds), *Proc. British Machine Vision Conference 1998, BMVC98*, **2**, pp. 680–689, 1998a

Cootes, T., Edwards, G. J. and Taylor, C. J, Active Appearance Models, In: H. Burkhardt and B. Neumann (Eds), *Proc. ECCV 98*, **2**, pp. 484–498, 1998b

Cootes, T. F., Edwards, G. J. and Taylor, C. J., Comparing Active Shape Models with Active Appearance Models, In: T. Pridmore and D. Elliman (Eds), *Proc. British Machine Vision Conference 1999, BMVC99*, **1**, pp. 173–182, 1999

Cremers, D., Tischhäuser, F., Weickert, J. and Schnörr, C., Diffusion Snakes: Introducing Statistical Shape Knowledge into the Mumford–Shah Functional, *Int. J. Comput. Vision*, **50**(3), pp. 295–313, 2002

Cremers, D., Rousson, M., and Deriche, R., A Review of Statistical Approaches to Level Set Segmentation: Integrating Color, Texture, Motion and Shape, *Int. J. Comput. Vision*, **72**(2), pp. 195–215, 2007

Duncan, J. S. and Ayache, N., Medical Image Analysis: Progress Over Two Decades and the Challenges Ahead, *IEEE Trans. PAMI*, **22**(1), pp. 85–106, 2000

Geiger, D., Gupta, A., Costa, L. A. and Vlontsos, J., Dynamical Programming for Detecting, Tracking and Matching Deformable Contours, *IEEE Trans. PAMI*, **17**(3), pp. 294–302, 1995

Goldberg, D., *Genetic Algorithms in Search, Optimization and Machine Learning*, Addison-Wesley, Reading, MA, 1988

Gross, A. D. and Boult, T. E., Analysing Skewed Symmetries, *Int. J. Comput. Vision*, **13**(1), pp. 91–111, 1994

Gross, R., Matthews, I. and Baker, S., Active Appearance Models with Occlusion, *Image Vision Comput.*, **24**(6), pp. 593–604, 2006

Gunn, S. R. and Nixon, M. S., A Robust Snake Implementation; A Dual Active Contour, *IEEE Trans. PAMI*, **19**(1), pp. 63–68, 1997

Gunn, S. R. and Nixon, M. S., Global and Local Active Contours for Head Boundary Extraction, *Int. J. Comput. Vision*, **30**(1), pp. 43–54, 1998

Hayfron-Acquah, J. B., Nixon, M. S. and Carter, J. N., Automatic Gait Recognition by Symmetry Analysis, *Pattern Recog. Lett.*, **24**(13), pp. 2175–2183, 2003

Hill, A., Cootes, T. F., Taylor, C. J. and Lindley, K., Medical Image Interpretation: A Generic Approach using Deformable Templates, *J. Med. Informatics*, **19**(1), pp. 47–59, 1994

Ivins, J. and Porrill, J., Active Region Models for Segmenting Textures and Colours, *Image Vision Comput.*, **13**(5), pp. 431–437, 1995

Jain, A. K., Zhong, Y. and Dubuisson-Jolly, M.-P., Deformable Template Models: A Review, *Signal Process.*, **71**, pp. 109–129, 1998

Kanatani, K., Comments on 'Symmetry as a Continuous Feature', *IEEE Trans. PAMI*, **19**(3), pp. 246–247, 1997

Kass, M., Witkin, A. and Terzopoulos, D., Snakes: Active Contour Models, *Int. J. Comput. Vision*, **1**(4), pp. 321–331, 1988

Katz, R. A. and Pizer, S. M., Untangling the Blum Medial Axis Transform, *Int. J. Comput. Vision*, **55**(2–3), pp. 139–153, 2003

Lai, K. F. and Chin, R. T., On Regularization, Extraction and Initialization of the Active Contour Model (Snakes), *Proc. 1st Asian Conference on Computer Vision*, pp. 542–545, 1994

Lai, K. F. and Chin, R. T., Deformable Contours – Modelling and Extraction, *IEEE Trans. PAMI*, **17**(11), pp. 1084–1090, 1995

Lanitis, A., Taylor, C. J. and Cootes, T., Automatic Interpretation and Coding of Face Images using Flexible Models, *IEEE Trans. PAMI*, **19**(7), pp. 743–755, 1997

Lee, D. T., Medial Axis Transformation of a Planar Shape, *IEEE Trans. PAMI*, **4**, pp. 363–369, 1982

McInerney, T. and Terzopolous, D., Deformable Models in Medical Image Analysis, a Survey, *Med. Image Analysis*, **1**(2), pp. 91–108, 1996

Malladi, R., Sethian, J. A. and Vemuri, B. C., Shape Modeling with Front Propagation: A Level Set Approach, *IEEE Trans. PAMI*, **17**(2), pp. 158–175, 1995

Matthews, I. and Baker, S., Active Appearance Models Revisited, *Int. J. Comput. Vision*, **60**(2), pp. 135–164, 2004

Mumford, D. and Shah, J., Optimal Approximation by Piecewise Smooth Functions and Associated Variational Problems, *Communs Pure Applied Math.*, **42**, pp. 577–685, 1989

Niblack, C. W., Gibbons, P. B. and Capson, D. W., Generating skeletons and centerlines from the distance transform, *CVGIP: Graphical Models and Image Processing*, **54**(5), pp. 420–437, 1992

Osher, S. J. and Paragios, N. (eds), *Geometric Level Set Methods in Imaging, Vision and Graphics*, Springer, New York, 2003

Osher, S. J. and Sethian, J., Eds., Fronts Propagating with Curvature Dependent Speed: Algorithms Based on the Hamilton–Jacobi Formulation, *J. Computational Physics*, **79**, pp. 12–49, 1988

van Otterloo, P. J., *A Contour-Oriented Approach to Shape Analysis*, Prentice Hall International (UK), Hemel Hempstead, 1991

Parsons, C. J. and Nixon, M. S., Introducing Focus in the Generalized Symmetry Operator, *IEEE Signal Process. Lett.*, **6**(1), 1999

Peterfreund, N., Robust Tracking of Position and Velocity, *IEEE Trans. PAMI*, **21**(6), pp. 564–569, 1999

Press, W. H., Teukolsky, S. A., Vettering, W. T. and Flannery, B. P., *Numerical Recipes in C – The Art of Scientific Computing*, 2nd edn, Cambridge University Press, Cambridge, 1992

Reisfeld, D., The Constrained Phase Congruency Feature Detector: Simultaneous Localization, Classification and Scale Determination, *Pattern Recog. Lett.*, **17**(11), pp. 1161–1169, 1996

Reisfeld, D., Wolfson, H. and Yeshurun, Y., Context-Free Attentional Operators: The Generalized Symmetry Transform, *Int. J. Comput. Vision*, **14**, pp. 119–130, 1995

Ronfard, R., Region-based Strategies for Active Contour Models, *Int. J. Comput. Vision*, **13**(2), pp. 229–251, 1994

Sethian, J. A., *Level Set Methods: Evolving Interfaces in Computational Geometry, Fluid Mechanics, Computer Vision, and Materials Science*, Cambridge University Press, Cambridge, 1996

Sethian, J., *Level Set Methods and Fast Marching Methods*, Cambridge University Press, New York, 1999

Siddiqi, K., Lauziere, Y., Tannenbaum, A. and Zucker, S., Area and Length Minimizing Flows for Shape Segmentation, *IEEE Trans. Image Process.*, **7**(3), pp. 433–443, 1998

Trier, O. D., Jain, A. K. and Taxt, T., Feature Extraction Methods for Character Recognition – A Survey, *Pattern Recog.*, **29**(4), pp. 641–662, 1996

Vangool, L., Moons, T., Ungureanu, D. and Oosterlinck, A., The Characterization and Detection of Skewed Symmetry, *Comput. Vision Image Understand.*, **61**(1), pp. 138–150, 1995

Waite, J. B. and Welsh, W. J., Head Boundary Location Using Snakes, *Br. Telecom J.*, **8**(3), pp. 127–136, 1990

Wang, Y. F. and Wang, J. F., Surface Reconstruction using Deformable Models with Interior and Boundary Constraints, *IEEE Trans. PAMI*, **14**(5), pp. 572–579, 1992

Weickert, J., Ter Haar Romeny, B. M. and Viergever, M. A., Efficient and Reliable Schemes for Nonlinear Diffusion Filtering, *IEEE Trans. Image Process.*, **7**(3), pp. 398–410, 1998

Williams, D. J. and Shah, M., A Fast Algorithm for Active Contours and Curvature Estimation, *CVGIP: Image Understand.*, **55**(1), pp. 14–26, 1992

Xie, X., and Mirmehdi, M., RAGS: Region-Aided Geometric Snake, *IEEE Trans. Image Process*, **13**(5), pp. 640–652, 2004

Xu, C. and Prince, J. L., Snakes, Shapes, and Gradient Vector Flow, *IEEE Trans. Image Process*, **7**(3), 359–369, 1998

Xu, C., Yezzi, A. and Prince, J. L., On the Relationship Between Parametric and Geometric Active Contours and its Applications, *Proc. 34th Asimolar Conf. Sig. Sys. Comput.*, Pacific Grove, CA, pp. 483–489, 2000

Xu, G., Segawa, E. and Tsuji, S., Robust Active Contours with Insensitive Parameters, *Pattern Recog.*, **27**(7), pp. 879–884, 1994

Yuille, L., Deformable Templates for Face Recognition, *J. Cognitive Neurosci.*, **3**(1), pp. 59–70, 1991

Zabrodsky, H., Peleg, S. and Avnir, D., Symmetry as a Continuous Feature, *IEEE Trans. PAMI*, **17**(12), pp. 1154–1166, 1995

7

Object description

7.1 Overview

Objects are represented as a collection of pixels in an image. Thus, for purposes of recognition we need to describe the properties of groups of pixels. The description is often just a set of numbers: the object's *descriptors*. From these, we can compare and recognize objects by simply matching the descriptors of objects in an image against the descriptors of known objects. However, to be useful for recognition, descriptors should have four important properties. First, they should define a *complete set*. That is, two objects must have the same descriptors if and only if they have the same shape. Secondly, they should be *congruent*. As such, we should be able to recognize *similar* objects when they have *similar* descriptors. Thirdly, it is convenient that they have *invariant* properties. For example, *rotation*-invariant descriptors will be useful for recognizing objects whatever their *orientation*. Other important invariance properties include scale and position and also invariance to affine and perspective changes. These last two properties are very important when recognizing objects observed from different viewpoints. In addition to these three properties, the descriptors should be a *compact* set. Namely, a descriptor should represent the essence of an object in an efficient way. That is, it should only contain information about what makes an object unique, or different from the other objects. The quantity of information used to describe this characterization should be less than the information necessary to have a complete description of the object itself. Unfortunately, there is no set of complete and compact descriptors to characterize general objects. Thus, the best recognition performance is obtained by carefully selected properties. As such, the process of recognition is strongly related to each particular application with a particular type of objects.

In this chapter, we present the characterization of objects by two forms of descriptors. These descriptors are summarized in Table 7.1. *Region* and *shape* descriptors characterize an arrangement of pixels within the *area* and the arrangement of pixels in the *perimeter* or *boundary*, respectively. This region versus perimeter kind of representation is common in image analysis. For example, edges can be located by *region growing* (to label area) or by *differentiation* (to label perimeter), as covered in Chapter 4. There are many techniques that can be used to obtain descriptors of an object's boundary. Here, we shall just concentrate on three forms of descriptors: *chain codes* and two forms based on *Fourier characterization*. For region descriptors we shall distinguish between basic descriptors and statistical descriptors defined by moments.

Table 7.1 Overview of Chapter 7

Main topic	Sub topics	Main points
Boundary descriptions	How to determine the boundary and the region it encloses. How to form a description of the boundary and necessary properties in that description. How we describe a curve/boundary by Fourier approaches.	Basic approach: chain codes. Fourier descriptors: discrete approximations; cumulative angular function and elliptic Fourier descriptors.
Region descriptors	How we describe the area of a shape. Basic shape measures: heuristics and properties. Describing area by statistical moments: need for invariance and more sophisticated descriptions. What moments describe, and reconstruction from the moments.	Basic shape measures: area; perimeter; compactness; dispersion. Moments: basic; centralized; invariant; Zernike. Properties and reconstruction.

7.2 Boundary descriptions

7.2.1 Boundary and region

A region usually describes *contents* (or interior points) that are surrounded by a *boundary* (or perimeter), which is often called the region's *contour*. The form of the contour is generally referred to as its *shape*. A point can be defined to be on the boundary (contour) if it is part of the region and there is at least one pixel in its neighbourhood that is not part of the region. The boundary itself is usually found by contour following: we first find one point on the contour and then progress round the contour either in a clockwise direction, or anticlockwise, finding the nearest (or next) contour point.

To define the interior points in a region and the points in the boundary, we need to consider neighbouring relationships between pixels. These relationships are described by means of *connectivity* rules. There are two common ways of defining connectivity: *four-way* (or four-neighbourhood) where only immediate *neighbours* are analysed for connectivity; or *eight-way* (or eight-neighbourhood) where all the eight pixels surrounding a chosen pixel are analysed for connectivity. These two types of connectivity are illustrated in Figure 7.1. In this figure, the pixel

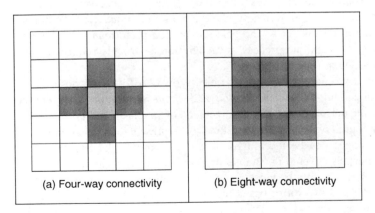

(a) Four-way connectivity   (b) Eight-way connectivity

Figure 7.1 Main types of connectivity analysis

is shown in light grey and its neighbours in dark grey. In four-way connectivity (Figure 7.1a), a pixel has four neighbours in the directions north, east, south and west, its immediate neighbours. The four extra neighbours in eight-way connectivity (Figure 7.1b) are those in the directions north-east, south-east, south-west and north-west, the points at the *corners*.

A boundary and a region can be defined using both types of connectivity and they are always *complementary*. That is, if the boundary pixels are connected in four-way, the region pixels will be connected in eight-way and vice versa. This relationship can be seen in the example shown in Figure 7.2. In this figure, the boundary is shown in dark grey and the region in light grey. We can observe that for a diagonal boundary, the four-way connectivity gives a staircase boundary, whereas eight-way connectivity gives a diagonal line formed from the points at the corners of the neighbourhood. Notice that all the pixels that form the region in Figure 7.2(b) have four-way connectivity, while the pixels in Figure 7.2(c) have eight-way connectivity. This is complementary to the pixels in the border.

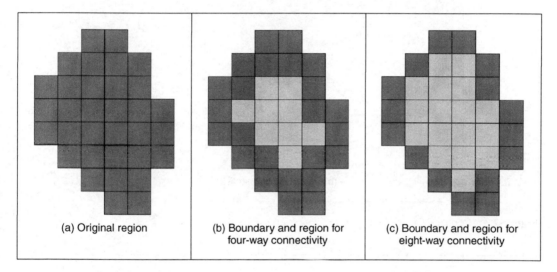

| (a) Original region | (b) Boundary and region for four-way connectivity | (c) Boundary and region for eight-way connectivity |

Figure 7.2 Boundaries and regions

7.2.2 Chain codes

To obtain a representation of a contour, we can simply store the coordinates of a sequence of pixels in the image. Alternatively, we can just store the relative position between consecutive pixels. This is the basic idea behind *chain codes*. Chain codes are one of the oldest techniques in computer vision, originally introduced in the 1960s (Freeman, 1961; an excellent review came later: Freeman, 1974). Essentially, the set of pixels in the border of a shape is translated into a set of connections between them. Given a complete border, one that is a set of connected points, then starting from one pixel we need to be able to determine the direction in which the next pixel is to be found. Namely, the next pixel is one of the adjacent points in one of the major compass directions. Thus, the chain code is formed by concatenating the number that designates the direction of the next pixel. That is, given a pixel, the successive direction from one pixel to the next pixel becomes an element in the final code. This is repeated for each point until the start point is reached when the (closed) shape is completely analysed.

Directions in four-way and eight-way connectivity can be assigned as shown in Figure 7.3. The chain codes for the example region in Figure 7.2(a) are shown in Figure 7.4. Figure 7.4(a) shows the chain code for the four-way connectivity. In this case, we have that the direction from the start point to the next is south (i.e. code 2), so the first element of the chain code describing the shape is 2. The direction from point P1 to the next, P2, is east (code 1), so the next element of the code is 1. The next point after P2 is P3, which is south, giving a code 2. This coding is repeated until P23, which is connected eastwards to the starting point, so the last element (the 12th element) of the code is 1. The code for eight-way connectivity shown in Figure 7.4(b) is obtained in an analogous way, but the directions are assigned according to the definition in Figure 7.3(b). Notice that the length of the code is shorter for this connectivity, given that the number of boundary points is smaller for eight-way connectivity than it is for four-way.

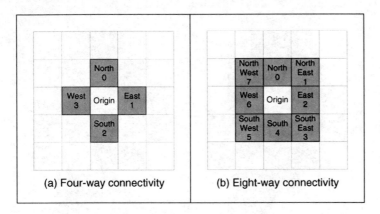

Figure 7.3   Connectivity in chain codes

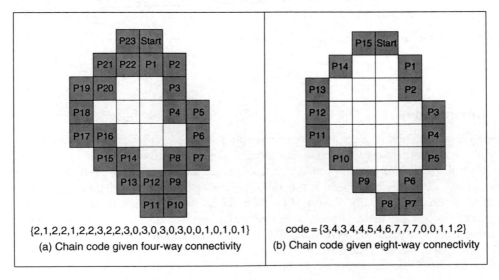

Figure 7.4   Chain codes by different connectivity

Clearly, this code will be different when the start point changes. Accordingly, we need *start point invariance*. This can be achieved by considering the elements of the code to constitute the digits in an integer. Then, we can shift the digits *cyclically* (replacing the least significant digit with the most significant one, and shifting all other digits left one place). The smallest integer is returned as the *start point invariant chain code description*. This is illustrated in Figure 7.5, where the initial chain code is that from the shape in Figure 7.4. Here, the result of the first shift is given in Figure 7.5(b); this is equivalent to the code that would have been derived by using point P1 as the starting point. The result of two shifts (Figure 7.5c) is the chain code equivalent to starting at point P2, but this is not a code corresponding to the minimum integer. The minimum integer code (Figure 7.5d) is the minimum of all the possible shifts and is the chain code that would have been derived by starting at point P11. That fact could not be used in application since we would need to find P11; it is much easier to shift to achieve a minimum integer.

code = {3,4,3,4,4,5,4,6,7,7,7,0,0,1,1,2}	code = {4,3,4,4,5,4,6,7,7,7,0,0,1,1,2,3}
(a) Initial chain code	(b) Result of one shift
code = {3,4,4,5,4,6,7,7,7,0,0,1,1,2,3,4}	code = {0,0,1,1,2,3,4,3,4,4,5,4,6,7,7,7}
(c) Result of two shifts	(d) Minimum integer chain code

Figure 7.5 Start point invariance in chain codes

In addition to starting point invariance, we can obtain a code that does not change with *rotation*. This can be achieved by expressing the code as a difference of chain code, since relative descriptions remove rotation dependence. Change of *scale* can complicate matters greatly, since we can end up with a set of points that is of different size to the original set. As such, the boundary needs to be *resampled* before coding. This is a tricky issue. Furthermore, *noise* can have drastic effects. If salt and pepper *noise* were to remove, or to add, some points the code would change. Such problems can lead to great difficulty with chain codes. However, their main virtue is their *simplicity* and as such they remain a popular technique for shape description. Further developments of chain codes have found application with *corner detectors* (Liu and Srinath, 1990; Seeger and Seeger, 1994). However, the need to be able to handle noise, the requirement of connectedness, and the local nature of description motivate alternative approaches. Noise can be reduced by *filtering*, which leads back to the *Fourier transform*, with the added advantage of a *global* description.

7.2.3 Fourier descriptors

Fourier descriptors, often attributed to early work by Cosgriff (1960), allow us to bring the power of Fourier theory to shape description. The main idea is to characterize a contour by a set of numbers that represent the frequency content of a whole shape. Based on frequency analysis, we can select a *small* set of numbers (the Fourier coefficients) that describe a shape rather than any noise (i.e. the noise affecting the spatial position of the boundary pixels). The general recipe to obtain a Fourier description of the curve involves two main steps. First, we have to define a representation of a curve. Secondly, we expand it using Fourier theory. We can obtain alternative flavours by combining different curve representations and different Fourier expansions. Here, we shall consider Fourier descriptors of angular and complex contour representations. However, Fourier expansions can be developed for other curve representations (van Otterloo, 1991).

In addition to the curve's definition, a factor that influences the development and properties of the description is the choice of Fourier expansion. If we consider that the trace of a curve defines a periodic function, we can opt to use a Fourier series expansion. However, we could also consider that the description is not periodic. Thus, we could develop a representation based on the Fourier transform. In this case, we could use alternative Fourier integral definitions. Here, we will develop the presentation based on expansion in Fourier series. This is the common way used to describe shapes in pattern recognition.

It is important to notice that although a curve in an image is composed of discrete pixels, Fourier descriptors are developed for continuous curves. This is convenient since it leads to a discrete set of Fourier descriptors. We should also remember that the pixels in the image are the sampled points of a continuous curve in the scene. However, the formulation leads to the definition of the integral of a continuous curve. In practice, we do not have a continuous curve, but a sampled version. Thus, the expansion is approximated by means of numerical integration.

7.2.3.1 *Basis of Fourier descriptors*

In the most basic form, the coordinates of boundary pixels are x and y point coordinates. A Fourier description of these essentially gives the set of spatial frequencies that fit the boundary points. The *first* element of the Fourier components (the d.c. component) is simply the average value of the x and y coordinates, giving the coordinates of the centre point of the boundary, expressed in complex form. The *second* component essentially gives the radius of the circle that best fits the points. Accordingly, a circle can be described by its zero-order and first order components (the d.c. component and first harmonic). The *higher* order components increasingly describe detail, as they are associated with higher frequencies.

This is illustrated in Figure 7.6. Here, the Fourier description of the ellipse in Figure 7.6(a) is the frequency components in Figure 7.6(b), depicted in logarithmic form for purposes of display. The Fourier description has been obtained by using the coordinates of the ellipse boundary points. Here we can see that the low-order components dominate the description, as to be

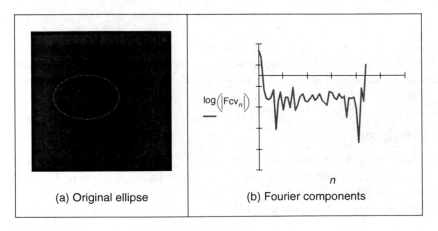

(a) Original ellipse   (b) Fourier components

Figure 7.6   An ellipse and its Fourier description

expected for such a smooth shape. In this way, we can derive a set a numbers that can be used to *recognize* the boundary of a shape: a similar ellipse should give a similar set of numbers, whereas a completely different shape will result in a completely different set of numbers.

We do, however, need to check the result. One way is to take the descriptors of a circle, since the first harmonic should be the circle's radius. A better way is to *reconstruct* the shape from its descriptors; if the reconstruction matches the original shape then the description would appear correct. We can reconstruct a shape from this Fourier description since the descriptors are *regenerative*. The zero-order component gives the position (or origin) of a shape. The ellipse can be reconstructed by adding in all spatial components, to extend and compact the shape along the *x*- and *y*-axes, respectively. By this inversion, we return to the original ellipse. When we include the zero and first descriptor, then we reconstruct a circle, as expected, shown in Figure 7.7(b). When we include all Fourier descriptors the reconstruction, Figure 7.7(c) is very close to the original Figure 7.7(a) with slight differences due to discretization effects.

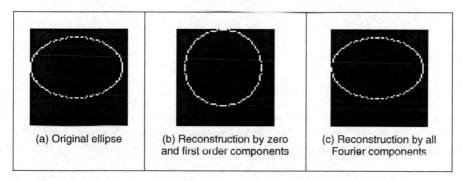

(a) Original ellipse

(b) Reconstruction by zero and first order components

(c) Reconstruction by all Fourier components

Figure 7.7 Reconstructing an ellipse from a Fourier description

This is only an outline of the basis to Fourier descriptors, since we have yet to consider descriptors that give the same description whatever an object's position, scale and rotation. Here we have just considered an object's description that is achieved in a manner that allows for reconstruction. To develop practically useful descriptors, we need to consider more basic properties. As such, we first turn to the use of Fourier theory for shape description.

7.2.3.2 *Fourier expansion*
To define a Fourier expansion, we can start by considering that a continuous curve $c(t)$ can be expressed as a summation of the form

$$c(t) = \sum_k c_k f_k(t) \tag{7.1}$$

where c_k defines the coefficients of the expansion, and the collection of functions $f_k(t)$ defines the basis functions. The expansion problem centres on finding the coefficients given a set of basis functions. This equation is very general and different basis functions can also be used. For example, $f_k(t)$ can be chosen such that the expansion defines a polynomial. Other bases define splines, Lagrange and Newton interpolant functions. A Fourier expansion represents periodic functions by a basis defined as a set of infinite complex exponentials. That is,

$$c(t) = \sum_{k=-\infty}^{\infty} c_k e^{jk\omega t} \tag{7.2}$$

Here, ω defines the fundamental frequency and it is equal to $T/2\pi$, where T is the period of the function. The main feature of the Fourier expansion is that it defines an orthogonal basis. This simply means that

$$\int_0^T f_k(t) f_j(t) dt = 0 \tag{7.3}$$

for $k \neq j$. This property is important for two main reasons. First, it ensures that the expansion does not contain redundant information (each coefficient is *unique* and contains no information about the other components). Secondly, it simplifies the computation of the coefficients. That is, to solve for c_k in Equation 7.1, we can simply multiply both sides by $f_k(t)$ and perform integration. Thus, the coefficients are given by

$$c_k = \int_0^T c(t) f_k(t) \Big/ \int_0^T f_k^2(t) \tag{7.4}$$

By considering the definition in Equation 7.2 we have:

$$c_k = \frac{1}{T} \int_0^T c(t) e^{-jk\omega t} \tag{7.5}$$

In addition to the exponential form given in Equation 7.2, the Fourier expansion can be expressed in trigonometric form. This form shows that the Fourier expansion corresponds to the summation of trigonometric functions that increase in frequency. It can be obtained by considering that

$$c(t) = c_0 + \sum_{k=1}^{\infty} \left(c_k e^{jk\omega t} + c_{-k} e^{-jk\omega t} \right) \tag{7.6}$$

In this equation the values of $e^{jk\omega t}$ and $e^{-jk\omega t}$ define a pairs of complex conjugate vectors. Thus, c_k and c_{-k} describe a complex number and its conjugate. Let us define these numbers as

$$c_k = c_{k,1} - jc_{k,2} \quad \text{and} \quad c_{-k} = c_{k,1} + jc_{k,2} \tag{7.7}$$

By substitution of this definition in Equation 7.6 we obtain

$$c(t) = c_0 + 2\sum_{k=1}^{\infty} \left(c_{k,1} \left(\frac{e^{jk\omega t} + e^{-jk\omega t}}{2} \right) + jc_{k,2} \left(\frac{-e^{jk\omega t} + e^{-jk\omega t}}{2} \right) \right) \tag{7.8}$$

That is,

$$c(t) = c_0 + 2\sum_{k=1}^{\infty} (c_{k,1} \cos(k\omega t) + c_{k,2} \sin(k\omega t)) \tag{7.9}$$

If we define

$$a_k = 2c_{k,1} \quad \text{and} \quad b_k = 2c_{k,2} \tag{7.10}$$

we obtain the standard trigonometric form given by

$$c(t) = \frac{a_0}{2} + \sum_{k=1}^{\infty} (a_k \cos(k\omega t) + b_k \sin(k\omega t)) \tag{7.11}$$

The coefficients of this expansion, a_k and b_k, are known as the *Fourier descriptors*. These control the amount of each frequency that contributes to make up the curve. Accordingly, these descriptors can be said to *describe* the curve, since they do not have the same values for different curves. Notice that according to Equations 7.7 and 7.10 the coefficients of the trigonometric and exponential form are related by

$$c_k = \frac{a_k - jb_k}{2} \quad \text{and} \quad c_{-k} = \frac{a_k + jb_k}{2} \tag{7.12}$$

The coefficients in Equation 7.11 can be obtained by considering the orthogonal property in Equation 7.3. Thus, one way to compute values for the descriptors is

$$a_k = \frac{2}{T} \int_0^T c(t) \cos(k\omega t) \, dt \quad \text{and} \quad b_k = \frac{2}{T} \int_0^T c(t) \sin(k\omega t) \, dt \tag{7.13}$$

To obtain the Fourier descriptors, a curve can be represented by the complex exponential form of Equation 7.2 or by the sin/cos relationship of Equation 7.11. The descriptors obtained by using either of the two definitions are equivalent, and they can be related by the definitions of Equation 7.12. In general, Equation 7.13 is used to compute the coefficients since it has a more intuitive form. However, some works have considered the complex form (e.g. Granlund, 1972). The complex form provides an elegant development of rotation analysis.

7.2.3.3 Shift invariance

Chain codes required special attention to give start point invariance. Let us see whether that is required here. The main question is whether the descriptors will change when the curve is shifted. In addition to Equations 7.2 and 7.11, a Fourier expansion can be written in another sinusoidal form. If we consider that

$$|c_k| = \sqrt{a_k^2 + b_k^2} \quad \text{and} \quad \varphi_k = a \tan^{-1}(b_k/a_k) \tag{7.14}$$

then the Fourier expansion can be written as

$$c(t) = \frac{a_0}{2} + \sum_{k=0}^{\infty} |c_k| \cos(k\omega t + \varphi_k) \tag{7.15}$$

Here, $|c_k|$ is the *amplitude* and φ_k is the *phase* of the Fourier coefficient. An important property of the Fourier expansion is that $|c_k|$ does not change when the function $c(t)$ is shifted (i.e. translated), as in Section 2.6.1. This can be observed by considering the definition of Equation 7.13 for a shifted curve $c(t + \alpha)$. Here, α represents the shift value. Thus,

$$a_k' = \frac{2}{T} \int_0^T c(t' + \alpha) \cos(k\omega t') \, dt' \quad \text{and} \quad b_k' = \frac{2}{T} \int_0^T c(t' + \alpha) \sin(k\omega t') \, dt' \tag{7.16}$$

By defining a change of variable by $t = t' + \alpha$, we have

$$a_k' = \frac{2}{T} \int_0^T c(t) \cos(k\omega t - k\omega \alpha) \, dt \quad \text{and} \quad b_k' = \frac{2}{T} \int_0^T c(t) \sin(k\omega t - k\omega \alpha) \, dt \tag{7.17}$$

After some algebraic manipulation we obtain

$$a_k' = a_k \cos(k\omega\alpha) + b_k \sin(k\omega\alpha) \quad \text{and} \quad b_k' = b_k \cos(k\omega\alpha) - a_k \sin(k\omega\alpha) \tag{7.18}$$

The amplitude $|c'_k|$ is given by

$$|c'_k| = \sqrt{(a_k \cos{(k\omega\alpha)} + b_k \sin{(k\omega\alpha)})^2 + (b_k \cos{(k\omega\alpha)} - a_k \sin{(k\omega\alpha)})^2} \quad (7.19)$$

That is,

$$|c'_k| = \sqrt{a_k^2 + b_k^2} \quad (7.20)$$

Thus, the amplitude is independent of the shift α. Although shift invariance could be incorrectly related to translation invariance, as we shall see, this property is related to rotation invariance in shape description.

7.2.3.4 Discrete computation

Before defining Fourier descriptors, we must consider the numerical procedure necessary to obtain the Fourier coefficients of a curve. The problem is that Equations 7.11 and 7.13 are defined for a *continuous* curve. However, given the discrete nature of the image, the curve $c(t)$ will be described by a collection of *points*. This discretization has two important effects. First, it limits the number of frequencies in the expansion. Secondly, it forces numerical approximation to the integral defining the coefficients.

Figure 7.8 shows an example of a discrete approximation of a curve. Figure 7.8(a) shows a continuous curve in a period, or interval, T. Figure 7.8(b) shows the approximation of the curve by a set of discrete points. If we try to obtain the curve from the sampled points, we will find that the sampling process reduces the amount of detail. According to the Nyquist theorem, the maximum frequency f_c in a function is related to the sample period τ by

$$\tau = \frac{1}{2f_c} \quad (7.21)$$

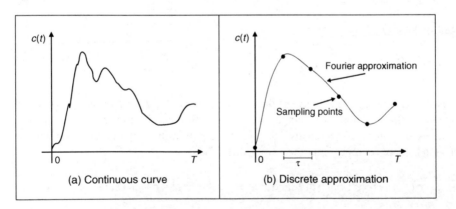

(a) Continuous curve   (b) Discrete approximation

Figure 7.8 Example of a discrete approximation

Thus, if we have m sampling points, then the sampling period is equal to $\tau = T/m$. Accordingly, the maximum frequency in the approximation is given by

$$f_c = \frac{m}{2T} \quad (7.22)$$

Each term in Equation 7.11 defines a trigonometric function at frequency $f_k = k/T$. By comparing this frequency with the relationship in Equation 7.15, we have that the maximum frequency is obtained when

$$k = \frac{m}{2} \tag{7.23}$$

Thus, to define a curve that passes through the m sampled points, we need to consider only $m/2$ coefficients. The other coefficients define frequencies higher than the maximum frequency. Accordingly, the Fourier expansion can be redefined as

$$c(t) = \frac{a_0}{2} + \sum_{k=1}^{m/2} (a_k \cos(k\omega t) + b_k \sin(k\omega t)) \tag{7.24}$$

In practice, Fourier descriptors are computed for fewer coefficients than the limit of $m/2$. This is because the low-frequency components provide most of the features of a shape. High frequencies are easily affected by noise and only represent detail that is of little value to recognition. We can interpret Equation 7.22 the other way around: if we know the maximum frequency in the curve, then we can determine the appropriate number of samples. However, the fact that we consider $c(t)$ to define a continuous curve implies that to obtain the coefficients in Equation 7.13, we need to evaluate an integral of a continuous curve. The approximation of the integral is improved by increasing the number of sampling points. Thus, as a practical rule, to improve accuracy, we must try to have a large number of samples even if it is theoretically limited by the Nyquist theorem.

Our curve is only a set of discrete points. We want to maintain a continuous curve analysis to obtain a set of discrete coefficients. Thus, the only alternative is to approximate the coefficients by approximating the value of the integrals in Equation 7.13. We can approximate the value of the integral in several ways. The most straightforward approach is to use a Riemann sum. Figure 7.9 illustrates this approach. In Figure 7.9(b), the integral is approximated as the summation of the rectangular areas. The middle point of each rectangle corresponds to each sampling point. Sampling points are defined at the points whose parameter is $t = i\tau$, where i is an integer between 1 and m. We consider that c_i defines the value of the function at the sampling point i. That is,

$$c_i = c(i\tau) \tag{7.25}$$

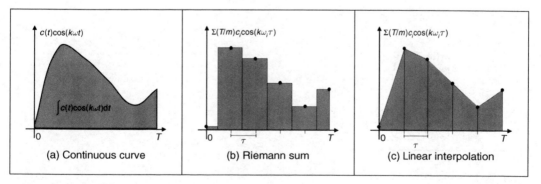

(a) Continuous curve   (b) Riemann sum   (c) Linear interpolation

Figure 7.9   Integral approximation

Thus, the height of the rectangle for each pair of coefficients is given by $c_i \cos(k\omega i\tau)$ and $c_i \sin(k\omega i\tau)$. Each interval has a length $\tau = T/m$. Thus,

$$\int_0^T c(t) \cos(k\omega t)dt \approx \sum_{i=1}^m \frac{T}{m} c_i \cos(k\omega i\tau)$$

and

$$\int_0^T c(t) \sin(k\omega t)dt \approx \sum_{i=1}^m \frac{T}{m} c_i \sin(k\omega i\tau) \tag{7.26}$$

Accordingly, the Fourier coefficients are given by

$$a_k = \frac{2}{m} \sum_{i=1}^m c_i \cos(k\omega i\tau) \quad \text{and} \quad b_k = \frac{2}{m} \sum_{i=1}^m c_i \sin(k\omega i\tau) \tag{7.27}$$

Here, the error due to the discrete computation will be reduced with increase in the number of points used to approximate the curve. These equations correspond to a linear approximation to the integral. This approximation is shown in Figure 7.9(c). In this case, the integral is given by the summation of the trapezoidal areas. The sum of these areas leads to Equation 7.26. Notice that b_0 is zero and a_0 is twice the average of the c_i values. Thus, the first term in Equation 7.24 is the average (or centre of gravity) of the curve.

7.2.3.5 *Cumulative angular function*

Fourier descriptors can be obtained by using many boundary representations. In a straightforward approach we could consider, for example, that t and $c(t)$ define the angle and modulus of a polar parameterization of the boundary. However, this representation is not very general. For some curves, the polar form does not define a single valued curve, and thus we cannot apply Fourier expansions. A more general description of curves can be obtained by using the angular function parameterization. This function was defined in Chapter 4 in the discussion about curvature.

The angular function $\varphi(s)$ measures the angular direction of the tangent line as a function of arc length. Figure 7.10 illustrates the angular direction at a point in a curve. In (Cosgriff, 1960) this angular function was used to obtain a set of Fourier descriptors. However, this first approach to Fourier characterization has some undesirable properties. The main problem is that the angular function has discontinuities even for smooth curves. This is because the angular direction is bounded from zero to 2π. Thus, the function has *discontinuities* when the angular

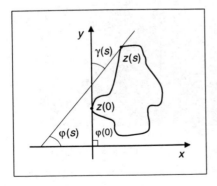

Figure 7.10   Angular direction

direction increases to a value of more than 2π or decreases to be less than zero (since it will change abruptly to remain within bounds). In Zahn and Roskies' (1972) approach, this problem is eliminated by considering a normalized form of the cumulative angular function.

The *cumulative angular function* at a point in the curve is defined as the amount of angular change from the starting point. It is called *cumulative*, since it represents the summation of the angular change to each point. Angular change is given by the derivative of the angular function $\varphi(s)$. We discussed in Chapter 4 that this derivative corresponds to the curvature $\kappa(s)$. Thus, the cumulative angular function at the point given by (s) can be defined as

$$\gamma(s) = \int_0^S \kappa(r)dr - \kappa(0) \qquad (7.28)$$

Here, the parameter s takes values from zero to L (i.e. the length of the curve). Thus, the initial and final values of the function are $\gamma(0) = 0$ and $\gamma(L) = -2\pi$, respectively. It is important to notice that to obtain the final value of -2π, the curve must be traced in a clockwise direction. Figure 7.10 illustrates the relation between the angular function and the cumulative angular function. In the figure, $z(0)$ defines the initial point in the curve. The value of $\gamma(s)$ is given by the angle formed by the inclination of the tangent to $z(0)$ and that of the tangent to the point $z(s)$. If we move the point $z(s)$ along the curve, this angle will change until it reaches the value of -2π. In Equation 7.28, the cumulative angle is obtained by adding the small angular increments for each point.

The cumulative angular function avoids the discontinuities of the angular function. However, it still has two problems. First, it has a discontinuity at the end. Secondly, its value depends on the length of curve analysed. These problems can be solved by defining the normalized function $\gamma^*(t)$, where

$$\gamma^*(t) = \gamma\left(\frac{L}{2\pi}t\right) + t \qquad (7.29)$$

Here, t takes values from 0 to 2π. The factor $L/2\pi$ normalizes the angular function such that it does not change when the curve is scaled. That is, when $t = 2\pi$, the function evaluates the final point of the function $\gamma(s)$. The term t is included to avoid discontinuities at the end of the function (remember that the function is periodic). That is, it makes that $\gamma^*(0) = \gamma^*(2\pi) = 0$. In addition, it causes the cumulative angle for a circle to be zero. This is consistent as a circle is generally considered the simplest curve and, intuitively, simple curves will have simple representations.

Figure 7.11 illustrates the definitions of the cumulative angular function with two examples. Figure 7.11(b)–(d) define the angular functions for a circle in Figure 7.11(a). Figure 7.11(f)–(h) define the angular functions for the rose in Figure 7.11(e). Figure 7.11(b) and (f) define the angular function $\varphi(s)$. We can observe the typical toroidal form. Once the curve is greater than 2π there is a discontinuity while its value returns to zero. The position of the discontinuity depends on the selection of the starting point. The cumulative function $\gamma(s)$ shown in Figure 7.11(c) and (g) inverts the function and eliminates discontinuities. However, the start and end points are not the same. If we consider that this function is periodic, there is a discontinuity at the end of each period. The normalized form $\gamma^*(t)$ shown in Figure 7.11(d) and (h) has no discontinuity and the period is normalized to 2π.

The normalized cumulative functions are very nice indeed. However, it is tricky to compute them from images. In addition, since they are based on measures of changes in angle, they are very sensitive to noise and difficult to compute at inflexion points (e.g. corners). Code 7.1 illustrates the computation of the angular functions for a curve given by a sequence of pixels.

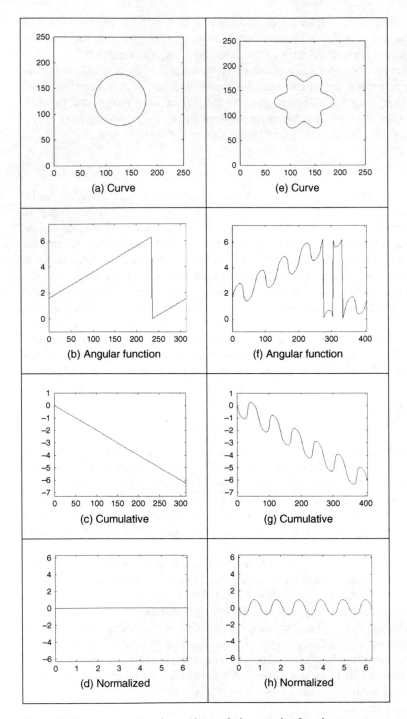

Figure 7.11 Angular function and cumulative angular function

```
%Angular function
 function AngFuncDescrp(curve)

%Function
 X=curve(1,:);   Y=curve(2,:);
 M=size(X,2);    %number points

%Arc length
 S=zeros(1,m);
 S(1)=sqrt((X(1)-X(m))^2+(Y(1)-Y(m))^2);
 for i=2:m
    S(i)=S(i-1)+sqrt((X(i)-X(i-1))^2+(Y(i)-Y(i-1))^2);
 End
 L=S(m);

%Normalized Parameter
 t=(2*pi*S)/L;

%Graph of the curve
 subplot(3,3,1);
 plot(X,Y);
 mx=max(max(X),max(Y))+10;
 axis([0,mx,0,mx]);   axis square;        %Aspect ratio

%Graph of the angular function y'/x'
 avrg=10;
 A=zeros(1,m);
 for i=1:m
    x1=0; x2=0; y1=0; y2=0;
    for j=1:avrg
        pa=i-j; pb=i+j;
        if(pa<1) pa=m+pa; end
        if(pb>m) pb=pb-m; end
        x1=x1+X(pa); y1=y1+Y(pa);
        x2=x2+X(pb); y2=y2+Y(pb);
    end
        x1=x1/avrg; y1=y1/avrg;
        x2=x2/avrg; y2=y2/avrg;
        dx=x2-x1;   dy=y2-y1;

        if(dx==0) dx=.00001; end
           if dx>0 & dy>0
           A(i)=atan(dy/dx);
           elseif dx>0 & dy<0
           A(i)=atan(dy/dx)+2*pi;
        else
           A(i)=atan(dy/dx)+pi;
        end
    end

    subplot(3,3,2);
```

```
    plot(S,A);
    axis([0,S(m),-1,2*pi+1]);

%Cumulative angular    G(s)=-2pi
 G=zeros(1,m);
 for i=2:m
    d=min(abs(A(i)-A(i-1)),abs(abs(A(i)-A(i-1))-2*pi));

    if d>.5
        G(i)=G(i-1);
    elseif (A(i)-A(i-1))<-pi
        G(i)=G(i-1)-(A(i)-A(i-1)+2*pi);
    elseif (A(i)-A(i-1))>pi
        G(i)=G(i-1)-(A(i)-A(i-1)-2*pi);
    else
        G(i)=G(i-1)-(A(i)-A(i-1));
    end
 end

 subplot(3,3,3);

 plot(S,G);
 axis([0,S(m),-2*pi-1,1]);

%Cumulative angular Normalized
 F=G+t;

 subplot(3,3,4);
 plot(t,F);
 axis([0,2*pi,-2*pi,2*pi]);
```

Code 7.1 Angular functions

The matrices X and Y store the coordinates of each pixel. The code has two important steps. First, the computation of the angular function stored in the matrix A. In general, if we use only the neighbouring points to compute the angular function, then the resulting function is useless owing to noise and discretization errors. Thus, it is necessary to include a procedure that can obtain accurate measures. For purposes of illustration, in the presented code we average the position of pixels to filter out noise; however, other techniques such as the fitting process discussed in Section 4.8.2 can provide a suitable alternative. The second important step is the computation of the cumulative function. In this case, the increment in the angle cannot be computed as the simple difference between the current and precedent angular values. This will produce as a result a discontinuous function. Thus, we need to consider the periodicity of the angles. In the code, this is achieved by checking the increment in the angle. If it is greater than a threshold, we consider that the angle has exceeded the limits of zero or 2π.

Figure 7.12 shows an example of the angular functions computed using Code 7.1, for a discrete curve. These are similar to those in Figure 7.11(a)–(d), but show noise due to discretization which

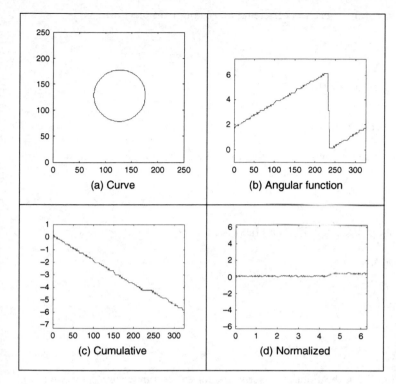

Figure 7.12 Discrete computation of the angular functions

produces a ragged effect on the computed values. The effects of noise will be reduced if we use more points to compute the average in the angular function. However, this reduces the level of detail in the curve. It also makes it more difficult to detect when the angle exceeds the limits of zero or 2π. In a Fourier expansion, noise will affect the coefficients of the high-frequency components, as seen in Figure 7.12(d).

To obtain a description of the curve we need to expand $\gamma^*(t)$ in Fourier series. In a straight-forward approach we can obtain $\gamma^*(t)$ from an image and apply the definition in Equation 7.27 for $c(t) = \gamma^*(t)$. However, we can obtain a computationally more attractive development with some algebraically simplifications. By considering the form of the integral in Equation 7.13 we have:

$$a_k^* = \frac{1}{\pi} \int_0^{2\pi} \gamma^*(t) \cos(kt)\mathrm{d}t \quad \text{and} \quad b_k^* = \frac{1}{\pi} \int_0^{2\pi} \gamma^*(t) \sin(kt)\mathrm{d}t \tag{7.30}$$

By substitution of Equation 7.29 we obtain

$$a_0^* = \frac{1}{\pi} \int_0^{2\pi} \gamma((L/2\pi)t)\mathrm{d}t + \frac{1}{\pi} \int_0^{2\pi} t\mathrm{d}t$$

$$a_k^* = \frac{1}{\pi} \int_0^{2\pi} \gamma((L/2\pi)t) \cos(kt)\mathrm{d}t + \frac{1}{\pi} \int_0^{2\pi} t \cos(kt)\mathrm{d}t \tag{7.31}$$

$$b_k^* = \frac{1}{\pi} \int_0^{2\pi} \gamma((L/2\pi)t) \sin(kt)\mathrm{d}t + \frac{1}{\pi} \int_0^{2\pi} t \sin(kt)\mathrm{d}t$$

By computing the second integrals of each coefficient, we obtain a simpler form as

$$a_0^* = 2\pi + \frac{1}{\pi} \int_0^{2\pi} \gamma((L/2\pi)t) dt$$

$$a_k^* = \frac{1}{\pi} \int_0^{2\pi} \gamma((L/2\pi)t) \cos(kt) dt \qquad (7.32)$$

$$b_k^* = -\frac{2}{k} + \frac{1}{\pi} \int_0^{2\pi} \gamma((L/2\pi)t) \sin(kt) dt$$

In an image, we measure distances, thus it is better to express these equations in arc-length form. For that, we know that $s = (L/2\pi)t$. Thus,

$$dt = \frac{2\pi}{L} ds \qquad (7.33)$$

Accordingly, the coefficients in Equation 7.32 can be rewritten as

$$a_0^* = 2\pi + \frac{2}{L} \int_0^L \gamma(s) ds$$

$$a_k^* = \frac{2}{L} \int_0^L \gamma(s) \cos\left(\frac{2\pi k}{L} s\right) ds \qquad (7.34)$$

$$b_k^* = -\frac{2}{k} + \frac{2}{L} \int_0^L \gamma(s) \sin\left(\frac{2\pi k}{L} s\right) ds$$

In a similar way to Equation 7.26, the Fourier descriptors can be computed by approximating the integral as a summation of rectangular areas. This is illustrated in Figure 7.13. Here, the discrete approximation is formed by rectangles of length τ_i and height γ_i. Thus,

$$a_0^* = 2\pi + \frac{2}{L} \sum_{i=1}^m \gamma_i \tau_i$$

$$a_k^* = \frac{2}{L} \sum_{i=1}^m \gamma_i \tau_i \cos\left(\frac{2\pi k}{L} s_i\right) \qquad (7.35)$$

$$b_k^* = -\frac{2}{k} + \frac{2}{L} \sum_{i=1}^m \gamma_i \tau_i \sin\left(\frac{2\pi k}{L} s_i\right)$$

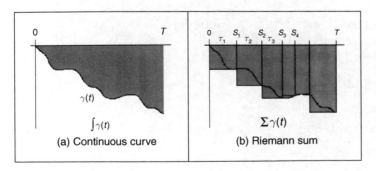

Figure 7.13 Integral approximations

where s_i is the arc-length at the ith point. Note that

$$s_i = \sum_{r=1}^{i} \tau_r \tag{7.36}$$

It is important to observe that although the definitions in Equation 7.35 only use the discrete values of $\gamma(t)$, they obtain a Fourier expansion of $\gamma^*(t)$. In the original formulation (Zahn and Roskies, 1972), an alternative form of the summations is obtained by rewriting the coefficients in terms of the increments of the angular function. In this case, the integrals in Equation 7.34 are evaluated for each interval. Thus, the coefficients are represented as a summation of integrals of constant values as

$$a_0^* = 2\pi + \frac{2}{L} \sum_{i=1}^{m} \int_{s_{i-1}}^{s_i} \gamma_i ds$$

$$a_k^* = \frac{2}{L} \sum_{i=1}^{m} \int_{s_{i-1}}^{s_i} \gamma_i \cos\left(\frac{2\pi k}{L}s\right) ds \tag{7.37}$$

$$b_k^* = -\frac{2}{k} + \frac{2}{L} \sum_{i=1}^{m} \int_{s_{i-1}}^{s_i} \gamma_i \sin\left(\frac{2\pi k}{L}s\right) ds$$

By evaluating the integral we obtain

$$a_0^* = 2\pi + \frac{2}{L} \sum_{i=1}^{m} \gamma_i (s_i - s_{i-1})$$

$$a_k^* = \frac{1}{\pi k} \sum_{i=1}^{m} \gamma_i \left(\sin\left(\frac{2\pi k}{L}s_i\right) - \sin\left(\frac{2\pi k}{L}s_{i-1}\right) \right) \tag{7.38}$$

$$b_k^* = -\frac{2}{k} + \frac{1}{\pi k} \sum_{i=1}^{m} \gamma_i \left(\cos\left(\frac{2\pi k}{L}s_i\right) - \cos\left(\frac{2\pi k}{L}s_{i-1}\right) \right)$$

A further simplification can be obtained by considering that Equation 7.28 can be expressed in discrete form as

$$\gamma_i = \sum_{r=1}^{i} \kappa_r \tau_r - \kappa_0 \tag{7.39}$$

where κ_r is the curvature (i.e. the difference of the angular function) at the rth point. Thus,

$$a_0^* = -2\pi - \frac{2}{L} \sum_{i=1}^{m} \kappa_i s_{i-1}$$

$$a_k^* = -\frac{1}{\pi k} \sum_{i=1}^{m} \kappa_i \tau_i \sin\left(\frac{2\pi k}{L}s_{i-1}\right) \tag{7.40}$$

$$b_k^* = -\frac{2}{k} - \frac{1}{\pi k} \sum_{i=1}^{m} \kappa_i \tau_i \cos\left(\frac{2\pi k}{L}s_{i-1}\right) + \frac{1}{\pi k} \sum_{i=1}^{m} \kappa_i \tau_i$$

Since

$$\sum_{i=1}^{m} \kappa_i \tau_i = 2\pi \tag{7.41}$$

thus,

$$a_0^* = -2\pi - \frac{2}{L} \sum_{i=1}^{m} \kappa_i s_{i-1}$$

$$a_k^* = -\frac{1}{\pi k} \sum_{i=1}^{m} \kappa_i \tau_i \sin\left(\frac{2\pi k}{L} s_{i-1}\right) \tag{7.42}$$

$$b_k^* = -\frac{1}{\pi k} \sum_{i=1}^{m} \kappa_i \tau_i \cos\left(\frac{2\pi k}{L} s_{i-1}\right)$$

These equations were originally presented in Zahn and Roskies (1972) and are algebraically equivalent to Equation 7.35. However, they express the Fourier coefficients in terms of increments in the angular function rather than in terms of the cumulative angular function. In practice, both implementations (Equations 7.35 and 7.40) produce equivalent Fourier descriptors.

It is important to notice that the parameterization in Equation 7.21 does not depend on the position of the pixels, but only on the change in angular information. That is, shapes in different position and with different scale will be represented by the same curve $\gamma^*(t)$. Thus, the Fourier descriptors obtained are *scale* and *translation* invariant. *Rotation*-invariant descriptors can be obtained by considering the shift-invariant property of the coefficients' amplitude. Rotating a curve in an image produces a shift in the angular function. This is because the rotation changes the starting point in the curve description. Thus, according to Section 7.2.3.2, the values

$$|c_k^*| = \sqrt{(a_k^*)^2 + (b_k^*)^2} \tag{7.43}$$

provide a rotation-, scale- and translation-invariant description. The function `AngFourier Descrp` in Code 7.2 computes the Fourier descriptors in this equation by using the definitions in Equation 7.35. This code uses the angular functions in Code 7.1.

```
%Fourier descriptors based on the Angular function
  function AngFuncDescrp(curve,n,scale)
                        %n=number coefficients
                        %if n=0 then n=m/2
                        %Scale amplitude output

%Angular functions
  AngFuncDescrp(curve);

%Fourier Descriptors
  if(n==0) n=floor(m/2); end;    %number of coefficients

  a=zeros(1,n); b=zeros(1,n);    %Fourier coefficients

for k=1:n
    a(k)=a(k)+G(1)*(S(1))*cos(2*pi*k*S(1)/L);
    b(k)=b(k)+G(1)*(S(1))*sin(2*pi*k*S(1)/L);
```

```
    for i=2:m
        a(k)=a(k)+G(i)*(S(i)-S(i-1))*cos(2*pi*k*S(i)/L);
        b(k)=b(k)+G(i)*(S(i)-S(i-1))*sin(2*pi*k*S(i)/L);
    end
    a(k)=a(k)*(2/L);
    b(k)=b(k)*(2/L)-2/k;
end

%Graphs
  subplot(3,3,7);
  bar(a);
  axis([0,n,-scale,scale]);

  subplot(3,3,8);
  bar(b);
  axis([0,n,-scale,scale]);

%Rotation invariant Fourier descriptors
    CA=zeros(1,n);
    for k=1:n
        CA(k)=sqrt(a(k)^2+b(k)^2);
    end

%Graph of the angular coefficients
  subplot(3,3,9);
  bar(CA);
  axis([0,n,-scale,scale]);
```

Code 7.2   Angular Fourier descriptors

Figure 7.14 shows three examples of the results obtained using the Code 7.2. In each example, we show the curve, the angular function, the cumulative normalized angular function and the Fourier descriptors. The curves in Figure 7.14(a) and (e) represent the same object (the contour of an F-14 fighter), but the curve in Figure 7.14(e) was scaled and rotated. We can see that the angular function changes significantly, while the normalized function is very similar but with a remarkable shift due to the rotation. The Fourier descriptors shown in Figure 7.14(d) and (h) are quite similar since they characterize the same object. We can see a clear difference between the normalized angular function for the object presented in Figure 7.14(i) (the contour of a different plane, a B1 bomber). These examples show that Fourier coefficients are indeed invariant to scale and rotation, and that they can be used to characterize different objects.

7.2.3.6   Elliptic Fourier descriptors

The cumulative angular function transforms the two-dimensional (2D) description of a curve into a one-dimensional periodic function suitable for Fourier analysis. In contrast, *elliptic Fourier descriptors* maintain the description of the curve in a 2D space (Granlund, 1972). This is achieved by considering that the image space defines the complex plane. That is, each pixel is

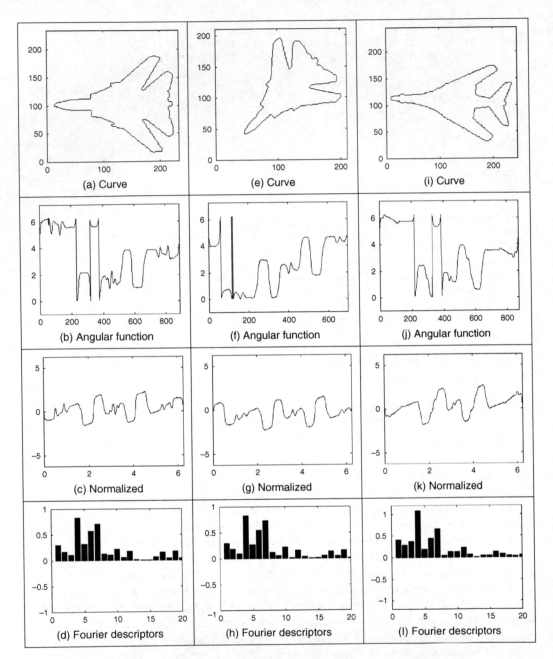

Figure 7.14 Example of angular Fourier descriptors

represented by a complex number. The first coordinate represents the real part, while the second coordinate represents the imaginary part. Thus, a curve is defined as

$$c(t) = x(t) + jy(t) \qquad (7.44)$$

Here, we will consider that the parameter t is given by the arc-length parameterization. Figure 7.15 shows an example of the complex representation of a curve. This example illustrates two periods of each component of the curve. In general, $T = 2\pi$, thus the fundamental frequency is $\omega = 1$. It is important to notice that this representation can be used to describe open curves. In this case, the curve is traced twice in opposite directions. In fact, this representation is very general and can be extended to obtain the elliptic Fourier description of irregular curves (i.e. those without derivative information) (Montiel et al., 1996, 1997).

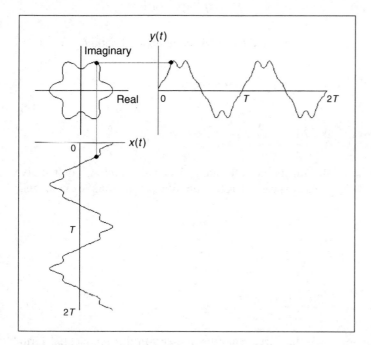

Figure 7.15 Example of complex curve representation

To obtain the elliptic Fourier descriptors of a curve, we need to obtain the Fourier expansion of the curve in Equation 7.44. The Fourier expansion can be performed by using the complex or trigonometric form. In the original work, in Granlund (1972), the expansion is expressed in the complex form. However, other works have used the trigonometric representation (Kuhl and Giardina, 1982). Here, we will pass from the complex form to the trigonometric representation. The trigonometric representation is more intuitive and easier to implement.

According to Equation 7.5, the elliptic coefficients are defined by

$$c_k = c_{xk} + jc_{yk} \tag{7.45}$$

where

$$c_{xk} = \frac{1}{T}\int_0^T x(t)e^{-jk\omega t} \quad \text{and} \quad c_{yk} = \frac{1}{T}\int_0^T y(t)e^{-jk\omega t} \tag{7.46}$$

By following Equation 7.12, we notice that each term in this expression can be defined by a pair of coefficients. That is,

$$c_{xk} = \frac{a_{xk} - jb_{xk}}{2} \quad c_{yk} = \frac{a_{yk} - jb_{yk}}{2}$$

$$c_{x-k} = \frac{a_{xk} - jb_{xk}}{2} \quad c_{y-k} = \frac{a_{yk} - jb_{yk}}{2} \tag{7.47}$$

Based on Equation 7.13, the trigonometric coefficients are defined as

$$a_{xk} = \frac{2}{T}\int_0^T x(t)\cos(k\omega t)dt \quad \text{and} \quad b_{xk} = \frac{2}{T}\int_0^T x(t)\sin(k\omega t)dt$$

$$a_{yk} = \frac{2}{T}\int_0^T y(t)\cos(k\omega t)dt \quad \text{and} \quad b_{yk} = \frac{2}{T}\int_0^T y(t)\sin(k\omega t)dt$$

(7.48)

which, according to Equation 7.27, can be computed by the discrete approximation given by

$$a_{xk} = \frac{2}{m}\sum_{i=1}^m x_i\cos(k\omega i\tau) \quad \text{and} \quad b_{xk} = \frac{2}{m}\sum_{i=1}^m x_i\sin(k\omega i\tau)$$

$$a_{yk} = \frac{2}{m}\sum_{i=1}^m y_i\cos(k\omega i\tau) \quad \text{and} \quad b_{yk} = \frac{2}{m}\sum_{i=1}^m y_i\sin(k\omega i\tau)$$

(7.49)

where x_i and y_i define the value of the functions $x(t)$ and $y(t)$ at the sampling point i. By considering Equations 7.45 and 7.47, we can express c_k as the sum of a pair of complex numbers. That is,

$$c_k = A_k - jB_k \quad \text{and} \quad c_{-k} = A_k + jB_k$$

(7.50)

where

$$A_k = \frac{a_{xk} + ja_{yk}}{2} \quad \text{and} \quad B_k = \frac{b_{xk} + jb_{yk}}{2}$$

(7.51)

Based on the definition in Equation 7.45, the curve can be expressed in the exponential form given in Equation 7.6 as

$$c(t) = c_0 + \sum_{k=1}^\infty (A_k - jB_k)e^{jk\omega t} + \sum_{k=-\infty}^{-1} (A_k + jB_k)e^{jk\omega t}$$

(7.52)

Alternatively, according to Equation 7.11 the curve can be expressed in trigonometric form as

$$c(t) = \frac{a_{x0}}{2} + \sum_{k=1}^\infty (a_{xk}\cos(k\omega t) + b_{xk}\sin(k\omega t))$$

$$+ j\left(\frac{a_{y0}}{2} + \sum_{k=1}^\infty (a_{yk}\cos(k\omega t) + b_{yk}\sin(k\omega t))\right)$$

(7.53)

In general, this equation is expressed in matrix form as

$$\begin{bmatrix} x(t) \\ y(t) \end{bmatrix} = \frac{1}{2}\begin{bmatrix} a_{x0} \\ a_{y0} \end{bmatrix} + \sum_{k=1}^\infty \begin{bmatrix} a_{xk} & b_{xk} \\ a_{yk} & b_{yk} \end{bmatrix}\begin{bmatrix} \cos(k\omega t) \\ \sin(k\omega t) \end{bmatrix}$$

(7.54)

Each term in this equation has an interesting geometric interpretation as an elliptic phasor (a rotating vector). That is, for a fixed value of k, the trigonometric summation defines the locus of an ellipse in the complex plane. We can imagine that as we change the parameter t the point traces ellipses moving at a speed proportional to the harmonic number k. This number indicates how many cycles (i.e. turns) give the point in the time interval from zero to T. Figure 7.16(a) illustrates this concept. Here, a point in the curve is given as the summation of three vectors that define three terms in Equation 7.54. As the parameter t changes, each vector defines an elliptic curve. In this interpretation, the values of $a_{x0}/2$ and $a_{y0}/2$ define the start point of the first vector (i.e. the location of the curve). The major axes of each ellipse are given by the values of $|A_k|$ and $|B_k|$. The definition of the ellipse locus for a frequency is determined by the coefficients, as shown in Figure 7.16(b).

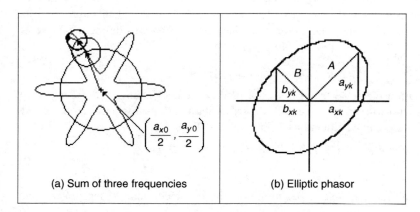

(a) Sum of three frequencies   (b) Elliptic phasor

Figure 7.16 Example of a contour defined by elliptic Fourier descriptors

7.2.3.7 *Invariance*

As in the case of angular Fourier descriptors, elliptic Fourier descriptors can be defined such that they remain invariant to geometric transformations. To show these definitions we must first study how geometric changes in a shape modify the form of the Fourier coefficients. Transformations can be formulated by using both the exponential or trigonometric form. We will consider changes in translation, rotation and scale using the trigonometric definition in Equation 7.54.

Let us denote $c'(t) = x'(t) + jy'(t)$ as the transformed contour. This contour is defined as

$$\begin{bmatrix} x'(t) \\ y'(t) \end{bmatrix} = \frac{1}{2}\begin{bmatrix} a'_{x0} \\ a'_{y0} \end{bmatrix} + \sum_{k=1}^{\infty}\begin{bmatrix} a'_{xk} & b'_{xk} \\ a'_{yk} & b'_{yk} \end{bmatrix}\begin{bmatrix} \cos(k\omega t) \\ \sin(k\omega t) \end{bmatrix} \tag{7.55}$$

If the contour is translated by t_x and t_y along the real and the imaginary axes, respectively, we have:

$$\begin{bmatrix} x'(t) \\ y'(t) \end{bmatrix} = \frac{1}{2}\begin{bmatrix} a_{x0} \\ a_{y0} \end{bmatrix} + \sum_{k=1}^{\infty}\begin{bmatrix} a_{xk} & b_{xk} \\ a_{yk} & b_{yk} \end{bmatrix}\begin{bmatrix} \cos(k\omega t) \\ \sin(k\omega t) \end{bmatrix} + \begin{bmatrix} t_x \\ t_y \end{bmatrix} \tag{7.56}$$

That is,

$$\begin{bmatrix} x'(t) \\ y'(t) \end{bmatrix} = \frac{1}{2}\begin{bmatrix} a_{x0} + 2t_x \\ a_{y0} + 2t_y \end{bmatrix} + \sum_{k=1}^{\infty}\begin{bmatrix} a_{xk} & b_{xk} \\ a_{yk} & b_{yk} \end{bmatrix}\begin{bmatrix} \cos(k\omega t) \\ \sin(k\omega t) \end{bmatrix} \tag{7.57}$$

Thus, by comparing Equations 7.55 and 7.57, the relationship between the coefficients of the transformed and original curves is given by

$$a'_{xk} = a_{xk} \quad b'_{xk} = b_{xk} \quad a'_{yk} = a_{yk} \quad b'_{yk} = b_{yk} \quad \text{for} \quad k \neq 0$$

$$a'_{x0} = a_{x0} + 2t_x \quad a'_{y0} = a_{y0} + 2t_y \tag{7.58}$$

Accordingly, all the coefficients remain invariant under translation except for a_{x0} and a_{y0}. This result can be intuitively derived by considering that these two coefficients represent the position of the centre of gravity of the contour of the shape and translation changes only the position of the curve.

The change in scale of a contour $c(t)$ can be modelled as the dilation from its centre of gravity. That is, we need to translate the curve to the origin, scale it and then return it to its original location. If s represents the scale factor, then these transformations define the curve as

$$\begin{bmatrix} x'(t) \\ y'(t) \end{bmatrix} = \frac{1}{2} \begin{bmatrix} a_{x0} \\ a_{y0} \end{bmatrix} + s \sum_{k=1}^{\infty} \begin{bmatrix} a_{xk} & b_{xk} \\ a_{yk} & b_{yk} \end{bmatrix} \begin{bmatrix} \cos(k\omega t) \\ \sin(k\omega t) \end{bmatrix} \tag{7.59}$$

Notice that in this equation the scale factor does not modify the coefficients a_{x0} and a_{y0} since the curve is expanded with respect to its centre. To define the relationships between the curve and its scaled version, we compare Equations 7.55 and 7.59. Thus,

$$a'_{xk} = sa_{xk} \quad b'_{xk} = sb_{xk} \quad a'_{yk} = sa_{yk} \quad b'_{yk} = sb_{yk} \quad \text{for} \quad k \neq 0$$

$$a'_{x0} = a_{x0} \quad a'_{y0} = a_{y0} \tag{7.60}$$

That is, under dilation, all the coefficients are multiplied by the scale factor except for a_{x0} and a_{y0}, which remain invariant.

Rotation can be defined in a similar way to Equation 7.59. If ρ represents the rotation angle, then we have:

$$\begin{bmatrix} x'(t) \\ y'(t) \end{bmatrix} = \frac{1}{2} \begin{bmatrix} a_{x0} \\ a_{y0} \end{bmatrix} + \begin{bmatrix} \cos(\rho) & \sin(\rho) \\ -\sin(\rho) & \cos(\rho) \end{bmatrix} \sum_{k=1}^{\infty} \begin{bmatrix} a_{xk} & b_{xk} \\ a_{yk} & b_{yk} \end{bmatrix} \begin{bmatrix} \cos(k\omega t) \\ \sin(k\omega t) \end{bmatrix} \tag{7.61}$$

This equation can be obtained by translating the curve to the origin, rotating it and then returning it to its original location. By comparing Equations 7.55 and 7.61, we have:

$$a'_{xk} = a_{xk} \cos(\rho) + a_{yk} \sin(\rho) \qquad b'_{xk} = b_{xk} \cos(\rho) + b_{yk} \sin(\rho)$$

$$a'_{yk} = -a_{xk} \sin(\rho) + a_{yk} \cos(\rho) \quad b'_{yk} = -b_{xk} \sin(\rho) + b_{yk} \cos(\rho) \tag{7.62}$$

$$a'_{x0} = a_{x0} \quad a'_{y0} = a_{y0}$$

That is, under translation, the coefficients are defined by a linear combination dependent on the rotation angle, except for a_{x0} and a_{y0}, which remain invariant. It is important to notice that rotation relationships are also applied for a change in the starting point of the curve.

Equations 7.58, 7.60 and 7.62 define how the elliptic Fourier coefficients change when the curve is translated, scaled or rotated, respectively. We can combine these results to define the changes when the curve undergoes the three transformations. In this case, transformations are applied in succession. Thus,

$$a'_{xk} = s(a_{xk} \cos(\rho) + a_{yk} \sin(\rho)) \qquad b'_{xk} = s(b_{xk} \cos(\rho) + b_{yk} \sin(\rho))$$

$$a'_{yk} = s(-a_{xk} \sin(\rho) + a_{yk} \cos(\rho)) \quad b'_{yk} = s(-b_{xk} \sin(\rho) + b_{yk} \cos(\rho)) \tag{7.63}$$

$$a'_{x0} = a_{x0} + 2t_x \quad a'_{y0} = a_{y0} + 2t_y$$

Based on this result we can define alternative invariant descriptors. To achieve invariance to translation, when defining the descriptors the coefficient for $k = 0$ is not used. In Granlund (1972), invariant descriptors are defined based on the complex form of the coefficients. Alternatively, invariant descriptors can be simply defined as

$$\frac{|A_k|}{|A_1|} + \frac{|B_k|}{|B_1|} \tag{7.64}$$

The advantage of these descriptors with respect to the definition in Granlund (1972) is that they do not involve negative frequencies and that we avoid multiplication by higher frequencies that are more prone to noise. By considering the definitions in Equations 7.51 and 7.63 we can prove that

$$\frac{|A'_k|}{|A'_1|} = \frac{\sqrt{a_{xk}^2 + a_{yk}^2}}{\sqrt{a_{x1}^2 + a_{y1}^2}} \quad \text{and} \quad \frac{|B'_k|}{|B'_1|} = \frac{\sqrt{b_{xk}^2 + b_{yk}^2}}{\sqrt{b_{x1}^2 + b_{y1}^2}} \tag{7.65}$$

These equations contain neither the *scale* factor, s, nor the *rotation*, ρ. Thus, they are *invariant*. Notice that if the square roots are removed, invariance properties are still maintained. However, high-order frequencies can have undesirable effects.

The function `EllipticDescrp` in Code 7.3 computes the elliptic Fourier descriptors of a curve. The code implements Equations 7.49 and 7.64 in a straightforward way. By default,

```
%Elliptic Fourier Descriptors
 function EllipticDescrp(curve,n,scale)
            %n=num coefficients
            %if n=0 then n=m/2
            %Scale amplitud output
%Function from image
 X=curve(1,:);
 Y=curve(2,:);
 m=size(X,2);

%Graph of the curve
          subplot(3,3,1);
          plot(X,Y);
          mx=max(max(X),max(Y))+10;
          axis([0,mx,0,mx]); %Axis of the graph pf the curve
          axis square;         %Aspect ratio

%Graph of X
 p=0:2*pi/m:2*pi-pi/m;          %Parameter
          subplot(3,3,2);
          plot(p,X);
          axis([0,2*pi,0,mx]); %Axis of the graph pf the curve

%Graph of Y
          subplot(3,3,3);
          plot(p,Y);
          axis([0,2*pi,0,mx]); %Axis of the graph pf the curve
```

```
%Elliptic Fourier Descriptors
 if(n==0) n=floor(m/2); end; %number of coefficients

%Fourier Coefficients
 ax=zeros(1,n); bx=zeros(1,n);
 ay=zeros(1,n); by=zeros(1,n);

t=2*pi/m;
   for k=1:n
     for i=1:m
       ax(k)=ax(k)+X(i)*cos(k*t*(i-1));
       bx(k)=bx(k)+X(i)*sin(k*t*(i-1));
       ay(k)=ay(k)+Y(i)*cos(k*t*(i-1));
       by(k)=by(k)+Y(i)*sin(k*t*(i-1);
     end
   ax(k)=ax(k)*(2/m);
   bx(k)=bx(k)*(2/m);
   ay(k)=ay(k)*(2/m);
   by(k)=by(k)*(2/m);
end

%Graph coefficient ax
 subplot(3,3,4);
 bar(ax);
 axis([0,n,-scale,scale]);

%Graph coefficient ay
 subplot(3,3,5);
 bar(ay);
 axis([0,n,-scale,scale]);

%Graph coefficient bx
 subplot(3,3,6);
 bar(bx);
 axis([0,n,-scale,scale]);

%Graph coefficient by
 subplot(3,3,7);
 bar(by);
 axis([0,n,-scale,scale]);

%Invariant
 CE=zeros(1,n);
 for k=1:n
   CE(k)=sqrt((ax(k)^2+ay(k)^2)/(ax(1)^2+ay(1)^2))
          +sqrt((bx(k)^2+by(k)^2)/(bx(1)^2+by(1)^2));
 end

%Graph of Elliptic descriptors
 subplot(3,3,8);
 bar(CE);
 axis([0,n,0,2.2]);
```

Code 7.3 Elliptic Fourier descriptors

the number of coefficients is half of the number of points that define the curve. However, the number of coefficients can be specified by the parameter n. The number of coefficients used defines the level of detail of the characterization. To illustrate this idea, we can consider the different curves that are obtained by using a different number of coefficients. Figure 7.17 shows an example of the reconstruction of a contour. In Figure 7.17(a) we can observe that the first coefficient represents an ellipse. When the second coefficient is considered (Figure 7.17b), the ellipse changes into a triangular shape. When adding more coefficients the contour is refined until the curve represents an accurate approximation of the original contour. In this example, the contour is represented by 100 points. Thus, the maximum number of coefficients is 50.

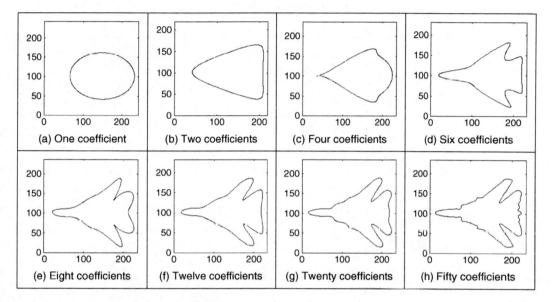

Figure 7.17 Fourier approximation

Figure 7.18 shows three examples of the results obtained using Code 7.3. Each example shows the original curve, the x and y coordinate functions and the Fourier descriptors defined in Equation 7.64. The maximum in Equation 7.64 is equal to two and is obtained when $k = 1$. In the figure we have scaled the Fourier descriptors to show the differences between higher order coefficients. In this example, we can see that the Fourier descriptors for the curves in Figure 7.18(a) and (e) (F-14 fighter) are very similar. Small differences can be explained by discretization errors. However, the coefficients remain the same after changing the location, orientation and scale. The descriptors of the curve in Figure 7.18(i) (B1 bomber) are clearly different, showing that elliptic Fourier descriptors truly characterize the shape of an object.

Fourier descriptors are one of the most popular boundary descriptions. As such, they have attracted considerable attention and there are many further aspects. We can use the descriptions for shape recognition (Aguado et al., 1998). It is important to mention that some work has suggested that there is some ambiguity in the Fourier characterization. Thus, an alternative set of descriptors has been designed specifically to reduce ambiguities (Crimmins, 1982). However, it is well known that Fourier expansions are unique. Thus, Fourier characterization should uniquely represent a curve. In addition, the mathematical opacity of the technique in Crimmins (1982) does not lend itself to tutorial type presentation. There has not been much

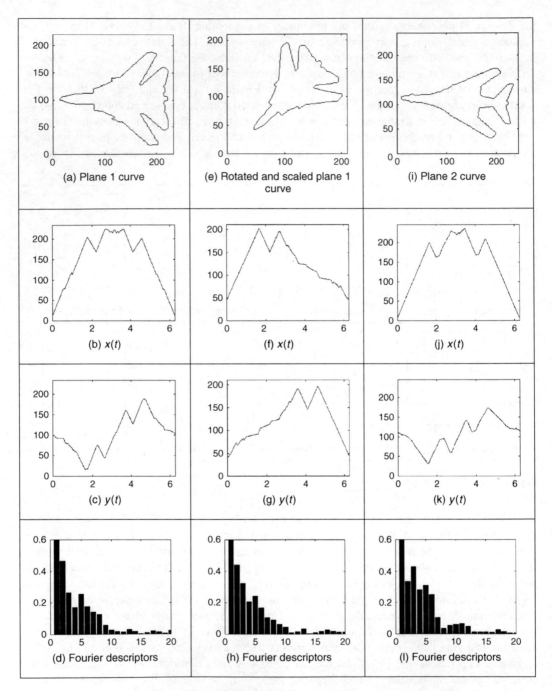

Figure 7.18 Example of elliptic Fourier descriptors

study on alternative decompositions to Fourier, although Walsh functions have been suggested for shape representation (Searle, 1970) and recently wavelets have been used (Kashi et al., 1996) (although these are not an orthonormal basis function). Three-dimensional Fourier descriptors were introduced for analysis of simple shapes (Staib and Duncan, 1992) and have been found to give good performance in application (Undrill et al., 1997). Fourier descriptors have also been used to model shapes in computer graphics (Aguado et al., 1999). Fourier descriptors cannot be used for occluded or mixed shapes, relying on extraction techniques with known indifference to occlusion (e.g. the Hough transform). However, there have been approaches aimed at classifying partial shapes using Fourier descriptors (Lin and Chellappa, 1987).

7.3 Region descriptors

So far, we have concentrated on descriptions of the perimeter, or boundary. The natural counterpart is to describe the *region*, or the *area*, by *regional shape descriptors*. Here, there are two main contenders that differ in focus: basic regional descriptors characterize the geometric properties of the region, whereas moments concentrate on the density of the region. First, though, we shall look at the simpler descriptors.

7.3.1 Basic region descriptors

A region can be described by considering scalar measures based on its geometric properties. The simplest property is given by its size or area. In general, the area of a region in the plane is defined as

$$A(S) = \int_x \int_y I(x, y) \mathrm{d}y \mathrm{d}x \tag{7.66}$$

where $I(x, y) = 1$ if the pixel is within a shape, $(x, y) \in S$, and 0 otherwise. In practice, integrals are approximated by summations. That is,

$$A(S) = \sum_x \sum_y I(x, y) \Delta A \tag{7.67}$$

where ΔA is the area of one pixel. Thus, if $\Delta A = 1$, then the area is measured in pixels. Area changes with changes in scale. However, it is invariant to image rotation. Small errors in the computation of the area will appear when applying a rotation transformation owing to discretization of the image.

Another simple property is defined by the perimeter of the region. If $x(t)$ and $y(t)$ denote the parametric coordinates of a curve enclosing a region S, then the perimeter of the region is defined as

$$P(S) = \int_t \sqrt{x^2(t) + y^2(t)} \mathrm{d}t \tag{7.68}$$

This equation corresponds to the sums all the infinitesimal arcs that define the curve. In the discrete case, $x(t)$ and $y(t)$ are defined by a set of pixels in the image. Thus, Equation 7.68 is approximated by

$$P(S) = \sum_i \sqrt{(x_i - x_{i-1})^2 + (y_i - y_{i-1})^2} \tag{7.69}$$

where x_i and y_i represent the coordinates of the ith pixel forming the curve. Since pixels are organized in a square grid, the terms in the summation can only take two values. When the pixels (x_i, y_i) and (x_{i-1}, y_{i-1}) are four-neighbours (as shown in Figure 7.1a), the summation term is unity. Otherwise, the summation term is equal to $\sqrt{2}$. Notice that the discrete approximation in Equation 7.69 produces small errors in the measured perimeter. As such, it is unlikely that an exact value of $2\pi r$ will be achieved for the perimeter of a circular region of radius r.

Based on the perimeter and area it is possible to characterize the compactness of a region. *Compactness* is an oft-expressed measure of shape given by the ratio of perimeter to area. That is,

$$C(S) = \frac{4\pi A(S)}{P^2(S)} \tag{7.70}$$

To show the meaning of this equation, we can rewrite it as

$$C(S) = \frac{A(S)}{P^2(S)/4\pi} \tag{7.71}$$

Here, the denominator represents the area of a circle whose perimeter is $P(S)$. Thus, compactness measures the ratio between the area of the shape and the circle that can be traced with the same perimeter. That is, compactness measures the efficiency with which a boundary encloses area. In mathematics, it is known as the *isoperimetric quotient*, which smacks rather of grandiloquency. For a perfectly circular region (Figure 7.19a) we have $C(circle) = 1$, which represents the maximum compactness value: a circle is the most compact shape. Figure 7.19(b) and (c) show two examples in which compactness is reduced. If we take the perimeter of these regions and draw a circle with the same perimeter, we can observe that the circle contains more area. This means that the shapes are not compact. A shape becomes more compact if we move region pixels far away from the centre of gravity of the shape to fill empty spaces closer to the centre of gravity. For a perfectly square region, $C(square) = \pi/4$. Note that for neither a perfect square nor a perfect circle does the measure include size (the width and radius, respectively). In this way, compactness is a measure of *shape* only. Note that compactness alone is not a good discriminator of a region; low values of C are associated with convoluted regions such as the one in Figure 7.19(b) and also with simple but highly elongated shapes. This ambiguity can be resolved by employing additional shape measures.

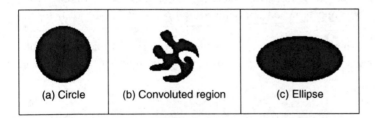

| (a) Circle | (b) Convoluted region | (c) Ellipse |

Figure 7.19   Examples of compactness

Another measure that can be used to characterize regions is *dispersion*. Dispersion (irregularity) has been measured as the ratio of major chord length to area (Chen et al., 1995). A simple version of this measure can be defined as *irregularity*:

$$I(S) = \frac{\pi \max\left((x_i - \bar{x})^2 + (y_i - \bar{y})^2\right)}{A(S)} \tag{7.72}$$

where $(\bar{x}, \bar{y})$ represent the coordinates of the centre of mass of the region. Notice that the numerator defines the area of the maximum circle enclosing the region. Thus, this measure describes the density of the region. An alternative measure of dispersion can also be expressed as the ratio of the maximum to the minimum radius. That is an alternative form of the irregularity

$$IR(S) = \frac{\max\left(\sqrt{(x_i - \bar{x})^2 + (y_i - \bar{y})^2}\right)}{\min\left(\sqrt{(x_i - \bar{x})^2 + (y_i - \bar{y})^2}\right)} \tag{7.73}$$

This measure defines the ratio between the radius of the maximum circle enclosing the region and the maximum circle that can be contained in the region. Thus, the measure will increase as the region spreads. In this way, the irregularity of a circle is unity, $IR(circle) = 1$; the irregularity of a square is $IR(square) = \sqrt{2}$, which is larger. As such the measure increases for irregular shapes, whereas the compactness measure decreases. Again, for perfect shapes the measure is irrespective of size and is a measure of shape only. One *disadvantage* of the irregularity measures is that they are insensitive to slight *discontinuity* in the shape, such as a thin crack in a disk. However, these discontinuities will be registered by the earlier measures of compactness since the perimeter will increase disproportionately with the area. This property might be desired and so irregularity is to be preferred when this property is required. In fact, the perimeter measures will vary with rotation owing to the nature of discrete images and are more likely to be affected by noise than the measures of area (since the area measures have inherent averaging properties). Since the irregularity is a ratio of distance measures and compactness is a ratio of area to distance, intuitively it would appear that irregularity will vary less with noise and rotation. Such factors should be explored in application, to check that desired properties have indeed been achieved.

Code 7.4 shows the implementation for the region descriptors. The code is a straightforward implementation of Equations 7.67, 7.69, 7.70, 7.72 and 7.73. A comparison of these measures for

```
%Region descriptors (compactness)

function RegionDescrp(inputimage)

%Image size
[rows,columns]=size(inputimage);

%area
A=0;
   for x=1:columns
      for y=1:rows
         if inputimage(y,x)==0 A=A+1; end
      end
   end

%Obtain Contour
C=Contour(inputimage);

%Perimeter & mean
X=C(1,:);    Y=C(2,:);    m=size(X,2);
```

```
mx=X(1); my=Y(1);
P=sqrt((X(1)-X(m))^2+(Y(1)-Y(m))^2);
for i=2:m
   P=P+sqrt((X(i)-X(i-1))^2+(Y(i)-Y(i-1))^2);
   mx=mx+X(i); my=my+Y(i);
end

mx=mx/m; my=my/m;

%Compactness
Cp=4*pi*A/P^2;

%Dispersion
max=0; min=99999;
   for i=1:m
      d=((X(i)-mx)^2+(Y(i)-my)^2);
      if (d>max) max=d; end
      if (d<min) min=d; end
   end
   I=pi*max/A;
   IR=sqrt(max/min);

%Results
disp('perimeter=');     disp(P);
disp('area=');          disp(A);
disp('Compactness=');   disp(Cp);
disp('Dispersion=');    disp(I);
disp('DispersionR=');   disp(IR);
```

Code 7.4   Evaluating basic region descriptors

the three regions shown in Figure 7.19 is shown in Figure 7.20. Clearly, for the circle the compactness and dispersion measures are close to unity. For the ellipse the compactness decreases while the dispersion increases. The convoluted region has the lowest compactness measure and the highest dispersion values. Clearly, these measurements can be used to characterize, and hence discriminate between areas of differing shape.

$A(S) = 4917$	$A(S) = 2316$	$A(S) = 6104$
$P(S) = 259.27$	$P(S) = 498.63$	$P(S) = 310.93$
$C(S) = 0.91$	$C(S) = 0.11$	$C(S) = 0.79$
$I(S) = 1.00$	$I(S) = 2.24$	$I(S) = 1.85$
$IR(S) = 1.03$	$IR(S) = 6.67$	$IR(S) = 1.91$
(a) Descriptors for the circle	(b) Descriptors for the convoluted region	(c) Descriptors for the ellipse

Figure 7.20   Basic region descriptors

Other measures, rather than focus on the geometric properties, characterize the structure of a region. This is the case of the *Poincarré measure* and the *Euler number*. The Poincarré measure concerns the number of holes within a region. Alternatively, the Euler number is the difference of the number of connected regions from the number of holes in them. There are many more potential measures for shape description in terms of structure and geometry. Recent interest has developed a measure (Rosin and Zunic, 2005) that can discriminate *rectilinear* regions, e.g. for discriminating buildings from within remotely sensed images. We could evaluate global or local *curvature* (convexity and concavity) as a further measure of geometry; we could investigate *proximity* and *disposition* as a further measure of structure. However, these do not have the advantages of a unified structure. We are simply suggesting measures with descriptive ability, but this ability is reduced by the correlation between different measures. We have already seen the link between the Poincarré measure and the Euler number; there is a natural link between circularity and irregularity. However, the region descriptors we have considered so far lack structure and are largely heuristic, although clearly they may have sufficient descriptive ability for some applications. As such, we shall now look at a *unified* basis for shape description which aims to reduce this correlation and provides a unified theoretical basis for region description, with some similarity to the advantages of the frequency selectivity in a Fourier transform description.

7.3.2 Moments

7.3.2.1 *Basic properties*

Moments describe a shape's *layout* (the arrangement of its pixels), a bit like combining area, compactness, irregularity and higher order descriptions together. Moments are a *global* description of a shape, accruing this same advantage as Fourier descriptors since there is selectivity, which is an in-built ability to discern, and filter, noise. Further, in image analysis, they are *statistical moments*, as opposed to *mechanical* ones, but the two are analogous. For example, the mechanical moment of inertia describes the rate of change in momentum; the statistical second order moment describes the rate of change in a shape's area. In this way, statistical moments can be considered as a global region description. Moments for image analysis were originally introduced in the 1960s (Hu, 1962) (an exciting time for computer vision researchers too!) and an excellent review is available (Prokop and Reeves, 1992).

Moments are often associated more with *statistical* pattern recognition than with *model-based* vision, since a major assumption is that there is an *unoccluded* view of the target shape. Target images are often derived by thresholding, usually one of the optimal forms that can require a single object in the field of view. More complex applications, including handling occlusion, could presuppose *feature extraction* by some means, with a model to in-fill for the missing parts. However, moments do provide a global description with invariance properties and with the advantages of a compact description aimed at avoiding the effects of noise. As such, they have proved popular and successful in many applications.

The *two-dimensional Cartesian moment* is associated with an order that starts from low (where the lowest is zero) up to higher orders. The moment of order p and q, m_{pq} of a function $I(x, y)$ is defined as

$$m_{pq} = \int_{-\infty}^{\infty} \int_{-\infty}^{\infty} x^p y^q I(x, y) \mathrm{d}x \mathrm{d}y \tag{7.74}$$

For discrete images, Equation 7.74 is usually approximated by

$$m_{pq} = \sum_x \sum_y x^p y^q I(x, y) \Delta A \tag{7.75}$$

where ΔA is again the area of a pixel. These descriptors have a *uniqueness* property in that it can be shown that if the function satisfies certain conditions, then moments of all orders exist. Also, and conversely, the set of descriptors uniquely determines the original function, in a manner similar to reconstruction via the inverse Fourier transform. However, these moments are descriptors, rather than a specification that can be used to reconstruct a shape. The *zero-order moment*, m_{00}, is

$$m_{00} = \sum_x \sum_y I(x, y)\Delta A \qquad (7.76)$$

which represents the total mass of a function. Notice that this equation is equal to Equation 7.67 when $I(x, y)$ takes values of zero and one. However, Equation 7.76 is more general since the function $I(x, y)$ can take a range of values. In the definition of moments, these values are generally related to density. The two *first order moments*, m_{01} and m_{10}, are given by

$$m_{10} = \sum_x \sum_y xI(x, y)\Delta A \qquad m_{01} = \sum_x \sum_y yI(x, y)\Delta A \qquad (7.77)$$

For binary images, these values are proportional to the shape's centre coordinates (the values merely require division by the shape's area). In general, the *centre of mass* $(\bar{x}, \bar{y})$ can be calculated from the ratio of the first order to the zero-order components as

$$\bar{x} = \frac{m_{10}}{m_{00}} \quad \bar{y} = \frac{m_{01}}{m_{00}} \qquad (7.78)$$

The first 10 x-axis moments of an ellipse are shown in Figure 7.21. The moments rise exponentially, so are plotted in logarithmic form. Evidently, the moments provide a set of descriptions of the shape: measures that can be collected together to differentiate between different shapes.

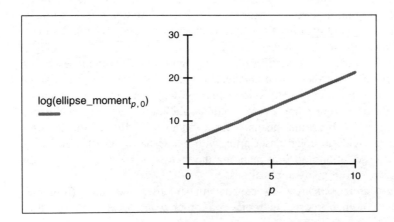

Figure 7.21 Horizontal axis ellipse moments

Should there be an intensity transformation that *scales* brightness by a particular factor, say α, such that a new image $I'(x, y)$ is a transformed version of the original one $I(x, y)$, given by

$$I'(x, y) = \alpha I(x, y) \qquad (7.79)$$

Then the transformed moment values m'_{pq} are related to those of the original shape m_{pq} by

$$m'_{pq} = \alpha m_{pq} \tag{7.80}$$

Should it be required to distinguish *mirror symmetry* (reflection of a shape about a chosen axis), then the rotation of a shape about the, say, the x-axis gives a new shape $I'(x, y)$, which is the reflection of the shape $I(x, y)$ given by

$$I'(x, y) = I(-x, y) \tag{7.81}$$

The transformed moment values can be given in terms of the original shape's moments as

$$m'_{pq} = (-1)^p m_{pq} \tag{7.82}$$

However, we are usually concerned with more basic invariants than mirror images, namely invariance to *position, size* and *rotation*. Given that we now have an estimate of a shape's centre (in fact, a reference point for that shape), the *centralized moments*, μ_{pq} which are invariant to *translation*, can be defined as

$$\mu_{pq} = \sum_x \sum_y (x - \bar{x})^p (y - \bar{y})^q I(x, y) \Delta A \tag{7.83}$$

Clearly, the zero-order *centralized* moment is again the shape's area. However, the first order centralized moment μ_{01} is given by

$$\mu_{01} = \sum_x \sum_y (y - \bar{y})^1 I(x, y) \Delta A$$

$$= \sum_x \sum_y y I(x, y) \Delta A - \sum_x \sum_y \bar{y} I(x, y) \Delta A \tag{7.84}$$

$$= m_{01} - \bar{y} \sum_x \sum_y I(x, y) \Delta A$$

From Equation 7.77, $m_{01} = \sum_x \sum_y y I(x, y) \Delta A$ and from Equation 7.78, $\bar{y} = m_{01}/m_{00}$, so

$$\mu_{01} = m_{01} - \frac{m_{01}}{m_{00}} m_{00}$$

$$= 0 \tag{7.85}$$

$$= \mu_{10}$$

Clearly, neither of the first order centralized moments has any description capability since they are both zero. Going to higher order, one of the second order moments, μ_{20}, is

$$\mu_{20} = \sum_x \sum_y (x - \bar{x})^2 I(x, y) \Delta A$$

$$= \sum_x \sum_y (x^2 - 2x\bar{x} + \bar{x}^2) I(x, y) \Delta A \tag{7.86}$$

$$= \sum_x \sum_y x^2 I(x, y) \Delta A - 2\bar{x} \sum_x \sum_y x I(x, y) \Delta A + \bar{x}^2 \sum_x \sum_y I(x, y) \Delta A$$

since $m_{10} = \sum_x \sum_y xI(x, y) \Delta A$ and since $\bar{x} = m_{10}/m_{00}$

$$\mu_{20} = m_{20} - 2\frac{m_{10}}{m_{00}}m_{10} + \left(\frac{m_{10}}{m_{00}}\right)^2 m_{00}$$

$$= m_{20} - \frac{m_{10}^2}{m_{00}}$$

(7.87)

and this has descriptive capability.

The use of moments to describe an ellipse is shown in Figure 7.22. Here, an original ellipse (Figure 7.22a) gives the second order moments in Figure 7.22(d). In all cases, the first order moments are zero, as expected. The moments (Figure 7.22e) of the translated ellipse (Figure 7.22b) are the same as those of the original ellipse. In fact, these moments show that the greatest rate of change in mass is around the horizontal axis, as consistent with the ellipse. The second order moments Figure 7.22(f) of the ellipse when rotated by 90° (Figure 7.22c) are simply swapped around, as expected: the rate of change of mass is now greatest around the vertical axis. This illustrates how centralized moments are invariant to translation, but not to rotation.

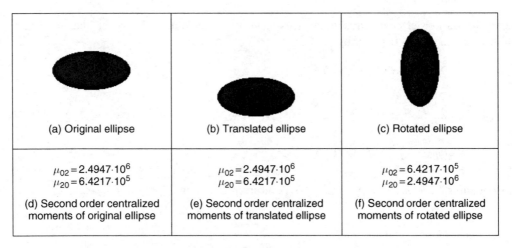

(a) Original ellipse	(b) Translated ellipse	(c) Rotated ellipse
$\mu_{02} = 2.4947 \cdot 10^6$ $\mu_{20} = 6.4217 \cdot 10^5$	$\mu_{02} = 2.4947 \cdot 10^6$ $\mu_{20} = 6.4217 \cdot 10^5$	$\mu_{02} = 6.4217 \cdot 10^5$ $\mu_{20} = 2.4947 \cdot 10^6$
(d) Second order centralized moments of original ellipse	(e) Second order centralized moments of translated ellipse	(f) Second order centralized moments of rotated ellipse

Figure 7.22   Describing a shape by centralized moments

7.3.2.2 Invariant moments

Centralized moments are only *translation* invariant: they are constant only with change in position, and no other appearance transformation. To accrue invariance to scale and rotation, we require *normalized central moments*, η_{pq}, defined as (Hu, 1962):

$$\eta_{pq} = \frac{\mu_{pq}}{\mu_{00}^{\gamma}}$$

(7.88)

where

$$\gamma = \frac{p+q}{2} + 1 \quad \forall p+q \geq 2$$

(7.89)

Seven *invariant moments* can be computed from these given by

$$M1 = \eta_{20} + \eta_{02}$$

$$M2 = (\eta_{20} - \eta_{02})^2 + 4\eta_{11}^2$$

$$M3 = (\eta_{30} - 3\eta_{12})^2 + (3\eta_{21} - \eta_{03})^2$$

$$M4 = (\eta_{30} + \eta_{12})^2 + (\eta_{21} + \eta_{03})^2$$

$$M5 = (\eta_{30} - 3\eta_{12})(\eta_{30} + \eta_{12}) + ((\eta_{30} + \eta_{12})^2 - 3(\eta_{21} - \eta_{03})^2) \qquad (7.90)$$

$$+ (3\eta_{21} - \eta_{03})(\eta_{21} + \eta_{03})(3(\eta_{30} + \eta_{12})^2 - (\eta_{21} + \eta_{03})^2)$$

$$M6 = (\eta_{20} - \eta_{02})((\eta_{30} + \eta_{12})^2 - (\eta_{21} + \eta_{03})^2) + 4\eta_{11}(\eta_{30} + \eta_{12})(\eta_{21} + \eta_{03})$$

$$M7 = (3\eta_{21} - \eta_{03})(\eta_{30} + \eta_{12})((\eta_{30} + \eta_{12})^2 - 3(\eta_{21} + \eta_{03})^2)$$

$$+ (3\eta_{12} - \eta_{30})(\eta_{21} + \eta_{03})(3(\eta_{12} + \eta_{30})^2 - (\eta_{21} + \eta_{03})^2)$$

The first of these, $M1$ and $M2$, are second order moments, those for which $p + q = 2$. Those remaining are third order moments, since $p + q = 3$. (The first order moments are of no consequence since they are zero.) The last moment, $M7$, is introduced as a skew invariant deigned to distinguish mirror images.

Code 7.5 shows the Mathcad implementation that computes the invariant moments $M1$, $M2$ and $M3$. The code computes the moments by straight implementation of Equations 7.83 and 7.90. The use of these invariant moments to describe three shapes is illustrated in Figure 7.23. Figure 7.23(b) corresponds to the same plane in Figure 7.23(a) but with a change of scale

```
μ(p,q,shape):= | cmom← 0
               |
               |      1            rows(shape)-1
               | xc← ──────────── ·    Σ          (shape_i)_0
               |     rows(shape)      i=0
               |
               |      1            rows(shape)-1
               | yc← ──────────── ·    Σ          (shape_i)_1
               |     rows(shape)      i=0
               |
               | for s∈0..rows(shape)-1
               |
               |     cmom←cmom+[(shape_s)_0-xc]^p·[(shape_s)_1-yc]^q·(shape_s)_2
               |
               | cmom

                   μ(p,q,im)
η(p,q,im):= ──────────────────
                       p+q
                       ─── +1
            μ(0,0,im)    2

M1(im):=η(2,0,im)+η(2,0,im)

M2(im):=(η(2,0,im)-η(0,2,im))^2+4·η(1,1,im)^2

M3(im):=(η(3,0,im)-3·η(1,2,im))^2+(3·η(2,1,im)-η(0,3,im))^2
```

Code 7.5 Computing $M1$, $M2$ and $M3$

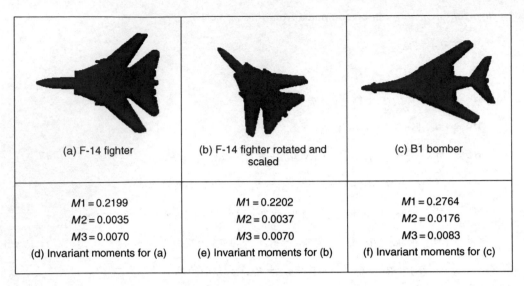

(a) F-14 fighter	(b) F-14 fighter rotated and scaled	(c) B1 bomber
M1 = 0.2199 M2 = 0.0035 M3 = 0.0070 (d) Invariant moments for (a)	M1 = 0.2202 M2 = 0.0037 M3 = 0.0070 (e) Invariant moments for (b)	M1 = 0.2764 M2 = 0.0176 M3 = 0.0083 (f) Invariant moments for (c)

Figure 7.23 Describing a shape by invariant moments

and a rotation. Thus, the invariant moments for these two shapes are very similar. In contrast, the invariant moments for the plane in Figure 7.23(c) differ. These invariant moments have the most important invariance properties. However, these moments are not *orthogonal*, and as such there is potential for *reducing* the size of the set of moments required to describe a shape accurately.

7.3.2.3 Zernike moments

Invariance can be achieved by using *Zernike moments* (Teague, 1980), which give an orthogonal set of rotation-invariant moments. These find greater deployment where *invariant* properties are required. *Rotation* invariance is achieved by using polar representation, as opposed to the Cartesian parameterization for centralized moments. The *complex* Zernike moment, Z_{pq}, is

$$Z_{pq} = \frac{p+1}{\pi} \int_0^{2\pi} \int_0^\infty V_{pq}(r, \theta)^* f(r, \theta) \, r \mathrm{d}r \mathrm{d}\theta \tag{7.91}$$

where p is now the radial magnitude and q is the radial direction and where * again denotes the complex conjugate (as in Section 5.3.2) of a *Zernike polynomial*, V_{pq}, given by

$$V_{pq}(r, \theta) = R_{pq}(r)e^{jq\theta} \quad \text{where} \quad p - q \text{ is } \mathbf{even} \quad \text{and} \quad 0 \le q \le |p| \tag{7.92}$$

where R_{pq} is a real-valued polynomial given by

$$R_{pq}(r) = \sum_{m=0}^{\frac{p-|q|}{2}} (-1)^m \frac{(p-m)!}{m! \left(\frac{p+|q|}{2} - m\right)! \left(\frac{p-|q|}{2} - m\right)!} r^{p-2m} \tag{7.93}$$

The order of the polynomial is denoted by p and the repetition by q. The repetition q can take negative values (since $q \le |p|$), so the radial polynomial uses its magnitude and thus the inverse relationship holds: $R_{p,q}(r) = R_{p,-q}(r)$ (changing the notation of the polynomial slightly by introducing a comma to make clear that the moment just has the sign of q inverted). The polynomials of lower degree are

$$R_{00}(r) = 1$$

$$R_{11}(r) = r$$

$$R_{22}(r) = r^2$$

$$R_{20}(r) = r^2 - 1 \qquad\qquad (7.94)$$

$$R_{31}(r) = 3r^2 - 2r$$

$$R_{40}(r) = 6r^4 - 6r^2 + 1$$

and some of these are plotted in Figure 7.24. In Figure 7.24(a) we can see that the frequency components increase with the order p and the functions approach unity as $r \to 1$. The frequency content reflects the level of detail that can be captured by the particular polynomial. The change between the different polynomials shows how together they can capture different aspects of an underlying signal, across the various values of r. The repetition controls the way in which the function approaches unity: the influence along the polynomial and the polynomials for different values of q are shown in Figure 7.24(b).

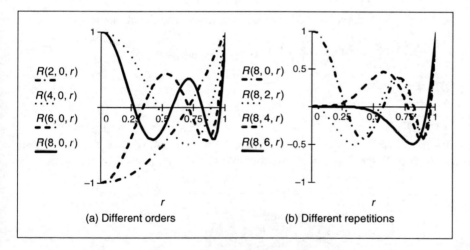

Figure 7.24 Zernike polynomials

These polynomials are orthogonal within the unit circle, so the analysed shape (the area of interest) has to be remapped to be of this size before calculation of its moments. This implies difficulty in mapping a unit circle to a Cartesian grid. As illustrated in Figure 7.25, the circle can be within the area of interest, losing corner information (but that is information rarely of interest) (Figure 7.25a); or around (encompassing) the area of interest, which then covers areas where there is no information, but ensures that all the information within the area of interest is included (Figure 7.25b).

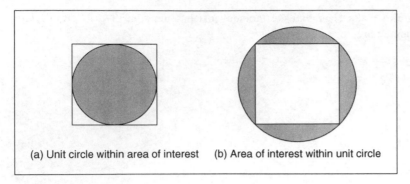

(a) Unit circle within area of interest   (b) Area of interest within unit circle

Figure 7.25 Mapping a unit circle to an area of interest

The *orthogonality* of these polynomials assures the *reduction* in the set of numbers used to describe a shape. More simply, the radial polynomials can be expressed as

$$R_{pq}(r) = \sum_{k=q}^{p} B_{pqk} r^k \tag{7.95}$$

where the Zernike coefficients are

$$B_{pqk} = (-1)^{\frac{p-k}{2}} \frac{((p+k)/2)!}{((p-k)/2)!((k+q)/2)!((k-q)/2)!} \tag{7.96}$$

for $p - k =$ even. The Zernike moments can be calculated from centralized moments as

$$Z_{pq} = \frac{p+1}{\pi} \sum_{k=q}^{p} \sum_{l=0}^{t} \sum_{m=0}^{q} (-j)^m \binom{t}{l} \binom{q}{m} B_{pqk} \mu_{(k-2l-q+m)(q+2l-m)} \tag{7.97}$$

where $t = (k-q)/2$ and where

$$\binom{t}{l} = \frac{t!}{l!(t-l)!} \tag{7.98}$$

A Zernike polynomial kernel is illustrated in Figure 7.26. This shows that the kernel can capture differing levels of shape detail (and that multiple kernels are needed to give a shape's

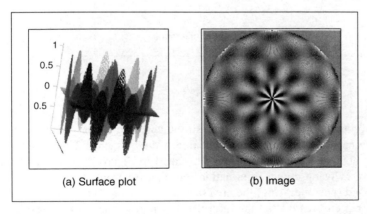

(a) Surface plot   (b) Image

Figure 7.26 Zernike polynomial kernel

description). This kernel is computed in radial form, which is how it is deployed in shape analysis. Note that differing sets of parameters such as order and repetition control the level of detail that is analysed by application of this kernel to a shape. The plot shows the real part of the kernel; the imaginary part is similar, but rotated.

Analysis (and by Equation 7.83), assuming that x and y are constrained to the interval $[-1, 1]$, gives

$$Z_{00} = \frac{\mu_{00}}{\pi}$$

$$Z_{11} = \frac{2}{\pi}(\mu_{01} - j\mu_{10}) = 0 \tag{7.99}$$

$$Z_{22} = \frac{3}{\pi}(\mu_{02} - j2\mu_{11} - \mu_{20})$$

which can be extended further (Teague, 1980), and with remarkable similarity to the Hu invariant moments (Equation 7.90).

The magnitude of these Zernike moments remains invariant to rotation, which affects only the phase; the Zernike moments can be made *scale* invariant by normalization. An additional advantage is that there is a *reconstruction* theorem. For Nm moments, the original shape f can be reconstructed from its moments and the Zernike polynomials as

$$f(x, y) \approx \sum_{p=0}^{Nm} \sum_{q} Z_{pq} V_{pq}(x, y) \tag{7.100}$$

This is illustrated in Figure 7.27 for reconstructing a simple binary object, the letter A, from different numbers of moments. When reconstructing this up to the 10th order of a Zernike moment description (this requires 66 moments) we achieve a grey-level image, which contains

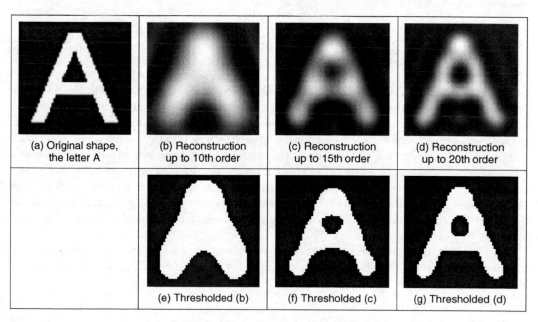

(a) Original shape, the letter A	(b) Reconstruction up to 10th order	(c) Reconstruction up to 15th order	(d) Reconstruction up to 20th order
	(e) Thresholded (b)	(f) Thresholded (c)	(g) Thresholded (d)

Figure 7.27   Reconstructing a shape from its moments (Prismall et al., 2002)

much of the overall shape (7.27b). This can be thresholded to give a binary image (Figure 7.27e), which shows the overall shape, without any corners. When we use more moments, we increase the detail in the reconstruction: Figure 7.27(c) is up to 15th order (136 moments) and Figure 7.27(d) is 20th order (231 moments). The latter of these is much closer to the original image, especially in its thresholded form (Figure 7.27d). This may sound like a lot of moments, but the compression from the original image is very high. Note also that even though we can achieve *recognition* from a smaller set of moments, these may not represent the hole in the shape, which is not present at the 10th order, which just shows the overall shape of the letter A. As such, reconstruction can give insight as to the shape contribution of selected moments: their *significance* can be assessed by this and other tests.

These Zernike descriptors have been shown to good effect in application by reconstructing a good approximation to a shape with only few descriptors (Boyce and Hossack, 1983) and in recognition (Khotanzad and Hong, 1990). As ever, fast computation has been of (continuing) interest (Mukundan and Ramakrishnan, 1995; Gu et al., 2002).

7.3.2.4 Other moments

Pseudo Zernike moments (Teh and Chin, 1988) aim to relieve the restriction on normalization to the unit circle. *Complex moments* (Abu-Mostafa and Psaltis, 1985) aim to provide a simpler moment description with invariance properties. In fact, since there is an infinite variety of functions that can be used as the basis function, we also have *Legendre* (Teague, 1980) and, more recently, *Tchebichef* (although this is sometimes spelt *Chebyshev*) moments (Mukundan, 2001). There is no detailed comparison yet available, but there are advantages and disadvantages to the differing moments, often exposed by application. As an extension into the time domain, Shutler and Nixon (2006) developed *velocity moments*, which can be used to recognize moving objects over a sequence of images, applied in that case to recognizing people by their gait. The moments sum over a sequence of I images as

$$vm_{pq\alpha\gamma} = N \sum_{i=2}^{I} \sum_{x \in \mathbf{P}} \sum_{y \in \mathbf{P}} U(i, \alpha, \gamma) S(i, p, q) \mathbf{P}_{i_{x,y}} \tag{7.101}$$

where N is a scaling coefficient, $\mathbf{P}_{i_{x,y}}$ is the ith image in the sequence, S are the moments describing a shape's structure (and can be Cartesian or Zernike), and U are moments that describe the movement of the shape's centre of mass between frames. Rotation was not included; the technique was shown to be capable for use in recognizing walking subjects, not gymnasts.

Finally, there are *affine invariant moments*, which do not change with position, rotation and different scales along the coordinate axes, as a result, say, of a camera not being normal to the object plane. Here, the earliest approach appears to be by Flusser and Suk (1993). One of the reviews (Teh and Chin, 1988) concentrates on information content (redundancy), noise sensitivity and representation ability, comparing the performance of several of the more popular moments in these respects.

It is possible to explore the link between moments and Fourier theory (Mukundan and Ramakrishnan, 1998). The discrete Fourier transform of an image (Equation 2.22), can be written as

$$\mathbf{FP}_{u,v} = \frac{1}{N} \sum_{x=0}^{N-1} \sum_{y=0}^{N-1} \mathbf{P}_{x,y} e^{-j\frac{2\pi}{N}ux} e^{-j\frac{2\pi}{N}vy} \tag{7.102}$$

By using the Taylor expansion of the exponential function

$$e^z = \sum_{p=0}^{\infty} \frac{z^p}{p!} \qquad (7.103)$$

we can substitute for the exponential functions as

$$\mathbf{FP}_{u,v} = \frac{1}{N} \sum_{x=0}^{N-1} \sum_{y=0}^{N-1} \mathbf{P}_{x,y} \sum_{p=0}^{\infty} \frac{\left(-j\frac{2\pi}{N} ux\right)^p}{p!} \sum_{q=0}^{\infty} \frac{\left(-j\frac{2\pi}{N} vy\right)^q}{q!} \qquad (7.104)$$

which, by collecting terms, gives

$$\mathbf{FP}_{u,v} = \frac{1}{N} \sum_{x=0}^{N-1} \sum_{y=0}^{N-1} x^p y^q \mathbf{P}_{x,y} \sum_{p=0}^{\infty} \sum_{q=0}^{\infty} \frac{\left(-j\frac{2\pi}{N}\right)^{p+q}}{p!q!} u^p v^q \qquad (7.105)$$

and by the definition of Cartesian moments, Equation 7.74, we have

$$\mathbf{FP}_{u,v} = \frac{1}{N} \sum_{p=0}^{\infty} \sum_{q=0}^{\infty} \frac{\left(-j\frac{2\pi}{N}\right)^{p+q}}{p!q!} u^p v^q m_{pq} \qquad (7.106)$$

This implies that the Fourier transform of an image can be derived from its moments. There is then a link between the Fourier decomposition and that by moments, showing the link between the two. But we can go further, since there is the inverse Fourier transform, Equation 2.23,

$$\mathbf{P}_{x,y} = \sum_{u=0}^{N-1} \sum_{v=0}^{N-1} \mathbf{FP}_{u,v} e^{j\frac{2\pi}{N} ux} e^{j\frac{2\pi}{N} vy} \qquad (7.107)$$

So the original image can be computed from the moments as

$$\mathbf{P}_{x,y} = \sum_{x=0}^{N-1} \sum_{y=0}^{N-1} e^{j\frac{2\pi}{N} ux} e^{j\frac{2\pi}{N} vy} \frac{1}{N} \sum_{p=0}^{\infty} \sum_{q=0}^{\infty} \frac{\left(-j\frac{2\pi}{N}\right)^{p+q}}{p!q!} u^p v^q m_{pq} \qquad (7.108)$$

and this shows that we can get back to the image from our moment description, although care must be exercised in the choice of windows from which data are selected. This is *reconstruction*: we can reconstruct an image from its moment description. There has not been much study on reconstruction from moments, despite its apparent importance in understanding the potency of the description that has been achieved. Potency is usually investigated in application by determining the best set of moment features to maximize recognition capability (and we shall turn to this in the next chapter). Essentially, reconstruction from basic geometric (Cartesian) moments is impractical (Teague, 1980) and the orthogonal bases functions such as the Zernike polynomials offer a simpler route to reconstruction, but these still require thesholding. More recently, Prismall et al. (2002) used (Zernike) moments for the reconstruction of moving objects.

7.4 Further reading

This chapter has essentially been based on unified techniques for border and region description. There is much more to contour and region analysis than indicated at the start of the chapter, for this is one the starting points of morphological analysis. There is an extensive review available (Loncaric, 1998) with many important references in this topic. The analysis neighbourhood can be extended to be larger (Marchand and Sharaiha, 1997) and there is consideration of appropriate distance metrics for this (Das and Chatterji, 1988). A much more detailed study of boundary-based representation and application can be found in van Otterloo's fine text

(1991). There are many other ways to describe features, although few have the unique attributes of moments and Fourier descriptors. There is an interrelation between boundary and region description: *curvature* can be computed from a chain code (Rosenfeld, 1974); Fourier descriptors can also be used to calculate region descriptions (Kiryati and Maydan, 1989). There have been many approaches to boundary approximation by fitting curves to the data. Some of these use polynomial approximation, and there are many *spline-based* techniques. A spline is a local function used to model a feature in sections. There are *quadratic* and *cubic* forms (for a good review of spline theory, try Ahlberg et al., 1967, or Dierckx, 1995); of interest, *snakes* are energy-minimizing splines. There are many methods for polygonal approximations to curves, and recently a new measure has been applied to compare performance on a suitable curve of techniques based on dominant point analysis (Rosin, 1997). To go with the earlier-mentioned review (Prokop and Reeves, 1992), there is a book available on moment theory (Mukundan and Ramakrishnan, 1998) showing the whole moment picture. As in the previous chapter, the skeleton of a shape can be used for recognition. This is a natural target for *thinning* techniques that have not been covered here. An excellent survey of these techniques, as used in character description following extraction, can be found in Trier et al. (1996), describing use of moments and Fourier descriptors.

7.5 References

Abu-Mostafa, Y. S. and Psaltis, D., Image Normalization by Complex Moments, *IEEE Trans. PAMI*, **7**, pp. 46–55, 1985

Aguado, A. S., Nixon, M. S. and Montiel, E., Parameterizing Arbitrary Shapes via Fourier Descriptors for Evidence-Gathering Extraction, *CVIU: Comput. Vision Image Understand.*, **69**(2), pp. 202–221, 1998

Aguado, A. S., Montiel, E. and Zaluska, E., Modelling Generalized Cylinders via Fourier Morphing, *ACM Trans. Graphics*, **18**(4), pp. 293–315, 1999

Ahlberg, J. H., Nilson, E. N. and Walsh, J. L., *The Theory of Splines and Their Applications*, Academic Press, New York, 1967

Boyce, J. F. and Hossack, W. J., Moment Invariants for Pattern Recognition, *Pattern Recog. Lett.*, **1**, pp. 451–456, 1983

Chen, Y. Q., Nixon, M. S. and Thomas, D. W., Texture Classification using Statistical Geometric Features, *Pattern Recog.*, **28**(4), pp. 537–552, 1995

Cosgriff, R. L., *Identification of Shape*, Rep. 820-11, ASTIA AD 254792, Ohio State University Research Foundation, Columbus, OH, 1960

Crimmins, T. R., A Complete Set of Fourier Descriptors for Two-Dimensional Shapes, *IEEE Trans. SMC*, **12**(6), pp. 848–855, 1982

Das, P. P. and Chatterji, B. N. Knight's Distances in Digital Geometry, *Pattern Recog. Lett.*, **7**, pp. 215–226, 1988

Dierckx, P., *Curve and Surface Fitting with Splines*, Oxford University Press, Oxford, 1995

Flusser, J. and Suk, T., Pattern Recognition by Affine Moment Invariants, *Pattern Recog.*, **26**(1), pp. 167–174, 1993

Freeman, H., On the Encoding of Arbitrary Geometric Configurations, *IRE Trans.*, **EC-10**(2), pp. 260–268, 1961

Freeman, H., Computer Processing of Line Drawing Images, *Comput. Surv.*, **6**(1), pp. 57–95, 1974

Granlund, G. H., Fourier Preprocessing for Hand Print Character Recognition, *IEEE Trans. Comput.*, **21**, pp. 195–201, 1972

Gu, J., Shua, H. Z., Toumoulinb, C. and Luoa, L. M., A Novel Algorithm for Fast Computation of Zernike Moments, *Pattern Recog.*, **35**(12), pp. 2905–2911, 2002

Hu, M. K., Visual Pattern Recognition by Moment Invariants, *IRE Trans. Inform. Theory*, **IT-8**, pp. 179–187, 1962

Kashi, R. S., Bhoj-Kavde, P., Nowakowski, R. S. and Papathomas, T. V., 2-D Shape Representation and Averaging using Normalized Wavelet Descriptors, *Simulation*, **66**(3), pp. 164–178, 1996

Khotanzad, A. and Hong, Y. H., Invariant Image Recognition by Zernike Moments, *IEEE Trans. PAMI*, **12**, pp. 489–498, 1990

Kiryati, N. and Maydan, D., Calculating Geometric Properties from Fourier Representation, *Pattern Recog.*, **22**(5), pp. 469–475, 1989

Kuhl, F. P. and Giardina, C. R., Elliptic Fourier Descriptors of a Closed Contour, *CVGIP*, **18**, pp. 236–258, 1982

Lin C. C. and Chellappa, R., Classification of Partial 2D Shapes using Fourier Descriptors, *IEEE Trans. PAMI*, **9**(5), pp. 686–690, 1987

Liu, H. C. and Srinath, M. D., Corner Detection from Chain-Coded Curves, *Pattern Recog.*, **23**(1), pp. 51–68, 1990

Loncaric, S., A Survey of Shape Analysis Techniques, *Pattern Recog.*, **31**(8), pp. 983–1001, 1998

Marchand, S. and Sharaiha, Y. M., Discrete Convexity, Straightness and the 16-Neighbourhood, *Comput. Vision Image Understand.*, **66**(3), pp. 416–429, 1997

Montiel, E., Aguado, A. S. and Zaluska, E., Topology in Fractals, *Chaos Solitons Fractals*, **7**(8), pp. 1187–1207, 1996

Montiel, E., Aguado, A. S. and Zaluska, E., Fourier Series Expansion of Irregular Curves, *Fractals*, **5**(1), pp. 105–199, 1997

Mukundan, R., Image Analysis by Tchebichef Moments, *IEEE Trans. Image Process.*, **10**(9), pp. 1357–1364, 2001

Mukundan, R. and Ramakrishnan, K. R., Fast Computation of Legendre and Zernike Moments, *Pattern Recog.*, **28**(9), pp. 1433–1442, 1995

Mukundan, R. and Ramakrishnan, K. R., *Moment Functions in Image Analysis: Theory and Applications*, World Scientific, Singapore, 1998

van Otterloo, P. J., *A Contour-Oriented Approach to Shape Analysis*, Prentice Hall International (UK), Hemel Hempstead, 1991

Persoon, E. and Fu, K.-S., Shape Description Using Fourier Descriptors, *IEEE Trans. SMC*, **3**, pp. 170–179, 1977

Prismall, S. P., Nixon, M. S. and Carter, J. N., On Moving Object Reconstruction by Moments, *Proc. BMVC 2002*, pp. 73–82, 2002

Prokop, R. J. and Reeves A. P., A Survey of Moment-Based Techniques for Unoccluded Object Representation and Recognition, *CVGIP: Graphical Models Image Process.*, **54**(5), pp. 438–460, 1992

Rosenfeld, A., Digital Straight Line Segments, *IEEE Trans. Comput.*, **23**, pp. 1264–1269, 1974

Rosin, P., Techniques for Assessing Polygonal Approximations to Curves, *IEEE Trans. PAMI*, **19**(6), pp. 659–666, 1997

Rosin, P. and Zunic, J., Measuring Rectilinearity, *Comput. Vision Image Understand.*, **99**(2), pp. 175–188, 2005

Searle, N. H., Shape Analysis by use of Walsh Functions, In: B. Meltzer and D. Mitchie (Eds), *Machine Intelligence 5*, Edinburgh University Press, Edinburgh, 1970

Seeger, U. and Seeger, R., Fast Corner Detection in Gray-Level Images, *Pattern Recog. Lett.*, **15**, pp. 669–675, 1994

Shutler, J. D. and Nixon, M. S., Zernike Velocity Moments for Sequence-Based Description of Moving Features, *Image Vision Comput.*, **24**(4), pp. 343–356, 2006

Staib, L. and Duncan, J., Boundary Finding with Parametrically Deformable Models, *IEEE Trans. PAMI*, **14**, pp. 1061–1075, 1992

Teague, M. R., Image Analysis by the General Theory of Moments, *J. Opt. Soc. Am.*, **70**, pp. 920–930, 1980

Teh, C. H. and Chin, R. T., On Image Analysis by the Method of Moments, *IEEE Trans. PAMI*, **10**, pp. 496–513, 1988

Trier, O. D., Jain, A. K. and Taxt, T., Feature Extraction Methods for Character Recognition – A Survey, *Pattern Recog.*, **29**(4), pp. 641–662, 1996

Undrill, P. E., Delibasis, K. and Cameron, G. G., An Application of Genetic Algorithms to Geometric Model-Guided Interpretation of Brain Anatomy, *Pattern Recog.*, **30**(2), pp. 217–227, 1997

Zahn, C. T. and Roskies, R. Z., Fourier Descriptors for Plane Closed Curves, *IEEE Trans. Comput.*, **C-21**(3), pp. 269–281, 1972

8

Introduction to texture description, segmentation and classification

8.1 Overview

This chapter is concerned with how we can use many of the feature extraction and description techniques presented earlier to characterize regions in an image. The aim here is to describe how we can collect together measurements for purposes of recognition, using texture by way of introduction and as a vehicle for using *feature extraction* in *recognition*.

We shall look first at what is meant by texture and then how we can use Fourier transform techniques, statistics and region measures to describe it. We shall then look at how the measurements provided by these techniques, the description of the texture, can be collected together to recognize it. Finally, we shall label an image according to the texture found within it, to give a segmentation into classes known to exist within the image. Since we could be recognizing shapes described by Fourier descriptors, region measures, or other feature extraction and description approaches, the material is general and could be applied for purposes of recognition to measures other than texture.

Table 8.1 Overview of Chapter 8

Main topic	Sub topics	Main points
Texture description	What is image texture and how do we determine sets of *numbers* that allow us to be able to *recognize* it.	Feature extraction: Fourier transform; co-occurrence; regions. Feature descriptions: energy; entropy and inertia.
Texture classification	How to *associate* the numbers we have derived with those that we have already *stored* for known examples.	k-Nearest neighbour rule; support vector machines and other classification approaches.
Texture segmentation	How to find *regions* of texture within images.	Convolution; tiling; thresholding.

8.2 What is texture?

Texture is a very nebulous concept, often attributed to human perception, as either the feel or the appearance of (woven) fabric. Everyone has their own interpretation as to the nature of texture; there is no mathematical definition of texture, it simply exists. By way of reference, let us consider one of the dictionary definitions (Oxford Dictionary, 1996):

> **texture** *n.*, & *v.t.* **1.** *n.* arrangement of threads etc. in textile fabric; characteristic feel due to this; arrangement of small constituent parts, perceived structure, (*of* skin, rock, soil, organic tissue, literary work, etc.); representation of structure and detail of objects in art;...

That covers quite a lot. If we change 'threads' for 'pixels', the definition could apply to images (except for the bit about artwork). Essentially, texture can be what we define it to be. Why might we want to do this? By way of example, analysis of remotely sensed images is now a major application of image processing techniques. In such analysis, pixels are labelled according to the categories of a required application, such as whether the ground is farmed or urban in land-use analysis, or water for estimation of surface analysis. An example remotely sensed image is given in Figure 8.1(a), which is of an urban area (in the top left) and some farmland. Here, the image resolution is low and each pixel corresponds to a large area of the ground. Square groups of pixels have then been labelled either as urban or as farmland, according to their texture properties as shown in Figure 8.1(b), where black represents the area classified as urban and white represents farmland. In this way we can assess the amount of area that urban areas occupy. As such, we have used *real* textures to label pixels, the perceived textures of the urban and farming areas.

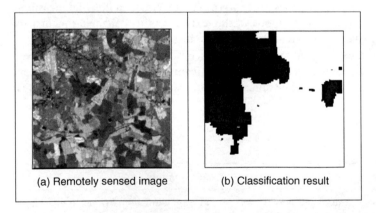

(a) Remotely sensed image   (b) Classification result

Figure 8.1   Example texture analysis

As an alternative definition of texture, we can consider it as a database of images that researchers use to test their algorithms. Many texture researchers have used a database of pictures of textures (Brodatz, 1968), produced for artists and designers, rather than for digital image analysis. Parts of three of the Brodatz texture images are given in Figure 8.2. Here, the French canvas (Brodatz index D20) in Figure 8.2(a) is a detail of Figure 8.2(b) (Brodatz index

D21), taken at four times the magnification. The beach sand in Figure 8.2(c), Brodatz index D29, is clearly of a different texture to that of cloth. Given the diversity of texture, there are now many databases available on the web, at the sites given in Chapter 1 or at this book's website.

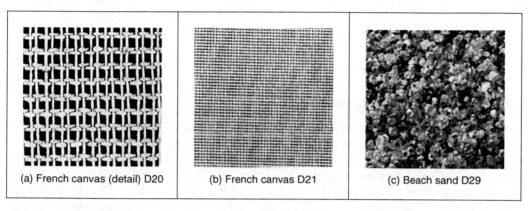

| (a) French canvas (detail) D20 | (b) French canvas D21 | (c) Beach sand D29 |

Figure 8.2 Three Brodatz textures

Alternatively, we can define texture as a quantity for which texture extraction algorithms provide meaningful results. One study (Karru et al., 1996) suggests

> The answer to the question 'is there any texture in the image?' depends not only on the input image, but also on the goal for which the image texture is used and the textural features that are extracted from the image.

As we shall find, texture analysis has a rich history in image processing and computer vision and there is now even a book devoted to texture analysis (Petrou and Sevilla, 2006). Despite this, approaches that synthesize texture are relatively recent. This is motivated also by graphics, and the need to include texture to improve the quality of the rendered scenes (Heckbert, 1986). By way of example, one well-known approach to texture synthesis is to use a Markov random field (Efros and Leung, 1999), but we shall not dwell on that here.

Essentially, there is no unique definition of texture and there are many ways to describe and extract it. It is a very large and exciting field of research and there continue to be many new developments.

Images will usually contain samples of more than one texture. Accordingly, we would like to be able to *describe* texture (*texture descriptions* are measurements that characterize a texture) and then to *classify* it (classification is attributing the correct class label to a set of measurements) and then, perhaps, to segment an image according to its texture content. We have used similar classification approaches to characterize the shape descriptions in the previous chapter. These are massive fields of research that move on to the broad subject of pattern recognition. We shall look at an introduction here; later references will point you to topics of particular interest and to some of the more recent developments. The main purpose of this introduction is to show how the measurements can be collected together to recognize objects. Texture is used as the vehicle for this since it is a region-based property that has not as yet been covered. Since texture itself is an enormous subject, you will find plenty of references to established approaches and to surveys of the field. First, we shall look at approaches to deriving the features (measurements) that can be used to describe textures. Broadly, these can be split into *structural* (transform-based), *statistical* and *combination* approaches. The frequency content of an image will reflect its texture; we shall start with Fourier. First, though, we shall consider some of the required properties of the descriptions.

8.3 Texture description

8.3.1 Performance requirements

The purpose of texture description is to derive some measurements that can be used to classify a particular texture. As such, there are *invariance* requirements on the measurements, as there were for shape description. The invariance requirements for feature extraction, namely invariance to position, scale and rotation, can apply equally to texture extraction. After all, texture is a feature, albeit a rather nebulous one as opposed to the definition of a shape. Clearly, we require *position* invariance: the measurements describing a texture should not vary with the position of the analysed section (of a larger image). We also require *rotation* invariance, but this is not as strong a requirement as position invariance; the definition of texture does not imply knowledge of orientation, but could be presumed to. The least strong requirement is that of *scale*, for this depends primarily on application. Consider using texture to analyse forests in remotely sensed images. Scale invariance would imply that closely spaced young trees should give the same measure as widely spaced mature trees. This should be satisfactory if the purpose was only to analyse foliage cover. It would be unsatisfactory if the purpose was to measure age for purposes of replenishment, since a scale-invariant measure would be of little use as it could not, in principle, distinguish between young and old trees.

Unlike feature extraction, texture description rarely depends on edge extraction, since one main purpose of edge extraction is to remove reliance on overall *illumination* level. The higher order invariants, such as perspective invariance, are rarely applied to texture description. This is perhaps because many applications are like remotely sensed imagery, or are in constrained industrial applications where the camera geometry can be controlled.

8.3.2 Structural approaches

The most basic approach to texture description is to generate the Fourier transform of the image and then to group the transform data in some way so as to obtain a set of measurements. The size of the set of measurements is smaller than the size of the image's transform. In Chapter 2 we saw how the transform of a set of horizontal lines was a set of vertical spatial frequencies (since the point spacing varies along the vertical axis). Here, we must remember that for display we rearrange the Fourier transform so that the d.c. component is at the centre of the presented image.

The transforms of the three Brodatz textures of Figure 8.2 are shown in Figure 8.3. Figure 8.3(a) shows a collection of frequency components which are then replicated with the same structure (consistent with the Fourier transform) in Figure 8.3(b). (Figure 8.3a and b also show the frequency scaling property of the Fourier transform: greater magnification reduces the high frequency content.) Figure 8.3(c) is clearly different in that the structure of the transform

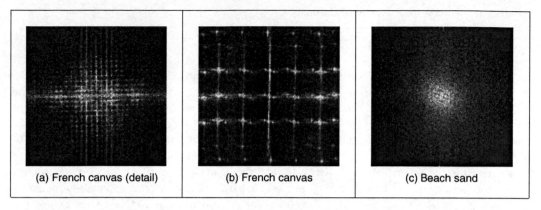

| (a) French canvas (detail) | (b) French canvas | (c) Beach sand |

Figure 8.3 Fourier transforms of the three Brodatz textures

data is spread a different manner to that of Figure 8.3(a) and (b). These images have been derived by application of the fast Fourier transform, which we shall denote as

$$\mathbf{FP} = \Im(\mathbf{P}) \tag{8.1}$$

where $\mathbf{FP}_{u,v}$ and $\mathbf{P}_{x,y}$ are the transform and pixel data, respectively. One clear advantage of the Fourier transform is that it possesses shift invariance (Section 2.6.1): the transform of a bit of (large and uniform) cloth will be the same, whatever segment we inspect. This is consistent with the observation that phase is of little use in Fourier-based texture systems (Pratt, 1992), so the modulus of the transform (its magnitude) is usually used. The transform is of the same size as the image, even though conjugate symmetry of the transform implies that we do not need to use all of its components as measurements. As such, we can *filter* the Fourier transform (Section 2.8) so as to select those frequency components deemed to be of interest to a particular application. Alternatively, it is convenient to collect the magnitude transform data in different ways to achieve a reduced set of measurements. First, though, the transform data can be normalized by the sum of the squared values of each magnitude component (excepting the zero-frequency components, those for $u = 0$ and $v = 0$), so that the magnitude data is invariant to linear shifts in illumination to obtain normalized Fourier coefficients **NFP** as

$$\mathbf{NFP}_{u,v} = \frac{\left|\mathbf{FP}_{u,v}\right|}{\sqrt{\displaystyle\sum_{(u\neq0)\wedge(v\neq0)}\left|\mathbf{FP}_{u,v}\right|^2}} \tag{8.2}$$

Alternatively, histogram equalization (Section 3.3.3) can provide such invariance, but is more complicated than using Equation 8.2. The spectral data can then be described by the *entropy*, h, as

$$h = \sum_{u=1}^{N}\sum_{v=1}^{N}\mathbf{NFP}_{u,v}\log\left(\mathbf{NFP}_{u,v}\right) \tag{8.3}$$

or by their *energy*, e, as

$$e = \sum_{u=1}^{N}\sum_{v=1}^{N}\left(\mathbf{NFP}_{u,v}\right)^2 \tag{8.4}$$

Introduction to texture description, segmentation and classification   333

Another measure is their *inertia*, i, defined as

$$i = \sum_{u=1}^{N} \sum_{v=1}^{N} (u-v)^2 \, \mathbf{NFP}_{u,v} \tag{8.5}$$

These measures are shown for the three Brodatz textures in Code 8.1. In a way, they are like the shape descriptions in the previous chapter: the measures should be the same for the same object and should differ for a different one. Here, the texture measures are different for each of the textures. Perhaps the detail in the French canvas (Code 8.1a) could be made to give a closer measure to that of the full resolution (Code 8.1b) by using the frequency scaling property of the Fourier transform, discussed in Section 2.6.3. The beach sand clearly gives a different set of measures from the other two (Code 8.1c). In fact, the beach sand in Code 8.1(c) would appear to be more similar to the French canvas in Code 8.1(b), since the inertia and energy measures are much closer than those for Code 8.1(a) (only the entropy measure in Code 8.1a is closest to Code 8.1b). This is consistent with the images: each of the beach sand and French canvas has a large proportion of higher frequency information, since each is a finer texture than that of the detail in the French canvas.

entropy(FD20)=-253.11 inertia(FD20)=5.55·10^5 energy(FD20)=5.41	entropy(FD21)=-196.84 inertia(FD21)=6.86·10^5 energy(FD21)=7.49	entropy(FD29)=-310.61 inertia(FD29)=6.38·10^5 energy(FD29)=12.37
(a) French canvas (detail)	(b) French canvas	(c) Beach sand

Code 8.1 Measures of the Fourier transforms of the three Brodatz textures

By Fourier analysis, the measures are inherently *position* invariant. Clearly, the entropy, inertia and energy are relatively immune to *rotation*, since order is not important in their calculation. The measures can also be made *scale* invariant, as a consequence of the frequency scaling property of the Fourier transform. Finally, the measurements (by virtue of the normalization process) are inherently invariant to linear changes in *illumination*. The descriptions will be subject to noise. To handle large datasets we need a larger set of measurements (larger than the three given here) to discriminate better between different textures. Other measures can include:

- the energy in the major peak
- the Laplacian of the major peak
- the largest horizontal frequency magnitude; and
- the largest vertical frequency magnitude.

Among others, these are elements of Liu's features (Liu and Jernigan, 1990) chosen in a way aimed to give Fourier transform-based measurements good performance in noisy conditions.

There are many other transforms and these can confer different attributes in analysis. The wavelet transform is very popular since it allows for localization in time and frequency (Laine and Fan, 1993; Lu et al., 1997). Other approaches use the Gabor wavelet (Bovik et al., 1990; Jain and Farrokhnia, 1991; Daugman, 1993; Dunn et al., 1994), as introduced in Section 2.7.3. One comparison between Gabor wavelets and tree- and pyramidal-structured wavelets suggested that Gabor has the greater descriptional ability, with the penalty of greater computational complexity

(Pichler et al., 1996), and more recent work is available (Grigorescu et al., 2002). There has also been renewed resurgence of interest in Markov random fields (Gimmel'farb and Jain, 1996; Wu and Wei, 1996). Others, such as the Walsh transform (where the basis functions are 1s and 0s) appear to await application in texture description, no doubt owing to basic properties. In fact, one survey (Randen and Husoy, 2000) includes the use of Fourier, wavelet and discrete cosine transforms (Section 2.7.1) for texture characterization. These approaches are structural in nature: an image is viewed in terms of a transform applied to a whole image, as such exposing its *structure*. This is like the dictionary definition of an arrangement of parts. Another part of the dictionary definition concerned *detail*; this can be exposed by analysis of the high-frequency components, but these can be prone to noise. An alternative way to analyse the detail is to consider the *statistics* of an image.

8.3.3 Statistical approaches

The most famous statistical approach is the *co-occurrence matrix*. This was the result of the first approach to describe, and then classify, image texture (Haralick et al., 1973). It remains popular today, by virtue of good performance. The co-occurrence matrix contains elements that are counts of the number of pixel pairs for specific brightness levels, when separated by some distance and at some relative inclination. For brightness levels $b1$ and $b2$ the co-occurrence matrix $\mathbf{C}$ is

$$\mathbf{C}_{b1,b2} = \sum_{x=1}^{N}\sum_{y=1}^{N} \left(\mathbf{P}_{x,y} = b1\right) \wedge \left(\mathbf{P}_{x',y'} = b2\right) \tag{8.6}$$

where the x coordinate x' is the offset given by the specified distance d and inclination θ by

$$x' = x + d\cos(\theta) \quad \forall \; (d \in 1, \max(d)) \wedge (\theta \subset 0, 2\pi) \tag{8.7}$$

and the y coordinate y' is

$$y' = y + d\sin(\theta) \quad \forall \; (d \in 1, \max(d)) \wedge (\theta \in 0, 2\pi) \tag{8.8}$$

When Equation 8.6 is applied to an image, we obtain a square, symmetric, matrix whose dimensions equal the number of grey levels in the picture. The co-occurrence matrices for the three Brodatz textures of Figure 8.2 are shown in Figure 8.4. In the co-occurrence matrix

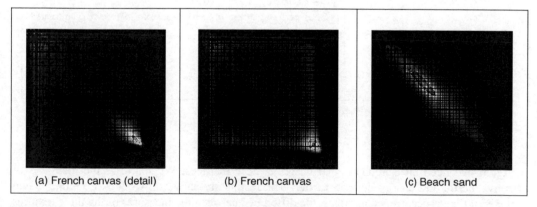

| (a) French canvas (detail) | (b) French canvas | (c) Beach sand |

Figure 8.4 Co-occurrence matrices of the three Brodatz textures

Introduction to texture description, segmentation and classification 335

generation, the maximum distance was 1 pixel and the directions were set to select the four nearest neighbours of each point. Now the results for the two samples of French canvas (Figure 8.4a and b) appear to be much more similar to each other, and quite different from the co-occurrence matrix for sand (Figure 8.4c). As such, the co-occurrence matrix looks like it can expose the underlying nature of texture better than the Fourier description. This is because the co-occurrence measures spatial relationships between brightness, as opposed to frequency content. This clearly gives alternative results. To generate results more quickly, the number of grey levels can be reduced by brightness scaling of the whole image, reducing the dimensions of the co-occurrence matrix, but this reduces discriminatory ability.

These matrices have been achieved by the implementation in Code 8.2. The subroutine `tex_cc` generates the co-occurrence matrix of an image `im` given a maximum distance `d` and a number of directions `dirs`. If `d` and `dirs` are set to 1 and 4, respectively (as was used to generate the results in Figure 8.4), then the co-occurrence will be evaluated from a point and its four nearest neighbours. First, the co-occurrence matrix is cleared. Then, for each point in the image and for each value of distance and relative inclination (and so long as the two points are within the image), the element of the co-occurrence matrix indexed by the brightnesses of the two points is incremented. There is a dummy operation after the incrementing process: this has been introduced for layout reasons (otherwise the Mathcad code would stretch out too far sideways). Finally, the completed co-occurrence matrix is returned. Note that even though the co-occurrence matrix is symmetric, this factor cannot be used to speed its production.

```
tex_cc(im,dist,dirs):=
            for x∈0..maxbri
              for y∈0..maxbri
                cocc_y,x←0
            for x∈0..cols(im)-1
              for y∈0..rows(im)-1
                for r∈1..dist
                                2·π
                  for θ∈0,─────..2·π
                                dirs
                      xc←floor(x+r·cos(θ))
                      yc←floor(y+r·sin(θ))
                      if(0≤yc)·(yc<rows(im))·(0≤xc)·(xc<cols(im))
                        cocc_{im_y,x, im_yc,xc}←cocc_{im_y,x, im_yc,xc}+1
                        I←1
            cocc
```

Code 8.2 Co-occurrence matrix generation

Again, we need measurements that describe these matrices. We shall use the measures of entropy, inertia and energy defined earlier. The results are shown in Code 8.3. Unlike visual analysis of the co-occurrence matrices, the difference between the measures of the three textures is less clear: classification from them will be discussed later. Clearly, the co-occurrence matrices have been reduced to only three different measures. In principle, these measurements are again invariant to linear shift in illumination (by virtue of brightness comparison) and to rotation (since order is of no consequence in their description and rotation only affects co-occurrence by discretization effects). As with Fourier, scale can affect the structure of the co-occurrence matrix, but the description can be made scale invariant.

entropy(CCD20)=7.052·10⁵ inertia(CCD20)=5.166·10⁸ energy(CCD20)=5.16·10⁸	entropy(CCD21)=5.339·10⁵ inertia(CCD21)=1.528·10⁹ energy(CCD21)=3.333·10⁷	entropy(CCD29)=6.445·10⁵ inertia(CCD29)=1.139·10⁸ energy(CCD29)=5.315·10⁷
(a) French canvas (detail)	(b) French canvas	(c) Beach sand

Code 8.3 Measures of co-occurrence matrices of the three Brodatz textures

Grey-level difference statistics (a first order measure) were later added to improve descriptional capability (Weszka et al., 1976). Other statistical approaches include the statistical feature matrix (Wu and Chen, 1992), with the advantage of faster generation.

8.3.4 Combination approaches

The previous approaches have assumed that we can represent textures by purely structural or purely statistical description, combined in some appropriate manner. Since texture is not an exact quantity, and is more a nebulous one, there are many alternative descriptions. One approach (Chen et al., 1995) suggested that texture combines geometric structures (e.g. in patterned cloth) with statistical ones (e.g. in carpet) and has been shown to give good performance in comparison with other techniques, and using the *whole* Brodatz dataset. The technique is called statistical geometric features (SGF), reflecting the basis of its texture description. This is not a dominant texture characterization: the interest here is that we shall now see the earlier *shape measures* in action, describing texture. Essentially, geometric features are derived from images, and then described by using statistics. The geometric quantities are derived from $NB - 1$ binary images **B** which are derived from the original image **P** (which has NB brightness levels). These binary images are given by

$$\mathbf{B}(\alpha)_{x,y} = \begin{vmatrix} 1 & \text{if } P_{x,y} = \alpha \\ 0 & \text{otherwise} \end{vmatrix} \quad \forall \alpha \in 1, NB \tag{8.9}$$

Then, the points in each binary region are connected into regions of 1s and 0s. Four geometric measures are made on these data. First, in each binary plane, the number of regions of 1s and 0s (the number of connected sets of 1s and 0s) is counted to give $NOC1$ and $NOC0$. Then, in each plane, each of the connected regions is described by its *irregularity*, which is a local shape measure of a region **R** of connected 1s giving irregularity $I1$ defined by

$$I1(\mathbf{R}) = \frac{1 + \sqrt{\pi} \max_{i \in \mathbf{R}} \sqrt{(x_i - \bar{x})^2 + (y_i - \bar{y})^2}}{\sqrt{N(\mathbf{R})}} - 1 \tag{8.10}$$

where x_i and y_i are coordinates of points within the region, $\bar{x}$ and $\bar{y}$ are the region's centroid (its mean x and y coordinates), and N is the number of points within (i.e. the area of) the region. The irregularity of the connected 0s, $I0(\mathbf{R})$, is similarly defined. When this is applied to the regions of 1s and 0s it gives two further geometric measures, $IRGL1(i)$ and $IRGL0(i)$, respectively. To balance the contributions from different regions, the irregularity of the regions of 1s in a particular plane is formed as a weighted sum $WI1(\alpha)$ as

$$WI1(\alpha) = \frac{\sum\limits_{\mathbf{R} \in B(\alpha)} N(\mathbf{R})I(\mathbf{R})}{\sum\limits_{\mathbf{R} \in P} N(\mathbf{R})} \qquad (8.11)$$

giving a single irregularity measure for each plane. Similarly, the weighted irregularity of the connected 0s is $WI0$. Together with the two counts of connected regions, $NOC1$ and $NOC0$, the weighted irregularities give the four geometric measures in SGF. The statistics are derived from these four measures. The derived statistics are the maximum value of each measure across all binary planes, M. Using $m(\alpha)$ to denote any of the four measures, the maximum is

$$M = \max_{\alpha i \in 1, NB} (m(\alpha)) \qquad (8.12)$$

the average, $\bar{m}$, is

$$\bar{m} = \frac{1}{255} \sum_{\alpha=1}^{NB} m(\alpha) \qquad (8.13)$$

the sample mean, $\bar{s}$, is

$$\bar{s} = \frac{1}{\sum\limits_{\alpha=1}^{NB} m(\alpha)} \sum_{\alpha=1}^{NB} \alpha m(\alpha) \qquad (8.14)$$

and the final statistic is the sample standard deviation, ssd, is

$$ssd = \sqrt{\frac{1}{\sum\limits_{\alpha=1}^{NB} m(\alpha)} \sum_{\alpha=1}^{NB} (\alpha - \bar{s})^2 m(\alpha)} \qquad (8.15)$$

The irregularity measure can be replaced by *compactness* (Section 7.3.1), but compactness varies with rotation, although this was not found to influence results much (Chen et al., 1995).

To implement these measures, we need to derive the sets of connected 1s and 0s in each of the binary planes. This can be achieved by using a version of the `connect` routine in hysteresis thresholding (Section 4.2.5). The reformulation is necessary because the `connect` routine just labels connected points, whereas the irregularity measures require a list of points in the connected region so that the centroid (and hence the maximum distance of a point from the centroid) can be calculated. The results for four of the measures (for the region of 1s, the maximum and average values of the number of connected regions and of the weighted irregularity) are shown in Code 8.4. Again, the set of measures is different for each texture. Note that the last measure, $\bar{m}(WI1)$, does not appear to offer much discriminatory capability here, whereas the measure $M(WI1)$ appears to be a much more potent descriptor. Classification, or *discrimination*, is used to select which class the measures refer to.

M(NOC1)=52.0	M(NOC1)=178	M(NOC1)=81
$\bar{m}$(NOC1)=8.75	$\bar{m}$(NOC1)=11.52	$\bar{m}$(NOC1)=22.14
M(WI1)=1.50	M(WI1)=1.42	M(WI1)=1.00
$\bar{m}$(WI1)=0.40	$\bar{m}$(WI1)=0.35	$\bar{m}$(WI1)=0.37
(a) French canvas (detail)	(b) French canvas	(c) Beach sand

Code 8.4 Four of the SGF measures of the three Brodatz textures

8.4 Classification

8.4.1 The *k*-nearest neighbour rule

In application, usually we have a description of a texture *sample* and we want to find which element of a database best matches that sample. This is *classification*: to associate the appropriate *class label* (type of texture) with the test sample by using the measurements that describe it. One way to make the association is by finding the member of the class (the sample of a known texture) with measurements that differ by the least amount from the test sample's measurements. In terms of *Euclidean* distance, the difference d between the M descriptions of a sample, $\mathbf{s}$, and the description of a known texture, $\mathbf{k}$, is

$$d = \sqrt{\sum_{i=1}^{M} (\mathbf{s}_i - \mathbf{k}_i)^2} \tag{8.16}$$

which is also called the L_2 *norm*. Alternative distance metrics include the L_1 *norm*, which is the sum of the modulus of the differences between the measurements

$$L_1 = \sum_{i=1}^{M} |\mathbf{s}_i - \mathbf{k}_i| \tag{8.17}$$

and the Bhattacharyya distance B

$$B = -\ln \sum_{i=1}^{M} \sqrt{\mathbf{s}_i \times \mathbf{k}_i} \tag{8.18}$$

but this appears to be used less, like other metrics such as the Matusita difference.

If we have M measurements of N known samples of textures and we have O samples of each, we have an M-dimensional *feature space* that contains the $N \times O$ points. If we select the point in the feature space that is closest to the current sample, then we have selected the sample's *nearest neighbour*. This is illustrated in Figure 8.5, where we have a two-dimensional feature space produced by the two measures made on each sample, measure 1 and measure 2. Each sample gives different values for these measures, but the samples of different classes give rise to clusters in the feature space where each cluster is associated with a single class. In Figure 8.5 we have seven samples of two known textures: class A and class B, depicted by X and O, respectively. We want to classify a test sample, depicted by +, as belonging either to class A or to class B (i.e. we assume that the training data contains representatives of all possible classes). Its nearest neighbour, the sample with least distance, is one of the samples of class A, so we could say that our test appears to be another sample of class A (i.e. the class label associated with it is class A). The clusters will be far apart for measures that have good discriminatory ability, whereas they will be overlap for measures that have poor discriminatory ability. That is how we can choose measures for particular tasks. Before that, let us look at how best to associate a class label with our test sample.

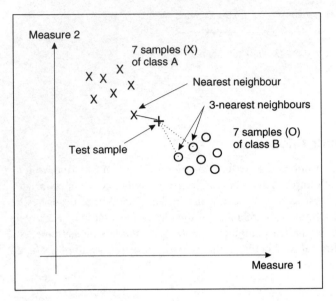

Figure 8.5 Feature space and classification

Classifying a test sample as the training sample to which it is closest in feature space is a specific case of a general classification rule known as the *k-nearest neighbour rule*. In this rule, the class selected is the *mode* of the sample's nearest k neighbours. By the k-nearest neighbour rule, for $k = 3$, we select the nearest three neighbours (those three with the least distance) and their mode, the maximally represented *class*, is attributed to the sample. In Figure 8.5, the 3-nearest neighbour is class B, since the three nearest samples contain one from class A (its nearest neighbour) and two from class B. Since there are two elements of class B, the sample is attributed to this class by the 3-nearest neighbour rule. As such, selection from more than one point introduces a form of feature space smoothing and allows the classification decision not to be affected by noisy *outlier* points. This smoothing has greater effect for larger values of k. (Further details concerning a modern view of the k-nearest neighbour rule can be found in Michie et al., 1994).

A Mathcad implementation of the k-nearest neighbour rule is given in Code 8.5. The arguments are `test` (the vector of measurements of the test sample), `data` (the list of vectors of measurements of all samples), `size` (the value of k) and `no`. The final parameter `no` dictates the structure of the presented data and is the number of classes within that data. The training data is presumed to have been arranged so that samples of each class are all stored together. For two classes in the training data, `no = 2`, where each occupies one half (the same situation as in Figure 8.5). If `no = 3` then there are three classes, each occupying one-third of the complete dataset; the first third contains the first class, the second third contains samples of another class and the remaining third contains samples of the final class. In application, first the distances between the current sample, `test`, and all other samples are evaluated by using the function distance. Then the k nearest neighbours are selected to form a vector of distances min; these are the k neighbours that are closest (in the feature space) to the sample test. The number of feature space splits fsp is the spacing between the classes in the data. The class that occurs the most number of times in the set of size nearest neighbours is then returned as the k-nearest neighbour, by incrementing the class number to which each of the k neighbours is associated. (If no such decision is possible, i.e. there is no maximally represented class, the technique can be arranged to return the class of the nearest neighbour, by default.)

```
k_nn(test,data,size,no):=
          for i∈0..rows(data)-1
              dist_i←0
              for j∈0..cols(data)-1
                dist_i←distance(test,data,i)
          for i∈0..size-1
              posmin←coord(min(dist),dist)
              dist_posmin←max(dist)+1
              min_i←posmin
          fsp← rows(data)
                ─────────
                   no
          for j∈1..no
            class_j←0
          for i∈0..size-1
            for j∈1..no
              class_j←class_j+1 if [min_i≥(j-1)·fsp]·(min_i<j·fsp)
          test_class←coord(max(class),class)
          test_class
```

Code 8.5 Implementing the k-nearest neighbour rule

The result of testing the k-nearest neighbour routine is illustrated on synthetic data in Code 8.6. Here, there are two different datasets. The first (Code 8.6a) has three classes of which there are three samples (each sample is a row of data, so this totals nine rows) and each sample is made up of three measurements (the three columns). As this is synthetic data, it can be seen that each class is quite distinct: the first class is for measurements around [1,2,3], the second class is around [4,6,8] and the third is around [8,6,3]. A small amount of noise has been added to the measurements. We then want to see the class associated with a test sample with measurements [4,6,8] (Code 8.6b). The 1-nearest nearest neighbour (Code 8.6c) associates it with the class with the closest measurements, which is class 2 as the test sample's nearest neighbour is the fourth row of data. (The result is class 1, class 2 or class 3.) The 3-nearest neighbour (Code 8.6d) is again class 2 as the nearest three neighbours are the fourth, fifth and sixth rows, and each of these is from class 2.

The second dataset (Code 8.6e) is two classes with three samples each made up of four measures. The test sample (Code 8.6f) is associated with class 1 by the 1-nearest neighbour (Code 8.6g), but with class 2 for the 3-nearest neighbour (Code 8.6h). This is because the test sample is closest to the sample in the third row. After the third row, the next two closest samples are in the fourth and sixth rows. The nearest neighbour is in a different class (class 1) to that

population1 := $\begin{bmatrix} 1 & 2 & 3 \\ 1.1 & 2 & 3.1 \\ 1 & 2.1 & 3 \\ 4 & 6 & 8 \\ 3.9 & 6.1 & 8.1 \\ 4.1 & 5.9 & 8.2 \\ 8.8 & 6.1 & 2.8 \\ 7.8 & 5.9 & 3.3 \\ 8.8 & 6.4 & 3.1 \end{bmatrix}$		population2 := $\begin{bmatrix} 2 & 4 & 6 & 8 \\ 2.1 & 3.9 & 6.2 & 7.8 \\ 2.3 & 3.6 & 5.8 & 8.3 \\ 2.5 & 4.5 & 6.5 & 8.5 \\ 3.4 & 4.4 & 6.6 & 8.6 \\ 2.3 & 4.6 & 6.4 & 8.5 \end{bmatrix}$	
(a) Three classes, three samples, three features		(e) Two classes, three samples, four features	
`test_point1:=(4 6 8)`		`test_point2:=(2.5 3.8 6.4 8.3)`	
(b) First test sample		(f) Second test sample	
`k_nn(test_point1,population1,1,3)=2`		`k_nn(test_point2,population2,1,2)=1`	
(c) 1-nearest neighbour		(g) 1-nearest neighbour	
`k_nn(test_point1,population1,3,3)=2`		`k_nn(test_point2,population2,3,2)=2`	
(d) 3-nearest neighbour		(h) 3-nearest neighbour	

Code 8.6   Applying the k-nearest neighbour rule to synthetic data

of the next two nearest neighbours (class 2); a different result has occurred when there is more *smoothing* in the feature space (when the value of k is increased).

The Brodatz database contains 112 textures, but few descriptions have been evaluated on the whole database, usually concentrating on a subset. It has been shown that the SGF description can afford better classification capability than the co-occurrence matrix and the Fourier transform features (described by Liu's features) (Chen et al., 1995). For experimental procedure, the Brodatz pictures were scanned into 256×256 images which were split into $16\ 64 \times 64$ subimages. Nine of the subimages were selected at random and results were classified using *leave-one-out cross-validation* (Lachenbruch and Mickey, 1968). *Leave-one-out* refers to a procedure where one of the samples is selected as the test sample and the others form the training data (this is the leave-one-out rule). *Cross-validation* is where the test is repeated for all samples: each sample becomes the test data once. In the comparison, the eight optimal Fourier transform features were used (Liu and Jernigan, 1990), and the five most popular measures from the co-occurrence matrix. The correct classification rate, the number of samples attributed to the correct class, showed better performance by the combination of statistical and geometric features (86%), as opposed to the use of single measures. The enduring capability of the co-occurrence approach was reflected by its performance (65%) in comparison with Fourier (33%; whose poor performance is rather surprising). An independent study (Walker and Jackway, 1996) has confirmed the experimental advantage of SGF over the co-occurrence matrix, based on a (larger) database of 117 cervical cell specimen images. Another study (Ohanian and Dubes, 1992) concerned the features that optimized classification rate and compared co-occurrence, fractal-based, Markov random field and Gabor-derived features. By analysis on synthetic and real imagery, via the *k*-nearest neighbour rule, the results suggested that co-occurrence offered the best overall performance. Wavelets (Porter and Canagarajah, 1996), Gabor wavelets and Gaussian Markov random fields have been compared (on a limited subset of the Brodatz database) to show that the wavelet-based approach had the best overall classification performance (in noise as well), together with the smallest computational demand.

8.4.2 Other classification approaches

Classification is the process by which we attribute a class label to a set of measurements. Essentially, this is the heart of pattern recognition: intuitively, there must be many approaches. These include *statistical* and *structural* approaches; a review can be found in Shalkoff (1992) and a more modern view in Cherkassky and Mulier (1998). One major approach is to use a *neural network*, which is a common alternative to using a classification rule. Essentially, modern approaches centre around using *multilayer perceptrons* with *artificial neural networks* in which the computing elements aim to mimic properties of neurons in the human brain. These networks require *training*, typically by error back-propagation, aimed to minimize classification error on the training data. At this point, the network should have learnt how to recognize the *test* data (they aim to learn its structure): the output of a neural network can be arranged to be class labels. Approaches using neural nets (Muhamad and Deravi, 1994) show how texture metrics can be used with neural nets as classifiers, while another uses cascaded neural nets for texture extraction (Shang and Brown, 1994). Neural networks lie within a research field that has shown immense growth in the past two decades. Further details may be found in Michie et al. (1994) and Bishop (1996; often a student favourite), and information more targeted at vision in Zhou and Chellappa (1992). *Support vector machines* (SVMs) (Vapnik, 1995) are one of the more popular approaches to data modelling and classification, more recently subsumed within *kernel methods* (Shawe-Taylor and Cristianini, 2004). Their advantages include excellent *generalization* capability, which concerns the ability to classify correctly samples that are not within feature space used for training. SVMs have found application in texture classification (Kim et al., 2002). Recently, interest in biometrics has focused on combining different classifiers, such as face and speech, and there are promising new approaches to accommodate this (Kittler, 1998; Kittler et al., 1998).

There are also methods aimed to improve classification capability by pruning the data to remove that which does not contribute to the classification decision. Guided ways that investigate the potency of measures for analysis are known as *feature (subset) selection*. *Principal components analysis* (Appendix 4) can reduce dimensionality, orthogonalize and remove redundant data. There is also *linear discriminant analysis* (also called *canonical analysis*) to improve class separability, while concurrently reducing cluster size (it is formulated concurrently to minimize the within-class distance and maximize the between-class distance). There are also algorithms aimed at choosing a reduced set of features for classification: feature selection for improved discriminatory ability; a comparison can be found in Jain and Zongker (1997). Alternatively, the basis functionals can be chosen in such a way as to improve classification capability.

8.5 Segmentation

To *segment* an image according to its texture, we can measure the texture in a chosen region and then classify it. This is equivalent to *template convolution*, but where the result applied to pixels is the class to which they belong, as opposed to the usual result of template convolution.

Here, we shall use a 7×7 template size: the texture measures will be derived from the 49 points within the template. First, though, we need data from which we can make a classification decision, the training data. This depends on a chosen application. Here, we shall consider the problem of segmenting the eye image into regions of *hair* and *skin*.

This is a two-class problem for which we need samples of each class, samples of skin and hair. We will take samples of each of the two classes; in this way, the classification decision is as illustrated in Figure 8.5. The texture measures are the energy, entropy and inertia of the co-occurrence matrix of the 7×7 region, so the feature space is three-dimensional. The training data is derived from regions of hair and from regions of skin, as shown in Figure 8.6(a) and (b), respectively. The first half of this data is the samples of hair, the other half is samples of the skin, as required for the k-nearest neighbour classifier of Code 8.5.

We can then segment the image by classifying each pixel according to the description obtained from its 7×7 region. Clearly, the training samples of each class should be classified correctly. The result is shown in Figure 8.7(a). Here, the top left corner is first (correctly) classified as hair, and the top row of the image is classified as hair until the skin commences (note that

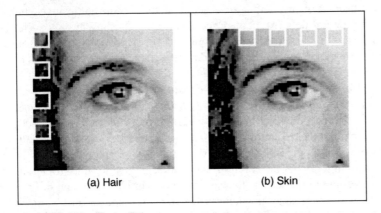

(a) Hair (b) Skin

Figure 8.6 Training regions for classification

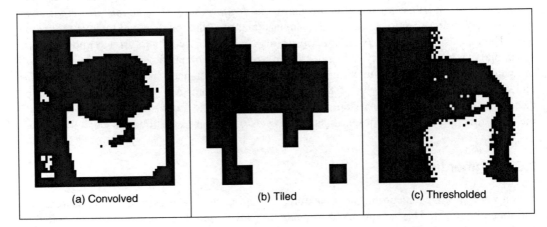

(a) Convolved (b) Tiled (c) Thresholded

Figure 8.7 Segmenting the eye image into two classes

344 *Feature Extraction and Image Processing*

the border inherent in template convolution reappears). In fact, much of the image appears to be classified as expected. The eye region is classified as hair, but this is a somewhat arbitrary decision; it is simply that hair is the closest texture feature. Some of the darker regions of skin are classified as hair, perhaps the result of training on regions of brighter skin.

This is a computationally demanding process. An alternative approach is simply to classify regions as opposed to pixels. This is the tiled approach, with the result shown in Figure 8.7(b). The resolution is very poor: the image has effectively been reduced to a set of 7×7 regions, but it is much *faster*, requiring only 2% of the computation of the convolution approach.

A comparison with the result achieved by uniform thresholding is given, for comparison, in Figure 8.7(c). This is equivalent to pixel segmentation by brightness alone. There are no regions where the hair and skin are mixed and in some ways the result appears superior. This is in part due to the simplicity in implementation of texture segmentation. However, the result of thresholding depends on *illumination* level and on appropriate choice of the threshold value. The texture segmentation method is completely *automatic* and the measures are known to have *invariance* properties to illumination, as well as other factors. In addition, in uniform thresholding there is no extension possible to separate *more* classes (except perhaps to threshold at differing brightness levels).

8.6 Further reading

There is much further reading in the area of texture description, classification and segmentation, as evidenced by the volume of published work in this area. The best place to start is Maria Petrou's book (Petrou and Sevilla, 2006) (the same author as in edge detection). There is one fairly comprehensive, but dated, survey (Reed and du Buf, 1993). An updated review has a wide bibliography (Tuceryan and Jain, 1998). Another (Zhang and Tan, 2002) offers a review of the approaches that are invariant to rotation, translation, and affine or projective transforms, but texture is a large field of work to survey with many applications. Even though it is a large body of work, it is still only a subset of the field of pattern recognition. In fact, reviews of pattern recognition give many pointers to this fascinating and extensive field (e.g. Jain et al., 2000). In this text, the general paradigm is to extract features that describe the target and then to classify them for purposes of recognition. In vision-based systems such approaches are used in *biometrics*: ways of recognizing a person's identity by some innate human properties. The biometrics of major recent interest are *signatures, speech, irises* and *faces*, although there is work in other areas including hand geometry (as used in US immigration) and gait. The first text on biometrics is not very old (Jain et al., 1999) and surveys all major biometric approaches. (It has just been updated.) There is much interest in automatic target recognition in both military and commercial applications. This translates to medical studies, where the interest is in either diagnosis or therapy. Here, researchers seek to be able to identify and recognize normal or abnormal features within one of the many medical imaging modalities, for surgical purposes. This is the world of image processing and computer vision. But all these operations depend on *feature extraction*, which is why this text has concentrated on these basic methods, for no practical vision-based system yet exists without them. We finish here; we hope you enjoyed the book and will find it useful in your career or study. Certainly have a look at our website, http://www.ecs.soton.ac.uk/~msn/book/, as you will find more material there. Don't hesitate to send us your comments or any suggestions. À bientôt!

8.7 References

Bishop, C. M., *Neural Networks for Pattern Recognition*, Oxford University Press, Oxford, 1996

Bovik, A. C., Clark, M. and Geisler, W. S., Multichannel Texture Analysis using Localized Spatial Filters, *IEEE Trans. PAMI*, **12**(1), pp. 55–73, 1990

Brodatz, P., *Textures: A Photographic Album for Artists and Designers*, Reinhold, New York, 1968

Chen, Y. Q., Nixon, M. S. and Thomas, D. W., Texture Classification using Statistical Geometric Features, *Pattern Recog.*, **28**(4), pp. 537–552, 1995

Cherkassky, V. and Mulier, F., *Learning from Data*, Wiley, New York, 1998

Daugman, J., G., High Confidence Visual Recognition of Persons using a Test of Statistical Independence, *IEEE Trans. PAMI*, **18**(8), pp. 1148–1161, 1993

Dunn, D., Higgins, W. E. and Wakely, J., Texture Segmentation using 2-D Gabor Elementary Functions, *IEEE Trans. PAMI*, **16**(2), pp. 130–149, 1994

Efros, A. and Leung, T., Texture Synthesis by Non-Parametric Sampling, *Proc. ICCV*, pp. 1033–1038, 1999

Gimmel'farb, G. L. and Jain, A. K., On Retrieving Textured Images from an Image Database, *Pattern Recog.*, **28**(12), pp. 1807–1817, 1996

Grigorescu, S. E., Petkov, N. and Kruizinga, P., Comparison of Texture Features based on Gabor Filters, *IEEE Trans. Image Process.*, pp. 1160–1167, **11**(10), 2002

Haralick, R. M., Shanmugam, K. and Dinstein, I., Textural Features for Image Classification, *IEEE Trans. SMC*, **2**, pp. 610–621, 1973

Heckbert, P. S., Survey of Texture Mapping, *IEEE Comput. Graphics Applic.*, pp. 56–67, 1986

Jain, A. K. and Farrokhnia, F., Unsupervised Texture Segmentation using Gabor Filters, *Pattern Recog.*, **24**(12), pp. 1186–1191, 1991

Jain, A. K. and Zongker, D., Feature Selection: Evaluation, Application and Small Sample Performance, *IEEE Trans. PAMI*, **19**(2), pp. 153–158, 1997

Jain, A. K., Bolle, R. and Pankanti, S. (Eds), *Biometrics – Personal Identification in Networked Society*, Kluwer Academic Publishers, Norwell, MA, 1999

Jain, A. K., Duin, R. P. W. and Mao, J., Statistical Pattern Recognition: A Review, *IEEE Trans. PAMI*, **22**(1), pp. 4–37, 2000

Karru, K., Jain, A. K. and Bolle, R., Is There any Texture in an Image?, *Pattern Recog*, **29**(9), pp. 1437–1446, 1996

Kim, K. I., Jung K., Park, S. H. and Kim, H. J., Support Vector Machines for Texture Classification, *IEEE Trans. PAMI*, **24**(11), pp. 1542–1550, 2002

Kittler, J., Hatef, M., Duin, R. P. W. and Matas, J., On Combining Classifiers, *IEEE Trans. PAMI*, **20**(3), pp. 226–239, 1998

Kittler, J., Combining Classifiers: A Theoretical Framework, *Pattern Analysis Applic.*, **1**(1), pp. 18–27, 1998

Lachenbruch, P. A. and Mickey, M. R., Estimation of Error Rates in Discriminant Analysis, *Technometrics*, **10**, pp. 1–11, 1968

Laine, A. and Fan, J., Texture Classification via Wavelet Pattern Signatures, *IEEE Trans. PAMI*, **15**(11), pp. 1186–1191, 1993

Liu, S. S. and Jernigan, M. E., Texture Analysis and Discrimination in Additive Noise, *CVGIP*, **49**, pp. 52–67, 1990

Lu, C. S., Chung, P. C. and Chen, C. F., Unsupervised Texture Segmentation via Wavelet Transform, *Pattern Recog.*, **30**(5), pp. 729–742, 1997

Michie, D., Spiegelhalter, D. J. and Taylor, C. C. (Eds), *Machine Learning, Neural and Statistical Classification*, Ellis Horwood, Hemel Hempstead, 1994

Muhamad, A. K., Deravi, F., Neural Networks for the Classification of Image Texture, *Engng Applic. Artif. Intell.*, **7**(4), pp. 381–393, 1994

Ohanian, P. P. and Dubes, R. C., Performance Evaluation for Four Classes of Textural Features, *Pattern Recognition*, **25**(8), pp. 819–833, 1992

Petrou, M. and Sevilla, O. G., *Image Processing: Dealing with Texture*, Wiley, New York, 2006

Pichler, O., Teuner, A. and Hosticka, B. J., A Comparison of Texture Feature Extraction using Adaptive Gabor Filtering, Pyramidal and Tree Structured Wavelet Transforms, *Pattern Recog.*, **29**(5), pp. 733–742, 1996

Porter, R. and Canagarajah, N., Robust Rotation-Invariant Texture Classification: Wavelet, Gabor Filter and GRMF Based Schemes, *IEE Proc. Vision Image Signal Process.*, **144**(3), pp. 180–188, 1997

Pratt, W. K., *Digital Image Processing*, Wiley, New York, 1992

Randen, T. and Husoy, J. H., Filtering for Texture Classification: A Comparative Study, *IEEE Trans. PAMI*, **21**(4), pp. 291–310, 2000

Reed, T. R. and du Buf, H., A Review of Recent Texture Segmentation and Feature Extraction Techniques, *CVGIP: Image Understanding*, **57**(3) pp. 359–372, 1993

Shalkoff, R. J., *Pattern Recognition – Statistical, Structural and Neural Approaches*, Wiley and Sons, New York, 1992

Shang, C. G. and Brown, K., Principal Features-Based Texture Classification with Neural Networks, *Pattern Recog.*, **27**(5), 675–687, 1994

Shawe-Taylor, J. and Cristianini, N., *Kernel Methods for Pattern Analysis*, Cambridge University Press, Cambridge, 2004

Tuceryan, M. and Jain, A. K., Texture Analysis, In: C. H. Chen, L. F. Pau and P. S. P. Wang (Eds), *The Handbook of Pattern Recognition and Computer Vision*, 2nd edn, pp. 207–248, World Scientific Publishing Co., Singapore, 1998

Vapnik, V., *The Nature of Statistical Learning Theory*, Springer, New York, 1995

Walker, R. F. and Jackway, P. T., Statistical Geometric Features – Extensions for Cytological Texture Analysis, *Proc. 13th ICPR*, Vienna, Vol. II (Track B), pp. 790–794, 1996

Weska, J. S., Dyer, C. R. and Rosenfeld, A., A Comparative Study of Texture Measures for Terrain Classification, *IEEE Trans. SMC*, **SMC-6**(4), pp. 269–285, 1976

Wu, C. M. and Chen, Y. C., Statistical Feature Matrix for Texture Analysis, *CVGIP: Graphical Models Image Process.*, **54**, pp. 407–419, 1992

Wu, W. and Wei, S., Rotation and Gray-Scale Transform-Invariant Texture Classification using Spiral Resampling, Subband Decomposition and Hidden Markov Model, *IEEE Trans. Image Process.*, **5**(10), pp. 1423–1434, 1996

Zhang, J. and Tan T., Brief review of invariant texture analysis methods, *Pattern Recog.*, **35**, pp. 735–747, 2002

Zhou, Y.-T. and Chellappa, R., *Artificial Neural Networks for Computer Vision*, Springer, New York, 1992

9

Appendix 1: Example worksheets

9.1 Example Mathcad worksheet for Chapter 3

The Mathcad worksheet has been typeset in this edition and the worksheets could be made to look (something) like this (but currently do not).

The appearance of the worksheets depends on the configuration of your system and of the Mathcad set up. To show you how they look, but as black and white only, here is part of a typeset version of the shortest worksheet. Note that the appearance of the real worksheet will depend largely on the setup of your machine.

Chapter 3 Basic image processing operations: CHAPTER3.MCD

This worksheet is the companion to Chapter 3 and implements the basic image processing operations described therein. The worksheet follows the text directly and allows you to process the eye image.

This chapter concerns basic image operations, essentially those which alter a pixel's *value* in a chosen way. We might want to make an image *brighter* (if it is too dark), or to remove contamination by *noise*. For these, we would need to make the pixel values *larger* (in some controlled way) or to change the pixel's value if we suspect it to be wrong, respectively. Let's start with images of pixels, by reading in the image of a human eye.

```
eye := READBMP(eye_orig)
```

We can view (part) of the image as a matrix of pixels

eye =

	0	1	2	3	4	5	6	7	8	9
0	115	117	130	155	155	146	146	135	115	132
1	135	130	139	155	141	146	146	115	115	135
2	139	146	146	152	152	155	117	117	117	139
3	139	144	146	155	155	146	115	114	117	139
4	139	146	146	152	150	136	117	115	135	139
5	146	146	146	155	149	130	115	137	135	145
6	147	146	142	150	136	115	132	146	146	146
7	146	141	155	152	130	115	139	139	146	146
8	136	145	160	141	115	129	139	147	146	141
9	117	146	155	130	115	115	137	149	141	139
10	132	152	150	130	115	115	142	149	141	118
11	137	149	136	130	130	114	135	139	141	139
12	137	145	130	117	115	115	117	117	132	132

or we can view it as an image (viewed using Mathcad's picture facility) as
This image is a 64 pixels wide and 64 pixels in height. Let's check:

    cols(eye)=64 rows(eye)=64

The gives us 4096 pixels. Each pixel is an *eight bit byte* (NB. it's stored in .BMP format), so this gives us 256 possible *intensity levels*, starting at zero and ending at 255. It is more common to use larger (say 256×256) images, but you won't be tempted to use much larger ones in Mathcad. It's very common to use 8 bits for pixels, as this is well suited to digitized video information.

We describe the occupation of intensity levels by a *histogram*. This is a *count* of all pixels with a specified brightness level, plotted against brightness level. As a function, we can calculate it by:

$$
\text{histogram(pic)} :=
\begin{array}{ll}
\text{for bright} \in 0..255 & \text{8 bits give 256 levels, 0..255} \\
\quad \text{pixels\_at\_level}_{\text{bright}} \leftarrow 0 & \text{Initialize histogram} \\
\text{for } x \in 0..\text{cols(pic)}-1 & \text{Cover whole picture} \\
\quad \text{for } y \in 0..\text{rows(pic)}-1 & \\
\quad\quad \text{level} \leftarrow \text{pic}_{y,x} & \text{Find level} \\
\quad\quad \text{pixels\_at\_level}_{\text{level}} & \text{Increment points at} \\
\quad\quad \leftarrow \text{pixels\_at\_level}_{\text{level}}+1 & \text{specified levels} \\
\text{pixels\_at\_level} & \text{Return histogram}
\end{array}
$$

So let's work out the histogram of our eye image:

    eye_histogram := histogram(eye)

To display it, we need a horizontal axis which gives the range of brightness levels

    bright := 0..255

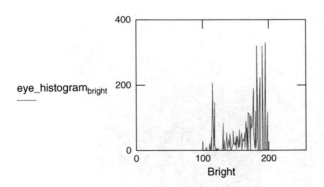

So here's the histogram of our picture of the eye image, p. The *bright* pixels relate mainly to the *skin*, the *darker* ones to the *hair*.

The most common *point operator* replaces each pixel by a scaled version of the original value. We therefore multiply each pixel by a number (like a *gain*), by specifying a function scale which is fed the picture and the gain, or a *level* shift (upwards or downwards). The function `scale` takes a picture `pic` and multiplies it by gain and adds a `level`

scale(pic,gain,level):= | for x∈0..cols(pic)-1   Address the whole picture
                        | for y∈0..rows(pic)-1
                        | newpic_y,x←floor   Multiply pixel
                        | (gain·pic_y,x+level)   by gain and add level
                        | newpic   Output the picture

So let's apply it: `brighter := scale(eye, 1.2, 10)`
You can change the settings of the parameters to see their effect; that's why you've got this electronic document. Try making it brighter and darker. What happens when the gain is too big (>1.23)?
So our new picture looks like (using Mathcad's picture display facility):

brighter =

	0	1	2	3	4	5	6	7	8	9
0	148	150	166	196	196	185	185	172	148	168
1	172	166	176	196	179	185	185	148	148	172
2	176	185	185	192	192	196	150	150	150	176
3	176	182	185	196	196	185	148	146	150	176
4	176	185	185	192	190	173	150	148	172	176
5	185	185	185	196	188	166	148	174	172	184
6	186	185	180	190	173	148	168	185	185	185
7	185	179	196	192	166	148	176	176	185	185
8	173	184	202	179	148	164	176	186	185	179
9	150	185	196	166	148	148	174	188	179	176
10	168	192	190	166	148	148	180	188	179	151
11	174	188	173	166	166	146	172	176	179	176
12	174	184	166	150	148	148	150	150	168	168

Processed

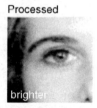

Original

The difference is clear in the magnitude of the pixels; those in the 'brighter' image are much larger than those in the original image, as well as by comparison of the processed with the original image. The difference between the images is much clearer when we look at the histogram of the brighter image.
So let's have a look at our scaled picture: `b_eye_hist := histogram(brighter)`

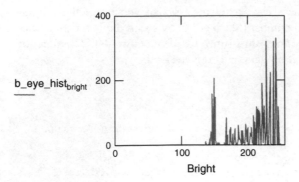

b_eye_hist_bright

Bright

Which is what we expect; it's just been moved *along* the brightness axis (it now starts well after 100), and reveals some detail in the histogram which was obscured earlier.

Do you want to read more for this and other Chapters? Then download the worksheets, and have a go!

9.2 Example Matlab worksheet for Chapter 4

This is part of the Matlab worksheet for Chapter 4. Essentially, the text is a Matlab script and the subroutines called and the images provided are set into figures.

```
%Chapter 4 Low-Level Feature Extraction and Edge Detection: CHAPTER4.M
%Written by: Mark S. Nixon

disp ('Welcome to the Chapter4 script')
disp ('This worksheet is the companion to Chapter 4 and is an
introduction.')
disp ('The worksheet follows the text directly and allows you to
process basic images.')

%Let's first empty the memory
clear

%Let's initialise the display colour
colormap(gray);

disp (' ')
disp ('Let us use the image of an eye.')
disp ('When you are ready to move on, press RETURN')
%read in the image
eye=imread('eye.jpg','jpg');
%images are stored as integers, so we need to double them for Matlab
%we also need to ensure we have a greyscale, not three colour planes
eye=double(eye(:,:,1));
%so let's display it
subplot(1,1,1), imagesc(eye);
plotedit on, title ('Image of an eye'), plotedit off
pause;
disp(' ')
```

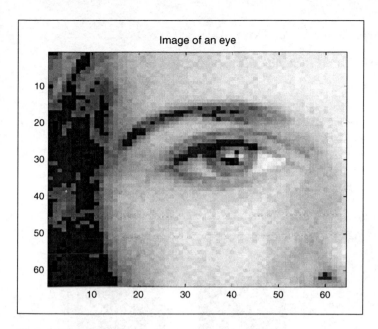

Figure 9.1 Image of an eye

```
disp ('We detect vertical edges by differencing horizontally adjacent')
disp ('points. Note how clearly the edge of the face appears')
%so we'll call the edge_x operator.

vertical=edge_x(eye);
imagesc(vertical);
plotedit on, title ('Vertical edges of an eye'), plotedit off
pause;
```

```
function vertical_edges=edge_x(image)
%Find edges by horizontal differencing
%
%Usage: [new image]=edge_x(image)
%
%Parameters: image-array of points
%
%Author: Mark S. Nixon

%get dimensions
[rows,cols]=size(image);

%set the output image to black
vertical_edges=zeros(rows,cols);
%this is equivalent to
vertical_edges(1:rows,1:cols)=0

%then form the difference between
horizontal successive points
for x=1:cols-1 %address all columns
                except border
  for y=1:rows %address all rows
    vertical_edges(y,x)=
        abs(image(y,x)-image(y,x+1));
  end
end
```

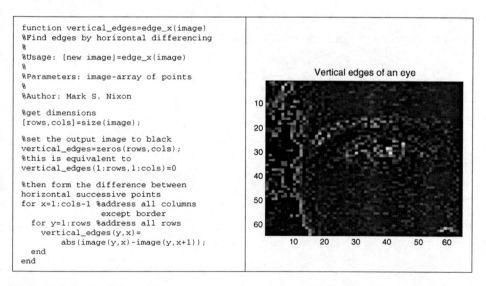

Figure 9.2 Vertical edge detection

```
disp (' ')
disp ('We detect horizontal edges by differencing vertically adjacent
points')
disp ('Notice how the side of the face now disappears, whereas the')
disp ('eyebrows appear')
%so we'll call the edge_y operator
subplot(1,2,2), horizontal=edge_y(eye);
subplot(1,2,1), imagesc(horizontal);
plotedit on, title ('Horizontal edges of an eye'), plotedit off
subplot(1,2,2), imagesc(vertical);
plotedit on, title ('Vertical edges of an eye'), plotedit off
pause;
```

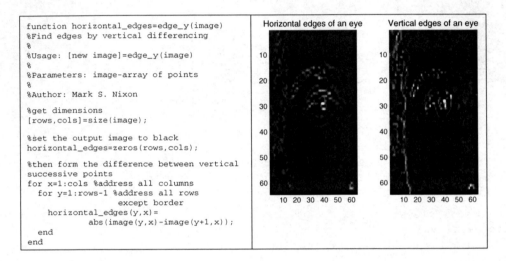

Figure 9.3   Vertical edge detection

Do you want to read more for this and other chapters? Then download the worksheets, and
have a go!

10

Appendix 2: Camera geometry fundamentals

10.1 Image geometry

This book has focused on techniques of image processing that use intensity or colour values of pixels to enhance and analyse images. Other image techniques include information about the geometry of image acquisition. These techniques are studied in the computer vision area and are mainly applied to three-dimensional (3D) scene analysis (Trucco and Verri, 1998; Hartley and Zisserman, 2001). This appendix does not cover computer vision techniques, but gives an introduction to the fundamental concepts of the geometry of computer vision. It aims to complement the concepts in Chapter 1 by increasing the background knowledge of how camera geometry is mathematically modelled.

As discussed in Chapter 1, an image is formed by a complex process involving optics, electronics and mechanical devices. This process maps information in a scene into pixels in an image. A camera model uses mathematical representations to describe this process. Different models include different aspects of the image formation and they are based on different assumptions or simplifications. This appendix explains basic aspects of common camera geometry models.

10.2 Perspective camera

Figure 10.1 shows the model of the *perspective* camera. This model is also known as the *pinhole* camera, since it describes the image formation process of a simple optical device with a small hole. This device is known as a *camera obscura* and it was developed in the sixteenth century as an artist's aid. Light going through a pinhole projects an image of a scene onto a back screen. The pinhole is called the centre of projection. Thus, a pixel is obtained by intersecting the image plane with the line between the 3D point and the centre of projection. In the projected image, parallel lines intersect at infinity, giving a correct perspective.

Although based on an ancient device, this model represents an accurate description of modern cameras where light is focused in a single point using lenses. In Figure 10.1, the centre of projection corresponds to the pinhole. Light passes through the point and is projected in the image plane. Figure 10.2 illustrates an alternative configuration where light is focused back to the image plane. The models are equivalent: the image is formed by projecting points through a single point; the point $\mathbf{x}_p$ is mapped into the point $\mathbf{x}_i$ in the image plane and the *focal length* determines the *zoom* distance.

355

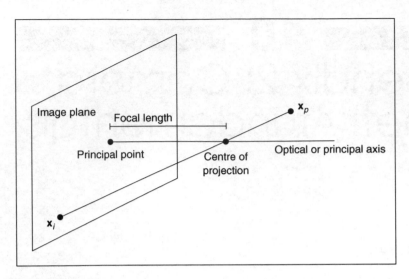

Figure 10.1 Pinhole model of perspective camera

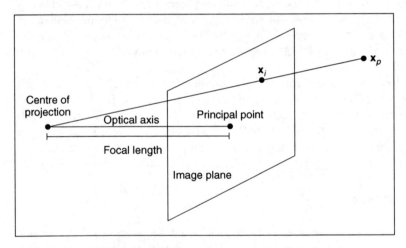

Figure 10.2 Perspective camera

The perspective camera model is formulated by an equation that describes how a point in space is mapped into an image, where the centre of projection is behind the image plane. This formulation can be developed using algebraic functions; nevertheless, the notation is greatly simplified by using matrix representations. In matrix form, points can be represented in Euclidean coordinates, yet a simpler notation is developed using *homogeneous coordinates*. Homogeneous coordinates simplify the formulation since translations and rotations are represented as matrix multiplications. In addition, homogeneous coordinates represent the projection of points and planes as a simple multiplication. Thus, before formulating the model of the perspective camera, we first review the basic concepts of homogeneous coordinates.

10.3 Perspective camera model

10.3.1 Homogeneous coordinates and projective geometry

Euclidean geometry is algebraically represented by the *Cartesian coordinate system*, in which points are defined by tuples of numbers. Each number is related to one axis and a set of axes determines the dimension. This representation is a very natural way of describing our 3D world and is very useful in image processing to describe pixels in two-dimensional (2D) images. Cartesian coordinates are convenient to describe angles and lengths and they are simply transformed by matrix algebra to represent translations, rotations and changes of scale. However, the relationship defined by projections cannot be described with the same algebraic simplicity.

Projective geometry is algebraically represented by the homogeneous coordinate system. This representation is a natural way of formulating how we relate camera coordinates to 'real-world' coordinates: the relation between image and physical space. Its major advantages are that image transformations such as rotations, change of scale and projections become matrix multiplications. Projections provide perspective, which corresponds to the distance of objects and affects their size in the image.

It is possible to map points from Cartesian coordinates into the homogeneous coordinates. The 2D point with Cartesian coordinates

$$\mathbf{x}_c = [x \quad y]^T \tag{10.1}$$

is mapped into homogeneous coordinates to the point

$$\mathbf{x}_h = [wx \quad wy \quad w]^T \tag{10.2}$$

where w is an arbitrary scalar. Notice that a point in Cartesian coordinates is mapped into several points in homogeneous coordinates; one point for any value of w. This is why homogeneous coordinates are also called redundant coordinates. We can use the definition in Equation 10.2 to obtain a mapping from homogeneous coordinates to Cartesian coordinates. That is,

$$x = wx/w \quad \text{and} \quad y = wy/w \tag{10.3}$$

The homogeneous representation can be extended to any dimension. For example, a 3D point in Cartesian coordinates

$$\mathbf{x}_c = [x \quad y \quad z]^T \tag{10.4}$$

is mapped into homogeneous form as

$$\mathbf{x}_h = [wx \quad wy \quad wz \quad w]^T \tag{10.5}$$

This point is mapped back to Cartesian coordinates by

$$x = wx/w, \quad y = wy/w \quad \text{and} \quad z = wz/w \tag{10.6}$$

Although it is possible to map points from Cartesian coordinates to homogeneous coordinates and vice versa, points in both systems define different geometric spaces. Cartesian coordinates define the Euclidean space and the points in homogeneous coordinates define the projective space. The projective space distinguishes a particular class of points defined when the last coordinate is zero. These are known as ideal points, and to understand them we need to understand how a line is represented in projective space. This is related to the concept of *duality*.

10.3.1.1 Representation of a line and duality

The homogeneous representation of points has a very interesting connotation that relates points and lines. Let us consider the equation of a 2D line in Cartesian coordinates:

$$Ax + By + C = 0 \qquad (10.7)$$

The same equation in homogeneous coordinates becomes

$$Ax + By + Cz = 0 \qquad (10.8)$$

What is interesting is that points and lines now become indistinguishable. Both a point $[x \ y \ z]^T$ and a line $[A \ B \ C]^T$ are represented by triplets and they can be interchanged in the homogeneous equation of a line. Similarly, in the 3D projective space, points are indistinguishable from planes. This symmetry is known as the duality of the projective space, which can be combined with the concept of concurrence and incidence to derive the principle of duality (Aguado et al., 2000). The principle of duality constitutes an important concept for understanding the geometric relationship in the projective space and the definition of the line can be used to derive the concept of ideal points.

10.3.1.2 Ideal points

We can use the algebra of homogeneous coordinates to find the intersection of parallel lines, planes and hyperplanes. For simplicity, let us consider lines in the 2D plane. In the Cartesian coordinates in Equation 10.7, two lines are parallel when their slopes $y' = -A/B$ are the same. Thus, to find the intersection between two parallel lines in the homogeneous form in Equation 10.8, we need to solve the following system of equations:

$$A_1 x + B_1 y + C_1 z = 0$$
$$A_2 x + B_2 y + C_2 z = 0 \qquad (10.9)$$

for $A_1/B_1 = A_2/B_2$. By dividing the first equation by B_1 and the second equation by B_2, and subtracting the second equation from the first, we have:

$$(C_2 - C_1) z = 0 \qquad (10.10)$$

Since we are considering different lines, $C_2 \neq C_1$ and consequently $z = 0$. That is, the intersection of parallel lines is defined by points of the form

$$\mathbf{x}_h = [x \quad y \quad 0]^T \qquad (10.11)$$

Similarly, in 3D, the intersection of parallel planes is defined by the points given by

$$\mathbf{x}_h = [x \quad y \quad z \quad 0]^T \qquad (10.12)$$

Since parallel lines are assumed to intersect at infinity, points with the last coordinate equal to zero are called points at infinity. They are also called ideal points and these points plus all the other homogeneous points form the projective space.

The points in the projective space can be visualized by extending the Euclidean space as shown in Figure 10.3. This figure illustrates the 2D projective space as a set of points in the 3D Euclidean space. According to Equation 10.3, points in the homogeneous space are mapped into the Euclidean space when $z = 1$. In the figure, this plane is called the Euclidean plane. Figure 10.3 shows two points in the Euclidean plane. These points define a line that is shown as a dotted line and it extends to infinity in the plane. In homogeneous coordinates, points in the Euclidean plane become rays from the origin in the projective space. Each point in the ray is given by a different value of z. The homogeneous coordinates of the line in the Euclidean plane define the plane between the two rays in the projective space. When two lines intersect in the Euclidean plane, they define a ray that passes through the intersection point in the Euclidean plane. However, if the lines are parallel, then they define an ideal point. That is, a point in the plane $z = 0$.

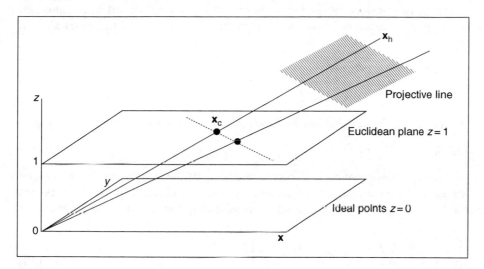

Figure 10.3 Model of the 2D projective space

Notice that the origin $[0\ 0\ 0]^\mathrm{T}$ is ambiguous since it can define any point in homogeneous coordinates or an ideal point. To avoid this ambiguity this point is not considered to be part of the projective space. Also remember that the concepts of point and line are indistinguishable, so it is possible to draw a dual diagram, where points become lines and vice versa.

10.3.1.3 *Transformations in the projective space*

In practice, perhaps the most relevant aspect of homogeneous coordinates is the way in which transformations are algebraically represented. *Transformations* in Cartesian coordinates are known as *similarity* or *rigid* transformations since they do not change angle values. They define *rotations*, changes in *scale* and *translations* (position), and they can be algebraically represented by matrix multiplications and additions. A 2D point $\mathbf{x}_1$ is transformed to a point $\mathbf{x}_2$ by a similarity transformation as

$$\begin{bmatrix} x_2 \\ y_2 \end{bmatrix} = \begin{bmatrix} \cos(\theta) & \sin(\theta) \\ -\sin(\theta) & \cos(\theta) \end{bmatrix} \begin{bmatrix} s_x \\ & s_y \end{bmatrix} \begin{bmatrix} x_1 \\ y_1 \end{bmatrix} + \begin{bmatrix} t_x \\ t_y \end{bmatrix} \tag{10.13}$$

where θ is a rotation angle, $\mathbf{S} = [s_x\ s_y]^\mathrm{T}$ defines the scale and $\mathbf{T} = [t_x\ t_y]^\mathrm{T}$ the translation along each axis. This transformation can be generalized to any dimension and it is written in short form as

$$\mathbf{x}_2 = \mathbf{R}\,\mathbf{S}\,\mathbf{x}_1 + \mathbf{T} \tag{10.14}$$

Notice that in these transformations $\mathbf{R}$ is an orthogonal matrix. That is, its transpose is equal to its inverse, or $\mathbf{R}^T = \mathbf{R}^{-1}$.

There is a more general type of transformation known as *affine transformations*, where the matrix $\mathbf{R}$ is replaced by a matrix $\mathbf{A}$ that is not necessarily orthogonal. That is,

$$\mathbf{x}_2 = \mathbf{A}\,\mathbf{S}\,\mathbf{x}_1 + \mathbf{T} \tag{10.15}$$

Affine transformations do not preserve the value of angles, but they preserve parallel lines. The principles and theorems studied under similarities define Euclidean geometry and the principles and theorems under affine transformations define affine geometry.

In the projective space, transformations are called *homographies*. They are more general than similarity and affine transformations; they only preserve collinearities and cross ratios and they are defined in homogeneous coordinates. A 2D point $\mathbf{x}_1$ is transformed to a point $\mathbf{x}_2$ by a homography as

$$\begin{bmatrix} x_2 \\ y_2 \\ w_2 \end{bmatrix} = \begin{bmatrix} h_{1,1} & h_{1,2} & h_{1,3} \\ h_{2,1} & h_{2,2} & h_{2,3} \\ h_{3,1} & h_{3,2} & h_{3,3} \end{bmatrix} \begin{bmatrix} x_1 \\ y_1 \\ w_1 \end{bmatrix} \tag{10.16}$$

This transformation can be generalized to other dimensions and it is written in short form as

$$\mathbf{x}_2 = H \, \mathbf{x}_1 \tag{10.17}$$

Notice that a similarity transformation is a special case of an affine transformation and that an affine transformation is a special case of a homography. Thus, rigid and affine transformations can be expressed as homographies. For example, a rigid transformation for a 2D point can be defined as

$$\begin{bmatrix} x_2 \\ y_2 \\ 1 \end{bmatrix} = \begin{bmatrix} s_x \cos(\theta) & s_x \sin(\theta) & t_x \\ -s_y \sin(\theta) & s_y \cos(\theta) & t_y \\ 0 & 0 & 1 \end{bmatrix} \begin{bmatrix} x_1 \\ y_2 \\ 1 \end{bmatrix} \tag{10.18}$$

Or in a more general form as

$$\mathbf{x}_2 = \begin{bmatrix} \mathbf{R}\,\mathbf{S} & \mathbf{T} \\ 0 & 1 \end{bmatrix} \mathbf{x}_1 \tag{10.19}$$

An affine transformation is defined as

$$\mathbf{x}_2 = \begin{bmatrix} \mathbf{A} & \mathbf{T} \\ 0 & 1 \end{bmatrix} \mathbf{x}_1 \tag{10.20}$$

The zeros in the last row define a transformation in a plane; the plane where $z = 1$. According to the discussion in Section 10.3.2, this plane defines the Euclidean plane. Thus these transformations are limited to Euclidean points.

10.3.2 Perspective camera model analysis

The *perspective camera model* uses the algebra of the projective space to describe the way in which space points are mapped into an image plane. The mapping can also be defined using Euclidean transformations, but the algebra becomes too elaborate. By using homogeneous coordinates, the geometry of image formation is simply defined by the projection of a 3D point into the plane by one special type of homography known as a projection. In a projection the matrix $\mathbf{H}$ is not square, so a point in a higher dimension is mapped into a lower dimension. The perspective camera model is defined by a projection transformation:

$$\begin{bmatrix} w_i x_i \\ w_i y_i \\ w_i \end{bmatrix} = \begin{bmatrix} p_{1,1} & p_{1,2} & p_{1,3} & p_{1,4} \\ p_{2,1} & p_{2,2} & p_{2,3} & p_{2,4} \\ p_{3,1} & p_{3,2} & p_{3,3} & p_{3,4} \end{bmatrix} \begin{bmatrix} x_p \\ y_p \\ z_p \\ 1 \end{bmatrix} \tag{10.21}$$

This equation can be written in short form as

$$\mathbf{x}_i = \mathbf{P}\mathbf{x}_p \tag{10.22}$$

Here, we have changed the elements from h to p to emphasize that we are using a projection. Also, we use $\mathbf{x}_i$ and $\mathbf{x}_p$ to denote the space and image points, as introduced in Figure 10.1. Notice that the point in the image is in homogeneous form, so the coordinates in the image are given by Equation 10.3.

The matrix $\mathbf{P}$ models three geometric transformations, so it can be factorized as

$$\mathbf{P} = \mathbf{V}\,\mathbf{Q}\,\mathbf{M} \tag{10.23}$$

The matrix $\mathbf{M}$ transforms the 3D coordinates of $\mathbf{x}_p$ to make them relative to the camera system. That is, it transforms world coordinates into camera coordinates. Notice that the point is not transformed, but we obtain its coordinates as if the camera were the origin of the coordinate system.

If the camera is posed in the world by a rotation $\mathbf{R}$ and a translation $\mathbf{T}$, then the transformation between world and camera coordinates is given by the inverse of rotation and translation. We define this matrix as

$$\mathbf{M} = [\mathbf{R} \quad \mathbf{T}] \tag{10.24}$$

or more explicitly as

$$\mathbf{M} = \begin{bmatrix} r_{1,1} & r_{1,2} & r_{1,3} & t_x \\ r_{2,1} & r_{2,2} & r_{2,3} & t_y \\ r_{3,1} & r_{3,2} & r_{3,3} & t_z \end{bmatrix} \tag{10.25}$$

The matrix $\mathbf{R}$ defines a rotation matrix and $\mathbf{T}$ a translation vector. The rotation matrix is composed by rotations along each axis. If α, β and γ are the rotation angles, then

$$\mathbf{R} = \begin{bmatrix} \cos(\alpha) & -\sin(\alpha) & 0 \\ \sin(\alpha) & \cos(\alpha) & 0 \\ 1 & 0 & 1 \end{bmatrix} \begin{bmatrix} \cos(\beta) & 0 & -\sin(\beta) \\ 0 & 1 & 0 \\ \sin(\beta) & 0 & \cos(\beta) \end{bmatrix} \begin{bmatrix} 1 & 0 & 0 \\ 0 & \cos(\gamma) & -\sin(\gamma) \\ 0 & \sin(\gamma) & \cos(\gamma) \end{bmatrix} \tag{10.26}$$

Once the points are made relative to the camera frame, the transformation $\mathbf{Q}$ obtains the coordinates of the point projected in the image. As illustrated in Figure 10.1, the focal length of a camera defines the distance between the centre of projection and the image plane. If f denotes the focal length of a camera, then

$$\mathbf{Q} = \begin{bmatrix} f & 0 & 0 \\ 0 & f & 0 \\ 0 & 0 & 1 \end{bmatrix} \tag{10.27}$$

To understand this projection, let us consider the way in which a point is mapped into the camera frame as shown in Figure 10.4. This figure illustrates the side view of the camera; to the right is the depth z-axis and from the top down is the y-axis. The image plane is shown as a dotted line. The point $\mathbf{x}_p$ is projected into $\mathbf{x}_i$ in the image plane. The tangent of the angle between the line from the centre of projection to $\mathbf{x}_p$ and the principal axis is given by

$$\frac{y_i}{f} = \frac{y_p}{z_p} \tag{10.28}$$

That is,

$$y_i = \frac{y_p}{z_p} f \tag{10.29}$$

Using a similar rationale we can obtain the value

$$\mathbf{x}_i = \frac{\mathbf{x}_p}{z_p} f \tag{10.30}$$

That is, the projection is obtained by multiplying by the focal length and by dividing by the depth of the point. Equation 10.27 multiplies each coordinate by the focal length and copies the depth value into the last coordinate of the point. However, since Equation 10.21 is in homogeneous coordinates, the depth value is used as a divisor when obtaining coordinates of the point according to Equation (10.3). Thus, projection can be simply defined by a matrix multiplication factor defined in Equation 10.27.

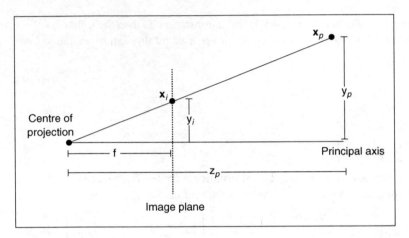

Figure 10.4 Projection of a point

The factors **M** and **Q** define the coordinates of a point in the image plane. However, the coordinates in an image are given in pixels. Thus, the last factor **V** is used to change from image coordinates to pixels. This transformation also includes a skew deformation to account for missed alignments that may occur in the camera system. The transformation **V** is defined as

$$\mathbf{V} = \begin{bmatrix} k_u & k_u \cot(\varphi) & u_0 \\ 0 & k_v \sin(\varphi) & v_0 \\ 0 & 0 & 1 \end{bmatrix} \tag{10.31}$$

The constants k_u and k_v define the number of pixels in a world unit, the angle φ defines the skew angle and (u_0, v_0) is the position of the principal point in the image.

Figure 10.5 illustrates the transformation in Equation 10.30. The image plane is shown as a dotted rectangle, but it actually extends to infinity. The image is delineated by the axes u and v. A point (x_1, y_1) in the image plane has coordinates (u_1, v_1) in the image frame. As previously discussed for Figure 10.1, the coordinates of (x_1, y_1) are relative to the principal point (u_0, v_0). As shown in Figure 10.5, the skew displaces the point form (u_0, v_0) by an amount given by

$$a_1 = y_1 \cot(\varphi) \quad \text{and} \quad c_1 = y_1/\sin(\varphi) \tag{10.32}$$

Thus, the new coordinates of the point after skew are

$$x_1 + y_1 \cot(\varphi) \quad \text{and} \quad y_1/\sin(\varphi) \tag{10.33}$$

To convert these coordinates to pixels, we need to multiply by the number of pixels to form a unit in the image plane. Finally, the point in pixel coordinates is obtained by adding the displacement (u_0, v_0) in pixels. That is,

$$u_1 = k_u x_1 + k_u y_1 \cot(\varphi) + u_0 \quad \text{and} \quad v_1 = k_v y_1/\sin(\varphi) + v_0 \tag{10.34}$$

These algebraic equations are expressed in matrix form in Equation 10.31.

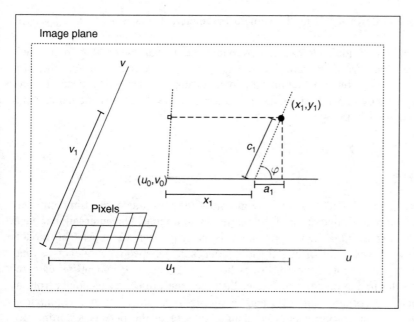

Figure 10.5 Image plane to pixels transformation

10.3.3 Parameters of the perspective camera model

The perspective camera model in Equation 10.21 has 12 elements. Thus, a particular camera model is completely defined by giving values to 12 unknowns. These unknowns are determined by the parameters of the transformations **M**, **Q** and **V**. The transformation **M** has three rotation angles (α, β, γ) and three translation parameters (t_x, t_y, t_z). The transformation **V** has a single parameter, f, while the transformation **Q** has the two translation parameters (u_0, v_0), two scale parameters (k_u, k_v) and one skew parameter φ. Thus, we need to set up 12 parameters to determine the elements of the projection matrix. However, one parameter can be eliminated by combining the matrices **V** and **Q**. That is, the projection matrix in Equation 10.23 can be written as

$$
\mathbf{P} = \begin{bmatrix} k_u & k_u \cot(\varphi) & u_0 \\ 0 & k_v \sin(\varphi) & v_0 \\ 0 & 0 & 1 \end{bmatrix} \begin{bmatrix} f & 0 & 0 \\ 0 & f & 0 \\ 0 & 0 & 1 \end{bmatrix} \begin{bmatrix} r_{1,1} & r_{1,2} & r_{1,3} & t_x \\ r_{2,1} & r_{2,2} & r_{2,3} & t_y \\ r_{3,1} & r_{3,2} & r_{3,3} & t_z \end{bmatrix}
\tag{10.35}
$$

or

$$
\mathbf{P} = \begin{bmatrix} s_u & s_u \cot(\varphi) & u_0 \\ 0 & s_v \sin(\varphi) & v_0 \\ 0 & 0 & 1 \end{bmatrix} \begin{bmatrix} r_{1,1} & r_{1,2} & r_{1,3} & t_x \\ r_{2,1} & r_{2,2} & r_{2,3} & t_y \\ r_{3,1} & r_{3,2} & r_{3,3} & t_z \end{bmatrix}
\tag{10.36}
$$

for

$$
s_u = fk_u \quad \text{and} \quad s_v = fk_v
\tag{10.37}
$$

Thus, the camera model is defined by the 11 camera parameters $(\alpha, \beta, \gamma, t_x, t_y, t_z, u_0, v_0, s_u, s_v, \varphi)$.

The camera parameters are divided into two groups to indicate the parameters that are internal or external to the camera. The intrinsic parameters are $(u_0, v_0, s_u, s_v, \varphi)$ and the extrinsic ones are $(\alpha, \beta, \gamma, t_x, t_y, t_z)$. In general, the intrinsic parameters do not change from scene to scene, so they are inherent to the system; they depend on the camera characteristics. The extrinsic parameters change by moving the camera in the world.

10.4 Affine camera

Although the perspective camera model is probably the most common model used in computer vision, there are alternative models that are useful in particular situations. One alternative model of reduced complexity that is useful in many applications is the *affine camera model*. This model is also called the *paraperspective* or *linear* model and it reduces the perspective model by setting the focal length f to *infinity*. Figure 10.6 illustrates how the perspective and affine camera models map points into the image plane. The figure illustrates the projection of points from a side view and it projects the corner points of a pair of objects represented by two rectangles. In the projective model, the projection produces changes of size in the objects according to their distance to the image plane; the far object is projected into a smaller area than the close object. The size and distance relationship is determined by the focal length f. As we increase the focal length, projection lines decrease their slope and become horizontal. As illustrated on the right of Figure 10.6, in the limit when the centre of projection is infinitely far away from the image plane, the lines not intersect and the objects have the same projected area in the image.

In spite of not accounting for changes in size due to distances, the affine camera provides a useful model when the depth position of objects in the scene with respect to the camera frame does not change significantly. This is the case in many indoor scenes and in many industrial applications where objects are aligned to a working plane. It is very useful to represent scenes on layers, that is, planes of objects with similar depth. In addition, affine models are simple and thus algorithms more stable, and an affine camera is linear since it does not include the projection division in Equations 10.28, 10.29 and 10.30.

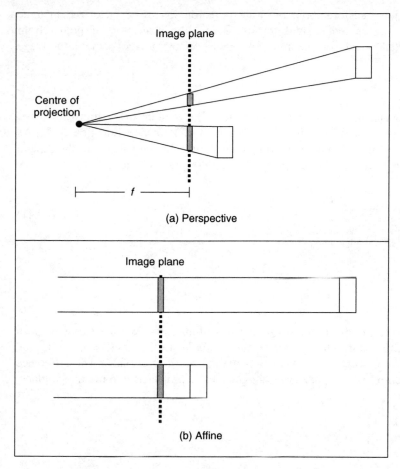

Figure 10.6   Perspective and affine camera models

10.4.1   Affine camera model

In the affine camera model, Equation 10.21 is changed to

$$
\begin{bmatrix} x_i \\ y_i \\ 1 \end{bmatrix} = \begin{bmatrix} p_{1,1} & p_{1,2} & p_{1,3} & p_{1,4} \\ p_{2,1} & p_{2,2} & p_{2,3} & p_{2,4} \\ 0 & 0 & 0 & 1 \end{bmatrix} \begin{bmatrix} x_p \\ y_p \\ z_p \\ 1 \end{bmatrix}
\tag{10.38}
$$

This equation can be written in short form as

$$
\mathbf{x}_i = \mathbf{P_A} \mathbf{x}_p
\tag{10.39}
$$

Here, we use the subindex $\mathbf{A}$ to indicate that the affine camera transformation is given by a special form of the projection $\mathbf{P}$. The last row in Equation 10.39 can be omitted. It is shown in the notation to emphasize that it is a special case of the perspective model. However, unlike the case for perspective camera, points in the image plane are in Euclidean coordinates. That is, the affine camera maps points from the projective space to the Euclidean plane.

Similar to the projection transformation, the transformation **A** can be factorized in three factors that account for the camera's rigid transformation, the projection of points from space into the image plane, and the mapping of points on the image plane into image pixels.

$$\mathbf{A} = \mathbf{V} \, \mathbf{Q_A} \, \mathbf{M_A} \tag{10.40}$$

Here, the subindex **A** indicates that these matrices are the affine versions of the transformations defined in Equation 10.23. We start by a rigid transformation as defined in Equation 10.25. As in the case of the perspective model, this transformation is defined by the position of the camera and makes the coordinates of a point in 3D space relative to the camera frame.

$$\mathbf{M_A} = \begin{bmatrix} r_{1,1} & r_{1,2} & r_{1,3} & t_x \\ r_{2,1} & r_{2,2} & r_{2,3} & t_y \\ r_{3,1} & r_{3,2} & r_{3,3} & t_z \\ 0 & 0 & 0 & 1 \end{bmatrix} \tag{10.41}$$

That is,

$$\mathbf{M}_A = \begin{bmatrix} \mathbf{R} & \mathbf{T} \\ 0 & 1 \end{bmatrix} \tag{10.42}$$

The last row is added so the transformation $\mathbf{Q_A}$ can have four rows. We need four rows in $\mathbf{Q_A}$ to define a parallel projection into the image plane. Similar to the transformation $\mathbf{Q}$, the transformation $\mathbf{Q_A}$ projects a point in the camera frame into the image plane. The difference is that in the affine model, points in space are orthographically projected into the image plane. This can be defined by

$$\mathbf{Q_A} = \begin{bmatrix} 1 & 0 & 0 & 0 \\ 0 & 1 & 0 & 0 \\ 0 & 0 & 0 & 1 \end{bmatrix} \tag{10.43}$$

This defines a projection when the focal length is set to infinity. Intuitively, you can see that when transforming a point $\mathbf{x}_p^T = \begin{bmatrix} x_p & y_p & z_p & 1 \end{bmatrix}$ by Equation 10.43, the x and y coordinates are copied and the depth z_p value does not change the projection. Thus, Equations 10.29 and 10.30 for the affine camera become

$$x_i = x_p \quad \text{and} \quad y_i = y_p \tag{10.44}$$

That is, the points in the camera frame are projected along the line $z_p = 0$. This is a line parallel to the image plane. The transformation **V** in Equation 10.40 provides the pixel coordinates of points in the image plane. This process is exactly the same in the perspective and affine models and is defined by Equation 10.31.

10.4.2   Affine camera model and the perspective projection

It is possible to show that the affine model is a particular case of the perspective model by considering the alternative camera representation illustrated in Figure 10.7. This figure is similar to the one used to illustrate Equation 10.27. The difference is that in the previous model, the centre of the camera frame was in the centre of projection, whereas in Figure 10.7 it is considered to be the principal point; that is, on the image plane. In general, the camera frame does not need to be located at a particular position in the camera, but can be arbitrarily set. When set in the image plane, as illustrated in Figure 10.7, the z camera coordinate of a point defines the depth in the image plane. Thus, Equation 10.28 is replaced by

$$\frac{y_i}{f} = \frac{h}{z_p} \tag{10.45}$$

From Figure 10.7, we can see that $y_p = y_i + h$. Thus,

$$y_p = y_i + z_p \frac{y_i}{f} \tag{10.46}$$

Solving for y_i, we have:

$$y_i = \frac{f\, y_p}{f + z_p} \tag{10.47}$$

We can use a similar development to find the $\mathbf{x}_i$ coordinate. That is,

$$\mathbf{x}_i = \frac{f\, \mathbf{x}_p}{f + z_p} \tag{10.48}$$

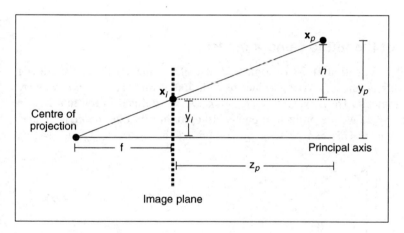

Figure 10.7 Projection of a point

Using homogeneous coordinates, Equations 10.47 and 10.48 can be written in matrix form as

$$\begin{bmatrix} x_i \\ y_i \\ z_i \end{bmatrix} = \begin{bmatrix} f & 0 & 0 & 0 \\ 0 & f & 0 & 0 \\ 0 & 0 & 1 & f \end{bmatrix} \begin{bmatrix} x_p \\ y_p \\ z_p \\ 1 \end{bmatrix} \tag{10.49}$$

This equation is an alternative to Equation 10.27; it represents a perspective projection. The difference is that in Equation 10.49 we assume that the camera axis is the located at the principal point of a camera. Using Equation 10.49 it is easy to see the projection in the affine camera model as a special case of projection in the perspective camera model. To show that Equation 10.29 becomes an affine model when f is set to be infinite, we define $B = 1/f$. Thus,

$$y_i = \frac{y_p}{1 + B\, z_p} \quad \text{and} \quad x_i = \frac{x_p}{1 + B\, z_p} \tag{10.50}$$

or

$$\begin{bmatrix} x_i \\ y_i \\ z_i \end{bmatrix} = \begin{bmatrix} 1 & 0 & 0 & 0 \\ 0 & 1 & 0 & 0 \\ 0 & 0 & B & 1 \end{bmatrix} \begin{bmatrix} x_p \\ y_p \\ z_p \\ 1 \end{bmatrix} \tag{10.51}$$

When f tends to infinity B tends to zero. Thus, the projection in Equation 10.51 for the affine camera becomes

$$\begin{bmatrix} x_i \\ y_i \\ z_i \end{bmatrix} = \begin{bmatrix} 1 & 0 & 0 & 0 \\ 0 & 1 & 0 & 0 \\ 0 & 0 & 0 & 1 \end{bmatrix} \begin{bmatrix} x_p \\ y_p \\ z_p \\ 1 \end{bmatrix} \tag{10.52}$$

The transformation in this equation is defined in Equation 10.43. Thus, the projection in the affine model is a special case of the projection in the perspective model obtained by setting the focal length to infinity.

10.4.3 Parameters of the affine camera model

The affine camera model in Equation 10.38 is composed of eight elements. Thus, a particular camera model is completely defined by giving values to eight unknowns. These unknowns are determined by the 11 parameters $(\alpha, \beta, \gamma, t_x, t_y, t_z, u_0, v_0, k_u, k_v, \varphi)$ defined in the matrices in Equation 10.40. However, since we are projecting points orthographically into the image plane, the translation in depth is lost. This can be seen by combining the matrices $\mathbf{Q_A}$ and $\mathbf{M_A}$ in Equation 10.40. That is,

$$\mathbf{G} = \begin{bmatrix} 1 & 0 & 0 & 0 \\ 0 & 1 & 0 & 0 \\ 0 & 0 & 0 & 1 \end{bmatrix} \begin{bmatrix} r_{1,1} & r_{1,2} & r_{1,3} & t_x \\ r_{2,1} & r_{2,2} & r_{2,3} & t_y \\ r_{3,1} & r_{3,2} & r_{3,3} & t_z \\ 0 & 0 & 0 & 1 \end{bmatrix} \tag{10.53}$$

or

$$\mathbf{G_A} = \begin{bmatrix} r_{1,1} & r_{1,2} & r_{1,3} & t_x \\ r_{2,1} & r_{2,2} & r_{2,3} & t_y \\ 0 & 0 & 0 & 1 \end{bmatrix} \tag{10.54}$$

Thus, Equation 10.40 becomes

$$\mathbf{A} = \mathbf{V}\, \mathbf{G_A} \tag{10.55}$$

Similar to Equation 10.42, the matrix $\mathbf{G_A}$ can be written as

$$\mathbf{G_A} = \begin{bmatrix} \mathbf{R_A} & \mathbf{T_A} \\ 0 & 1 \end{bmatrix} \tag{10.56}$$

and it defines the orthographic projection of the rigid transformation $\mathbf{M_A}$ into the image plane. According to Equation 10.53,

$$\mathbf{T_A} = \begin{bmatrix} t_x \\ t_y \\ 1 \end{bmatrix} \tag{10.57}$$

Since we do not have t_z we cannot determine whether objects are far away or close to the camera. Just because an object is small does not mean that it is far away. According to Equation 10.53, we also have that

$$\mathbf{R_A} = \begin{bmatrix} \cos(\alpha) & -\sin(\alpha) & 0 \\ \sin(\alpha) & \cos(\alpha) & 0 \\ 0 & 0 & 0 \end{bmatrix} \begin{bmatrix} \cos(\beta) & 0 & -\sin(\beta) \\ 0 & 1 & 0 \\ 0 & 0 & 0 \end{bmatrix} \begin{bmatrix} 1 & 0 & 0 \\ 0 & \cos(\gamma) & -\sin(\gamma) \\ 0 & 0 & 0 \end{bmatrix} \tag{10.58}$$

Thus, the eight elements of the affine camera projection matrix are determined by the intrinsic parameters $(u_0, v_0, s_u, s_v, \varphi)$ and the extrinsic parameters $(\alpha, \beta, \gamma, t_x, t_y)$.

10.5 Weak perspective model

The *weak perspective model* defines a geometric mapping that stands between the perspective and affine models. This model considers that the distance between points in the scene is small relative to the focal length. Thus, Equations 10.29 and 10.30 are approximated by

$$y_i = \frac{y_p}{\mu_z} f \quad \text{and} \quad x_i = \frac{x_p}{\mu_z} f \tag{10.59}$$

where μ_z is the average z coordinate of all the points in a scene.

Figure 10.8 illustrates two possible geometric interpretations for the relationships defined in Equation 10.59. Figure 10.8(a) illustrates a two-step process wherein, first, all points are affine projected to a plane orthogonal to the image plane and at a distance μ_z. Points on this plane are then mapped into the image plane by a perspective projection. The projection on the plane $z = \mu_z$ simply replaces the z coordinates of the points by μ_z. Since points are assumed to be close, this projection is a good approximation of the scene. Thus, the weak perspective model corresponds to a perspective model for scenes approximated by planes parallels to the image plane.

A second geometric interpretation of Equation 10.59 is illustrated in Figure 10.8(b). In Equation 10.59, we can combine the values f and μ_z into a single constant. Thus, Equation 10.59 corresponds to a scaled version of Equation 10.44. In Figure 10.8(b), first, objects in the scene are mapped into the image plane by an affine projection and then the image is rescaled by a value f/μ_z. Thus, the affine model can be seen as a particular case of the weak perspective model when $f/\mu_z = 1$.

By following the two geometric interpretations discussed above, the weak perspective model can be formulated by changing the projection equations of the perspective or the affine models. It can also be formulated by considering the camera model presented in Section 10.3.2. For simplicity, we consider the weak perspective from the affine model. Thus, Equation 10.43 should include a change in scale. That is,

$$\mathbf{Q_A} = \begin{bmatrix} f/\mu_z & 0 & 0 & 0 \\ 0 & f/\mu_z & 0 & 0 \\ 0 & 0 & 0 & 1 \end{bmatrix} \tag{10.60}$$

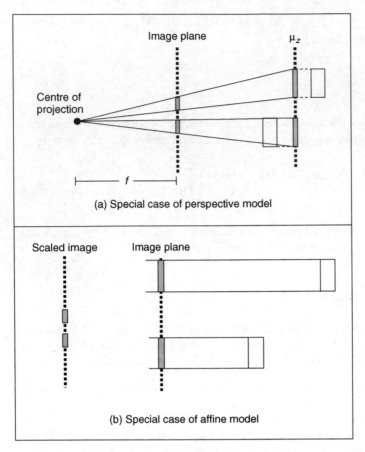

Figure 10.8 Weak perspective camera model

By considering the definition in Equation 10.40, we can move the scale factor in this matrix to the matrix **V**. Thus, the model for the weak perspective model can be expressed as

$$
\mathbf{P} = \begin{bmatrix} s_u & s_u \cot(\varphi) & u_0 \\ 0 & s_v \sin(\varphi) & v_0 \\ 0 & 0 & 1 \end{bmatrix} \begin{bmatrix} 1 & 0 & 0 & 0 \\ 0 & 1 & 0 & 0 \\ 0 & 0 & 0 & 1 \end{bmatrix} \begin{bmatrix} r_{1,1} & r_{1,2} & r_{1,3} & t_x \\ r_{2,1} & r_{2,2} & r_{2,3} & t_y \\ r_{3,1} & r_{3,2} & r_{3,3} & t_z \\ 0 & 0 & 0 & 1 \end{bmatrix} \tag{10.61}
$$

for

$$
s_u = f k_u / \mu_z \quad \text{and} \quad s_v = f k_v / \mu_z \tag{10.62}
$$

Thus, the weak perspective is a scaled version of the affine model. The scale is a function of the f that defines the distance from the centre of the camera to the image plane and the average distance μ_z.

10.6 Example of camera models

This section illustrates the mapping of points into an image frame for the perspective and affine camera models. Code 10.1 contains the functions used to create

```matlab
%************************************************
%Draw a set of points
%-----------------------------------------------
function DrawPoints(P,colour)
%-----------------------------------------------

[r,c]=size(P);
plot3(P(1,1:c),P(2,1:c),P(3,1:c),'+','color',colour);

%************************************************
%Draw a set of Image points
%-----------------------------------------------
function DrawImagePoints(P,C,colour)
%-----------------------------------------------

[r,c]=size(P);

for column=1:c
   P(1:r-1,column)=P(1:r-1,column)/P(3,column);
end

plot(P(1,1:c),P(2,1:c),'+','color',colour);
axis([0 C(10) 0 C(11)]);

%************************************************
%Draw a coordinate frame
%-----------------------------------------------
function DrawWorldFrame(x0,x1,y0,y1,z0,z1);
%-----------------------------------------------

axis equal; %same aspect ratio
axis([x0,x1,y0,y1,z0,z1]);

xlabel('X','FontSize',14);
ylabel('Y','FontSize',14);
zlabel('Z','FontSize',14);

grid on;
hold on;

%************************************************
%Draw an image plane
%-----------------------------------------------
function DrawImagePlane(C,dx,dy);
%-----------------------------------------------
```

```
%Draw camera origin
plot3(C(1),C(2),C(3),'o');   %optic centre
plot3(C(1),C(2),C(3),'+');   %optic centre


%coordinates of 4 points on the
%image plane (to draw a rectangle)
p(1,1)=-dx/2; p(2,1)=-dy/2; p(3,1)=C(7); p(4,1)=1;
p(1,2)=-dx/2; p(2,2)=+dy/2; p(3,2)=C(7); p(4,2)=1;
p(1,3)=+dx/2; p(2,3)=+dy/2; p(3,3)=C(7); p(4,3)=1;
p(1,4)=+dx/2; p(2,4)=-dy/2; p(3,4)=C(7); p(4,4)=1;


%CW: Camera to world transformation
CW=CameraToWorld(C);


%transform coordinates to world coordinates
P(:,1)=CW*p(:,1);
P(:,2)=CW*p(:,2);
P(:,3)=CW*p(:,3);
P(:,4)=CW*p(:,4);


%Draw image plane
patch(P(1,:),P(2,:),P(3,:),[.9,.9,1]);


%***************************************************
%Draw a line between optical centre and 3D points
%
%---------------------------------------------------------
function DrawPerspectiveProjectionLines (o,P,colour);
%---------------------------------------------------------


[r,c]=size(P);
for i=1:c
  plot3([o(1) P(1,i)],[o(2) P(2,i)],[o(3) P(3,i)],'color',colour);
%optic centre
end;


%**************************************************
%Draw a line between image plane and 3D points
%
%--------------------------------------------------
function DrawAffineProjectionLines(z,P,colour);
%--------------------------------------------------


[r,c]=size(P);
for column=1:c
  plot3([P(1,column) P(1,column)],[P(2,column) P(2,column)],[z(3)
P(3,column)],'color',colour);   %optic centre
end;
```

Code 10.1   Drawing functions

figures: DrawPoints, DrawImagePoints, DrawWorldFrame, DrawImagePlane, DrawPerspectiveProjectionLines and DrawAffineProjectionLines.

The function DrawPoints draws a set of points given its three world coordinates. In our example, it is used to draw points in the world. The function DrawImagePoints draws points on the image plane. It uses homogeneous coordinates, so it implements Equation 10.3. The function DrawWorldFrame draws the three axes given the location of the origin. This is used to illustrate the world frame and the origin is always set to zero. The function DrawImagePlane draws a rectangle that represents the image plane. It also draws a point to illustrate the location of the centre of the camera. In our examples, we assume that the centre of the camera is at the position of the focal point, for both the perspective and the affine model.

The functions DrawPerspectiveProjectionLines and DrawAffine ProjectionLines are used to illustrate the projection of points into the image plane for the perspective and affine projections, respectively. These functions take a vector of points and trace a line to the image plane that represent the projection. For perspective, the lines intersects the focal point and for affine they are projected parallel to the image plane.

In the example implementation, we group the camera parameters in the vector $C = [x0, y0, z0, a, b, g, f, u0, v0, kx, ky]$. The first six elements define the location and rotation parameters. The value of f defines the focal length. The remaining parameters define the location of the optical centre and the pixel size. For simplicity, we assume that there is no skew. Code 10.2 contains the functions that compute camera transformations from camera parameters. The function CameraToWorld computes a matrix that defines the position of the camera.

```
%*************************************************
%Compute matrix that transforms coordinates
%in the camera frame to the world frame%
%-------------------------------------------------
  function CW=CameraToWorld(C);
%-------------------------------------------------

%rotation
Rx=[cos(C(6))   sin(C(6))   0
   -sin(C(6))   cos(C(6))   0
        0           0       1];

Ry=[cos(C(5))      0       sin(C(5))
        0          1          0
   -sin(C(5))      0       cos(C(5))];

Rz=[1              0              0
       0        cos(C(4))      sin(C(4))
       0       -sin(C(4))      cos(C(4))];

%translation
T=[C(1) C(2) C(3)]';

%transformation
CW=(Rz*Ry*Rx);
CW(:,4)=T;
```

```
%*************************************************
%Compute matrix that transforms coordinates
%in the world frame to the camera frame
%------------------------------------------------
  function WC=WorldToCamera(c);
%------------------------------------------------

%translation T'=-R'T
%for R'=inverse rotation
Rx=[cos(C(6))  -sin(C(6))   0
    sin(C(6))   cos(C(6))   0
        0           0       1];

Ry=[cos(C(5))       0   -sin(C(5))
        0           1       0
    sin(C(5))       0    cos(C(5))];

Rz=[1           0           0
    0       cos(C(4))   -sin(C(4))
    0       sin(C(4))    cos(C(4))];

T=[-c(1)  -c(2)  -c(3)]';

%transformation
WC=(Rz*Ry*Rx); %rotation inverse
Tp=WC*T;        %translation
WC(:,4)=Tp;     %compose homogeneous form

%*************************************************
%Convert from homogeneous coordinates in pixels
%to distance coordinates in the camera frame
%------------------------------------------------
function p=ImageToCamera(P,C);
%------------------------------------------------

%inverse of K
Ki=[1/C(10)         0       -C(8)/C(10)
        0       1/C(11)     -C(9)/C(11)
        0           0           1  ];

%coordinate in distance units
p=Ki*P;

%coordinates in the image plane
p(1,:)=p(1,:)./p(3,:);
p(2,:)=p(2,:)./p(3,:);
p(3,:)=p(3,:)./p(3,:);

%the third coordinate gives the depth
%the focal length C(7) defines depth
p(3,:)=p(3,:).*C(7);

%include homogeneous coordinates
p(4,:)=p(1,:)/p(1,:);
```

Code 10.2   Transformation function

It poses the camera in the world. Its inverse is computed in the function `WorldToCamera` and it defines the matrix in Equation 10.25. The inverse is simply obtained by the transpose of the rotation and by changing the signs of the translation. The matrix obtained from `WorldToCamera` can be used to obtain the coordinates of world points in the camera frame.

The function `ImageToCamera` in Code 10.2 obtains the inverse of the transformation defined in Equation 10.31. This is used to draw points in pixel coordinates. The pixel's coordinates are converted into world coordinates, which are then drawn to show its position.

Code 10.3 contains the two functions that compute the projection matrices for the perspective and affine camera models. Both functions start by computing the matrix that transforms the world points into the camera frame. For the affine model, the dimensions of the matrix transformation are augmented according to Equation 10.42. For the perspective model, the world to camera

```
%*******************************************
%Obtain the projection matrix parameters
%from the camera position
%------------------------------------------
  function M=PerspectiveProjectionMatrix(C);
%------------------------------------------

%World to camera
WC=WorldToCamera(C);

%Project point in the image
F=[ C(7)    0     0
      0    C(7)   0
      0     0    1 ];

%Distance units to pixels
K=[ C(10)    0      C(8)
      0    C(11)    C(9)
      0      0       1 ];

%Projection matrix
M=K*F*WC;

%*******************************************
%Obtain the projection matrix parameters
%from the camera position
%------------------------------------------
  function M=AffineProjectionMatrix(C);
%------------------------------------------

%world to camera
WC=WorldToCamera(C);
WC(4,1:4)=[0 0 0 1];

%project point in the image
F=[ 1    0    0    0
    0    1    0    0
    0    0    0    1];
```

```
%distance units to pixels
K=[ C(10)      0        C(8)
       0      C(11)     C(9)
       0       0         1 ];

%projection matrix
M=K*F*WC;
```

Code 10.3 Camera models

matrix is multiplied by the projection defined in Equation 10.27. The affine model implements the projection defined in Equation 10.43. In the perspective and affine functions, coordinates are transformed to pixels by the transformation defined in Equation 10.31.

Code 10.4 uses the previous functions to generate figures that illustrate the projection of a pair of points in the image plane for the perspective and affine models. The camera is defined with a translation of 1 in y and the focal length is 0.5 from the camera plane. The image is defined to be 100×100 pixels and the principal point is in the middle of the image. That is, at a pixel of coordinates (50, 50). After the definition of the camera, the code defines two 3D

```
%***************************************************
%Example of the computation projection of points
%for the perspective and affine camera models
%-------------------------------------------------
  function ProjectionExample();
%-------------------------------------------------

%C=[x0,y0,z0,a,b,g,f,u0,v0,kx,ky]
%
%Camera parameters:
%x0,y0,z0: location
%a,b,g    : orientation
%f        : focal length
%u0,v0    : optical centre
%kx,ky    : pixel size

C=[0,1,0,0,0,0,0.5,50,50,100,100];

%3D points in homogeneous form
XYZ=[ 0,     .2      %x
      1,     .6      %y
      1.7   2        %z
      1     1];

%Perspective example
figure(1);
clf;

%Draw world frame
DrawWorldFrame(-.5,2,-.5,2,-.5,2);
```

```
%Draw camera
DrawImagePlane(C,1,1);

%Draw world points
DrawPoints(XYZ,[0,0,0]);

%Perspective projection matrix
P=PerspectiveProjectionMatrix(C);

%Project into camera frame, in pixels
UV=P*XYZ;

%Convert to camera coordinates
PC=ImageToCamera(UV,C);

%Convert to world frame
MI=CameraToWorld(C);
PW=MI*PC;

%Draw Projected points in world frame
DrawPoints(PW,[1,0,0]);

%Draw projection lines
DrawPerspectiveProjectionLines([C(1),C(2),C(3)],XYZ,[.3,.3,.3]);

%Draw image points
figure(2);
clf;
DrawImagePoints(UV,C,[0,0,0]);

%Affine example
figure(3);
clf;

%Draw world frame
DrawWorldFrame(-.5,2,-.5,2,-.5,2);

%Draw camera
DrawImagePlane(C,1,1);

%3D points in homogeneous form
DrawPoints(XYZ,[0,1,0]);

%Affine projection matrix
P=AffineProjectionMatrix(C);

%Project into camera frame, in pixels
UV=P*XYZ;

%Convert to camara coordinates
PC=ImageToCamera(UV,C);
```

```
%Convert to world frame
PW=MI*PC;

%Draw Projected points in world frame
DrawPoints(PW,[0,0,0]);

%Draw projection lines
DrawAffineProjectionLines([C(1),C(2),C(3)],XYZ,[.3,.3,.3]);

%Draw image points
figure(4);
clf;
DrawImagePoints(UV,C,[0,0,0]);
```

Code 10.4   Main example

points in homogeneous form. These points will be used to illustrate how camera models map world points into images.

The example first draws a frame to represent the world frame. The parameters are chosen to show the image plane and the world points. These are drawn using the functions DrawWorldFrame and DrawImagePlane defined in Code 10.1. After the drawing, the code computes the projection matrix by calling the function PerspectiveProjectionMatrix discussed in Code 10.3. This transformation is used to project the 3D points into the image. In the code, the matrix **UV** contains the coordinates of the points in pixels. To draw these points in the 3D space, first they are converted to the camera coordinates by calling ImageToCamera and then they are converted to the world frame. The function DrawPerspectiveProjectionLines draws the lines from the world points to the centre of projection.

The result of the perspective projection example is shown in Figure 10.9. Here, we can see the projection lines pass through the points obtained by the projection matrix. The image

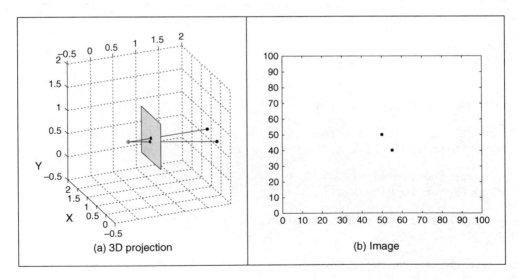

(a) 3D projection   (b) Image

Figure 10.9   Perspective camera example

shown in Figure 10.9(b) was obtained by calling the function `DrawImagePoints` defined in Code 10.1. This function draws the points obtained by the projection matrix. One of the points is projected into the centre of the image. This is because its x and y coordinates are the same as those of the principal point.

The last two figures created in Code 10.4 illustrate the projection for the affine matrix. The process is similar to the perspective example, but they use the projection obtained by the function `AffineProjectionMatrix` defined in Code 10.3. The resulting figures are shown in Figure 10.9. Here, we can see that the projection matrix transforms the points by following rays perpendicular to the image plane. As such, the points in Figure 10.9(b) are further apart than the points in Figure 10.10(b). In the perspective model the distance between the points depends on the distance from the image plane, while in the affine model this information is lost.

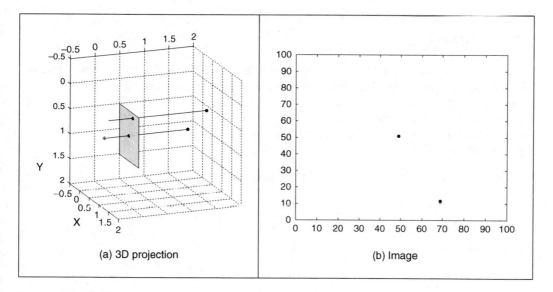

(a) 3D projection   (b) Image

Figure 10.10   Affine camera example

10.7 Discussion

In this appendix we have formulated the most common models of camera geometry. However, in addition to perspective and affine camera models there exist other models that consider different camera properties. For example, cameras built from a linear array of sensors can be modelled by particular versions of the perspective and affine models obtained by considering a one-dimensional (1D) image plane. These 1D camera models can also be used to represent strips of pixels obtained by cameras with 2D image planes and they have found an important application in mosaic construction from video images.

Besides image plane dimensionality, perhaps the most evident extension of camera models is in considering lens distortions. Small geometric distortions are generally ignored or dealt with as noise in computer vision techniques. Strong geometric distortions such as those produced by wide-angle or fish-eye lenses can be modelled by considering a spherical image plane or by non-linear projections. The model of wide-angle cameras has found applications in environmental map capture and panoramic mosaics.

The formulation of camera models is the basis of two central problems of computer vision. The first problem is known as camera calibration and it centres on computing the camera parameters from image data. There are many camera calibration techniques based on the camera model and different types of data. However, camera calibration techniques are grouped into two main classes. Strong camera calibration assumes knowledge of the 3D coordinates of image points. Weak calibration techniques do not know 3D coordinates, but they assume knowledge of the type of motion of a camera. In addition, some techniques focus on intrinsic or extrinsic parameters. The second central problem in computer vision is called scene reconstruction and centres on recovering the coordinates of points in the 3D scene from image data. Techniques have been developed for each camera model.

10.8 References

Aguado, A. S., Montiel, E. and Nixon, M. S., On the Intimate Relationship Between the Principle of Duality and the Hough Transform, *Proc. R. Soc. Lond. A*, **456**, pp. 503–526, 2000

Hartley, R. and Zisserman, A., *Multiple View Geometry in Computer Vision*, Cambridge University Press, Cambridge, 2001

Trucco, E. and Verri, A., *Introductory Techniques for 3-D Computer Vision*, Prentice Hall, New Jersey, 1998

11

Appendix 3: Least squares analysis

11.1 The least squares criterion

The *least squares criterion* is one of the foundations of *estimation theory*. This theory concerns extracting the *true* value of signals from *noisy* measurements. Estimation theory techniques have been used to guide Exocet missiles and astronauts on moon missions (where navigation data was derived using sextants!), all based on techniques that use the least squares criterion. The least squares criterion was originally developed by Gauss when he was confronted by the problem of measuring the six parameters of the orbits of planets, given astronomical measurements. These measurements were subject to error, and Gauss realized that they could be combined together in some way to reduce a best estimate of the six parameters of interest.

Gauss assumed that the noise corrupting the measurements would have a *normal distribution*; indeed, such distributions are often now called Gaussian to honour his great insight. As a consequence of the *central limit theorem*, it may be assumed that many real random noise sources are normally distributed. In cases where this assumption is not valid, the mathematical advantages that accrue from its use generally offset any resulting loss of accuracy. In addition, the assumption of normality is particularly invaluable in view of the fact that the output of a system excited by Gaussian-distributed noise is also Gaussian distributed (as seen in Fourier analysis, Chapter 2). A Gaussian probability distribution of a variable x is defined by

$$p(x) = \frac{1}{\sigma\sqrt{2\pi}} e^{\frac{-(x-\bar{x})^2}{\sigma^2}} \tag{11.1}$$

where $\bar{x}$ is the mean (loosely the average) of the distribution and σ^2 is the second moment or variance of the distribution. Given many measurements of a single unknown quantity, when that quantity is subject to errors of a zero-mean (symmetric) normal distribution, it is well known that the best estimate of the unknown quantity is the average of the measurements. In the case of two or more unknown quantities, the requirement is to combine the measurements in such a way that the error in the estimates of the unknown quantities is minimized. Clearly, direct averaging will not suffice when measurements are a function of two or more unknown quantities.

Consider the case where N equally precise measurements, $f_1, f_2 \ldots f_N$, are made on a linear function $f(a)$ of a single parameter a. The measurements are subject to zero-mean additive Gaussian noise $v_i(t)$; as such, the measurements are given by

$$f_i = f(a) + v_i(t) \quad \forall i \in 1, N \tag{11.2}$$

The differences $\tilde{f}$ between the true value of the function and the noisy measurements of it are then

$$\tilde{f}_i = f(a) - f_i \quad \forall i \in 1, N \tag{11.3}$$

By Equation 11.1, the probability distribution of these errors is

$$p(\tilde{f}_i) = \frac{1}{\sigma \sqrt{2\pi}} e^{\frac{-(\tilde{f}_i)^2}{\sigma^2}} \quad \forall i \in 1, N \tag{11.4}$$

Since the errors are *independent*, the *compound* distribution of these errors is the product of their distributions, and is given by

$$p(\tilde{f}) = \frac{1}{\sigma \sqrt{2\pi}} e^{\frac{-\left((\tilde{f}_1)^2 + (\tilde{f}_2)^2 + (\tilde{f}_3)^2 + \cdots + (\tilde{f}_N)^2\right)}{\sigma^2}} \tag{11.5}$$

Each of the errors is a function of the unknown quantity, a, which is to be estimated. Different estimates of a will give different values for $p(\tilde{f})$. The most probable system of errors will be that for which $p(\tilde{f})$ is a *maximum*, and this corresponds to the *best* estimate of the unknown quantity. Thus, to maximize $p(\tilde{f})$,

$$\max\{p(\tilde{f})\} = \max\left\{ \frac{1}{\sigma \sqrt{2\pi}} e^{\frac{-\left((\tilde{f}_1)^2 + (\tilde{f}_2)^2 + (\tilde{f}_3)^2 + \cdots + (\tilde{f}_N)^2\right)}{\sigma^2}} \right\}$$

$$= \max\left\{ e^{\frac{-\left((\tilde{f}_1)^2 + (\tilde{f}_2)^2 + (\tilde{f}_3)^2 + \cdots + (\tilde{f}_N)^2\right)}{\sigma^2}} \right\} \tag{11.6}$$

$$= \max\left\{ -\left((\tilde{f}_1)^2 + (\tilde{f}_2)^2 + (\tilde{f}_3)^2 + \cdots + (\tilde{f}_N)^2\right) \right\}$$

$$= \min\left\{ (\tilde{f}_1)^2 + (\tilde{f}_2)^2 + (\tilde{f}_3)^2 + \cdots + (\tilde{f}_N)^2 \right\}$$

Thus, the required estimate is that which *minimizes* the sum of the differences squared and this estimate is the one that is *optimal* by the least squares criterion.

This criterion leads on to the method of least squares, which follows in the next section. This is a method commonly used to fit curves to measured data. It concerns estimating the values of parameters from a complete set of measurements. There are also techniques that provide estimate of parameters at time instants, based on a set of previous measurements. These techniques include the Weiner filter and the Kalman filter. The Kalman filter was the algorithm chosen for guiding Exocet missiles and moon missions (an extended square root Kalman filter, no less).

11.2 Curve fitting by least squares

Curve fitting by the method of least squares concerns combining a set of measurements to derive *estimates* of the parameters which specify the curve that best fits the data. By the least squares criterion, given a set of N (noisy) measurements $f_i \ i \in 1, N$ which are to be fitted to a curve $f(\mathbf{a})$, where $\mathbf{a}$ is a vector of parameter values, we seek to minimize the square of the difference between the measurements and the values of the curve to give an estimate of the parameters $\hat{\mathbf{a}}$ according to

$$\hat{\mathbf{a}} = \min \sum_{i=1}^{N} (f_i - f(x_i, y_i, \mathbf{a}))^2 \tag{11.7}$$

Since we seek a minimum, by differentiation we obtain

$$\frac{\partial \sum_{i=1}^{N} (f_i - f(x_i, y_i, \mathbf{a}))^2}{\partial \mathbf{a}} = 0 \tag{11.8}$$

which implies that

$$2 \sum_{i=1}^{N} (f_i - f(x_i, y_i, \mathbf{a})) \frac{\partial f(\mathbf{a})}{\partial \mathbf{a}} = 0 \tag{11.9}$$

The solution is usually of the form

$$\mathbf{Ma} = \mathbf{F} \tag{11.10}$$

where $\mathbf{M}$ is a matrix of summations of products of the index i and $\mathbf{F}$ is a vector of summations of products of the measurements and i. The solution, the best estimate of the values of $\mathbf{a}$, is then given by

$$\hat{\mathbf{a}} = \mathbf{M}^{-1} \mathbf{F} \tag{11.11}$$

By way of example, let us consider the problem of fitting a two-dimensional surface to a set of data points. The surface is given by

$$f(x, y, \mathbf{a}) = a + bx + cy + dxy \tag{11.12}$$

where the vector of parameters $\mathbf{a} = [a \, b \, c \, d]^{\mathrm{T}}$ controls the shape of the surface, and (x, y) are the coordinates of a point on the surface. Given a set of (noisy) measurements of the value of the surface at points with coordinates (x, y), $f_i = f(x, y) + v_i$, we seek to estimate values for the parameters using the method of least squares. By Equation 11.7 we seek

$$\hat{\mathbf{a}} = \begin{bmatrix} \hat{a} & \hat{b} & \hat{c} & \hat{d} \end{bmatrix}^T = \min \sum_{i=1}^{N} (f_i - f(x_i, y_i, \mathbf{a}))^2 \tag{11.13}$$

By Equation 11.9 we require

$$2 \sum_{i=1}^{N} (f_i - (a + bx_i + cy_i + dx_i y_i)) \frac{\partial f(x_i, y_i, \mathbf{a})}{\partial \mathbf{a}} = 0 \tag{11.14}$$

By differentiating $f(x, y, \mathbf{a})$ with respect to each parameter we have

$$\frac{\partial f(x_i, y_i)}{\partial a} = 1 \tag{11.15}$$

$$\frac{\partial f(x_i, y_i)}{\partial b} = x \tag{11.16}$$

$$\frac{\partial f(x_i, y_i)}{\partial c} = y \tag{11.17}$$

and

$$\frac{\partial f(x_i, y_i)}{\partial d} = xy \tag{11.18}$$

and by substitution of Equations 11.15–11.18 in Equation 11.14, we obtain four simultaneous equations:

$$\sum_{i=1}^{N}(f_i - (a + bx_i + cy_i + dx_iy_i)) \times 1 = 0 \tag{11.19}$$

$$\sum_{i=1}^{N}(f_i - (a + bx_i + cy_i + dx_iy_i)) \times x_i = 0 \tag{11.20}$$

$$\sum_{i=1}^{N}(f_i - (a + bx_i + cy_i + dx_iy_i)) \times y_i = 0 \tag{11.21}$$

and

$$\sum_{i=1}^{N}(f_i - (a + bx_i + cy_i + dx_iy_i)) \times x_iy_i = 0 \tag{11.22}$$

Since $\sum_{i=1}^{N}a = Na$, Equation 11.19 can be reformulated as

$$\sum_{i=1}^{N}f_i - Na - b\sum_{i=1}^{N}x_i - c\sum_{i=1}^{N}y_i - d\sum_{i=1}^{N}x_iy_i = 0 \tag{11.23}$$

and Equations 11.20–11.22 can be reformulated likewise. The simultaneous equations can be expressed in matrix form:

$$\begin{bmatrix} N & \sum_{i=1}^{N}x_i & \sum_{i=1}^{N}y_i & \sum_{i=1}^{N}x_iy_i \\ \sum_{i=1}^{N}x_i & \sum_{i=1}^{N}(x_i)^2 & \sum_{i=1}^{N}x_iy_i & \sum_{i=1}^{N}(x_i)^2y_i \\ \sum_{i=1}^{N}y_i & \sum_{i=1}^{N}x_iy_i & \sum_{i=1}^{N}(y_i)^2 & \sum_{i=1}^{N}x_i(y_i)^2 \\ \sum_{i=1}^{N}x_iy_i & \sum_{i=1}^{N}(x_i)^2y_i & \sum_{i=1}^{N}x_i(y_i)^2 & \sum_{i=1}^{N}(x_i)^2(y_i)^2 \end{bmatrix} \begin{bmatrix} a \\ b \\ c \\ d \end{bmatrix} = \begin{bmatrix} \sum_{i=1}^{N}f_i \\ \sum_{i=1}^{N}f_ix_i \\ \sum_{i=1}^{N}f_iy_i \\ \sum_{i=1}^{N}f_ix_iy_i \end{bmatrix} \tag{11.24}$$

This is the same form as Equation 11.10 and can be solved by inversion, as in Equation 11.11. Note that the matrix is *symmetric* and its inversion, or solution, does not impose such a great computational penalty as appears. Given a set of data points, the values need to be entered in the summations, thus completing the matrices from which the solution is found. This technique can replace the one used in the zero-crossing detector within the Marr–Hildreth edge detection operator (Section 4.3.3), but appears to offer no significant advantage over the (much simpler) function implemented there.

12

Appendix 4: Principal components analysis

12.1 Introduction

This appendix introduces *principal components analysis* (PCA). This technique is also known as the *Karhunen Loeve* (KL) transform or the *Hotelling transform*. It is based on factorization techniques developed in linear algebra. Factorization is commonly used to diagonalize a matrix, so its inverse can be easily obtained. PCA uses factorization to transform data according to its statistical properties. The data transformation is particularly useful for classification and compression.

Here we will give an introduction to the mathematical concepts and give examples and simple implementations, so you should be able to understand the basic ideas of PCA, to develop your own implementation and to apply the technique to your own data. We use simple matrix notations to develop the main ideas of PCA. If you want to have a more rigorous mathematical understanding of the technique, you should review concepts of eigenvalues and eigenvectors in more detail (Anton, 2005).

You can think of PCA as a technique that takes a collection of data and transforms it such that the new data has given statistical properties. The statistical properties are chosen such that the transformation highlights the importance of data elements. Thus, the transformed data can be used for classification by observing important components of the data. Data can also be reduced or compressed by eliminating (filtering out) the less important elements. The data elements can be seen as features, but in a mathematical sense they define the axes in the coordinate system.

Before defining the data transformation process defined by PCA, we need to understand how data is represented and also have a clear understanding of the statistical measure known as the covariance.

12.2 Data

Generally, data is represented by a set of m vectors:

$$\mathbf{X} = \{\mathbf{x}_1, \mathbf{x}_2, \ldots, \mathbf{x}_m\} \tag{12.1}$$

Each vector $\mathbf{x}_i$ has n elements or features. That is,

$$\mathbf{x}_i = \left\{ x_{i,1}, x_{i,2}, \ldots, x_{i,n} \right\} \tag{12.2}$$

The way you interpret each vector $\mathbf{x}_i$ depends on your application. For example, in pattern classification, each vector can represent a measure and each component of the vector a feature such as colour, size or edge magnitude.

We can group features by taking the elements of each vector. That is, the feature column vector k for the set $\mathbf{X}$ can be defined as

$$\mathbf{c}_{\mathbf{X},k} = \begin{bmatrix} x_{1,k} \\ x_{2,k} \\ \vdots \\ x_{m,k} \end{bmatrix} \tag{12.3}$$

for k ranging from 1 to n. The subindex $\mathbf{X}$ may seem unnecessary now; however, this will help us to distinguish features of the original set and of the transformed data.

We can group all the features in the feature matrix by considering each vector $\mathbf{c}_{\mathbf{X},k}$ to be a column in a matrix. That is,

$$\mathbf{c}_{\mathbf{X}} = \begin{bmatrix} \mathbf{c}_{\mathbf{X},1} \ \mathbf{c}_{\mathbf{X},2} \cdots \mathbf{c}_{\mathbf{X},n} \end{bmatrix} \tag{12.4}$$

The PCA technique transforms the feature vectors $\mathbf{c}_{\mathbf{X},k}$ to define new vectors defining components with better classification capabilities. Thus, the new vectors can be grouped by clustering according distance criteria on the more important elements, that is, the elements that define important variations in the data. PCA ensures that we highlight the data that accounts for the maxima variation measured by the covariance.

12.3 Covariance

Broadly speaking, the covariance measures the linear dependence between two random variables (DeGroot and Schervish, 2001). So, by computing the covariance, we can determine whether there is a relationship between two sets of data. If we consider that the data defined in the previous section has only two components, then the covariance between features can be defined by considering the component of each vector. That is, if $\mathbf{x}_i = \{x_{i,1}, x_{i,2}\}$, then the covariance is

$$\sigma_{\mathbf{X},1,2} = E\left[\left(\mathbf{c}_{\mathbf{X},1} - \boldsymbol{\mu}_{\mathbf{X},1} \right)\left(\mathbf{c}_{\mathbf{X},2} - \boldsymbol{\mu}_{\mathbf{X},2} \right) \right] \tag{12.5}$$

Here, the multiplication is assumed to be element by element and $E[]$ denotes the expectation, which is loosely the average value of the elements of the vector. We denote as $\boldsymbol{\mu}_{\mathbf{X},k}$ a column vector obtained by multiplying the scalar value $E\left[\mathbf{c}_{\mathbf{X},k} \right]$ by a unitary vector. That is, $\boldsymbol{\mu}_{\mathbf{X},k}$ is a vector that has the mean value on each element. Thus, according to Equation 12.5, we first subtract the mean value for each feature and then compute the mean of the multiplication of each element.

The definition of covariance can be expressed in matrix form as

$$\sigma_{\mathbf{X},1,2} = \frac{1}{m}\left(\left(\mathbf{c}_{\mathbf{X},1} - \boldsymbol{\mu}_{\mathbf{X},1} \right)^T \left(\mathbf{c}_{\mathbf{X},2} - \boldsymbol{\mu}_{\mathbf{X},2} \right) \right) \tag{12.6}$$

Here, T denotes the matrix transpose. Sometimes, features are represented as rows, so you can find the transpose operating on the second factor rather than on the first. Notice that the covariance is symmetric, thus $\sigma_{\mathbf{X},1,2} = \sigma_{\mathbf{X},2,1}$.

In addition to Equations 12.5 and 12.6, there is a third alternative definition of covariance that is obtained by developing the products in Equation 12.6. That is,

$$\sigma_{X,1,2} = \frac{1}{m}\left(\mathbf{c}_{X,1}{}^{T}\mathbf{c}_{X,2} - \boldsymbol{\mu}_{X,1}{}^{T}\mathbf{c}_{X,2} - \mathbf{c}_{X,1}{}^{T}\boldsymbol{\mu}_{X,2} + \boldsymbol{\mu}_{X,1}{}^{T}\boldsymbol{\mu}_{X,2}\right) \tag{12.7}$$

Since

$$\boldsymbol{\mu}_{X,1}{}^{T}\mathbf{c}_{X,2} = \mathbf{c}_{X,1}{}^{T}\boldsymbol{\mu}_{X,2} = \boldsymbol{\mu}_{X,1}{}^{T}\boldsymbol{\mu}_{X,2} \tag{12.8}$$

then

$$\sigma_{X,1,2} = \frac{1}{m}\left(\mathbf{c}_{X,1}{}^{T}\mathbf{c}_{X,2} - \boldsymbol{\mu}_{X,1}{}^{T}\boldsymbol{\mu}_{X,2}\right) \tag{12.9}$$

This can be written in short form as

$$\sigma_{X,1,2} = E\lfloor\mathbf{c}_{X,1},\mathbf{c}_{X,2}\rfloor - E\lfloor\mathbf{c}_{X,1}\rfloor E\lfloor\mathbf{c}_{X,2}\rfloor \tag{12.10}$$

for

$$E\left[\mathbf{c}_{X,1},\mathbf{c}_{X,2}\right] = \frac{1}{m}\left(\mathbf{c}_{X,1}^{T}\mathbf{c}_{X,2}\right) \tag{12.11}$$

Equations 12.5, 12.6 and 12.11 are alternative ways to compute the covariance. They are obtained by expressing products and averages in algebraic equivalent definitions.

As a simple example of the covariance, you can think of one variable representing the value of a spectral band of an aerial image, while the other the amount of vegetation in the ground region covered by the pixel. If you measure the covariance and you get a positive value, then for new data you should expect that an increase in the pixel intensity means an increase in vegetation. If the covariance value is negative, then you should expect that an increase in the pixel intensity means a decrease in vegetation. When the values are zero or very small, the values are uncorrelated; the pixel intensity and vegetation are independent and we cannot tell whether the change in intensity is related to any change in vegetation. Recall that the probability of two independent events happening together is equal to the product of the probability of each event. Thus, $E\lfloor\mathbf{c}_{X,1},\mathbf{c}_{X,2}\rfloor = E\lfloor\mathbf{c}_{X,1}\rfloor E\lfloor\mathbf{c}_{X,2}\rfloor$ is characteristic of independent events.

The covariance value ranges from 0 to indicate no relationship, to large positive and negative values that reflect strong dependences. The maximum and minimum values are obtained by using the Cauchy–Schwarz inequality and they are given by

$$\left|\sigma_{X,1,2}\right| \le \sigma_{X,1}\sigma_{X,2} \tag{12.12}$$

Here, | | denotes the absolute value and $\sigma_{X,1}^{2} = E\left[\mathbf{c}_{X,1},\mathbf{c}_{X,1}\right] - E\left[\mathbf{c}_{X,1}\right]E\left[\mathbf{c}_{X,1}\right]$ defines the variance of $\mathbf{c}_{X,1}$. Remember that the variance is a measure of dispersion; therefore, this inequality indicates that the covariance will be large if the data has large ranges. When the sets are totally dependent, $\left|\sigma_{X,1,2}\right| = \sigma_{X,1}\sigma_{X,2}$.

It is important to stress that the covariance measures a linear relationship. In general, data can be related to each other in different ways. For example, the colour of a pixel can increase exponentially as the heat of a surface increases, or the area of a region can increase in square proportion to its radius. However, the covariance only measures the degree of linear dependence. If features are related by another relationship, for example quadratic, then the covariance will produce a low value, even if there is perfect relationship. Linearity is generally considered to be the main limitation of PCA; however, PCA has proved to give a simple and effective solution in many applications; linear modelling is a very common model for many data and covariance is particularly good if you are using some form of linear classification.

To understand the linearity in the covariance definition, we can consider that features $\mathbf{c}_{\mathbf{X},2}$ are a linear function of $\mathbf{c}_{\mathbf{X},1}$. That is, $\mathbf{c}_{\mathbf{X},2} = A\mathbf{c}_{\mathbf{X},1} + \mathbf{B}$ for A, an arbitrary constant, and $\mathbf{B}$, an arbitrary column vector. Thus, according to Equation 12.11,

$$E\left[\mathbf{c}_{\mathbf{X},1}, \mathbf{c}_{\mathbf{X},2}\right] = E\left[A\mathbf{c}_{\mathbf{X},1}^{T}\mathbf{c}_{\mathbf{X},1} + \mathbf{c}_{\mathbf{X},1}^{T}\mathbf{B}\right] \tag{12.13}$$

We also have that

$$E\left[\mathbf{c}_{\mathbf{X},1}\right]E\left[\mathbf{c}_{\mathbf{X},2}\right] = AE\left[\mathbf{c}_{\mathbf{X},1}\right]^{2} + E\left[\mathbf{B}\right]E\left[\mathbf{c}_{\mathbf{X},1}\right] \tag{12.14}$$

By substitution of these equations in the definition of covariance in Equation 12.10, we have:

$$\sigma_{\mathbf{X},1,2} = A\left(E\left[\mathbf{c}_{\mathbf{X},1}^{T}\mathbf{c}_{\mathbf{X},1}\right] - E\left[\mathbf{c}_{\mathbf{X},1}\right]^{2}\right) \tag{12.15}$$

That is,

$$\sigma_{\mathbf{X},1,2} = A\sigma_{\mathbf{X},1}^{2} \tag{12.16}$$

That is, when features are related by a linear function, the covariance is a scalar value of the variance. We can follow a similar development to find the covariance as a function of $\sigma_{\mathbf{X},2}^{2}$. If we consider that $\mathbf{c}_{\mathbf{X},1} = \frac{1}{A}\mathbf{c}_{\mathbf{X},2} - \frac{\mathbf{B}}{A}$, then

$$\sigma_{\mathbf{X},1,2} = \frac{1}{A}\sigma_{\mathbf{X},2}^{2} \tag{12.17}$$

Thus, we can use Equations 12.16 and 12.17 to solve for A. That is,

$$A = \frac{\sigma_{\mathbf{X},2}}{\sigma_{\mathbf{X},1}} \tag{12.18}$$

By substitution in Equation 12.16, we have:

$$\sigma_{\mathbf{X}1,2} = \sigma_{\mathbf{X},1}\sigma_{\mathbf{X},2} \tag{12.19}$$

That is, the covariance value takes its maximum value given in Equation 12.12 when the features are related by a linear relationship.

12.4 Covariance matrix

When data has more than two dimensions, the covariance can be defined by considering every pair of components. These components are generally represented in a matrix that is called the covariance matrix. This matrix is defined as

$$\Sigma_{\mathbf{X}} = \begin{bmatrix} \sigma_{\mathbf{X},1,1} & \sigma_{\mathbf{X},1,2} & \cdots & \sigma_{\mathbf{X},1,n} \\ \sigma_{\mathbf{X},2,1} & \sigma_{\mathbf{X},2,2} & \cdots & \sigma_{\mathbf{X},2,n} \\ \vdots & \vdots & \ddots & \vdots \\ \sigma_{\mathbf{X},n,1} & \sigma_{\mathbf{X},n,2} & \cdots & \sigma_{\mathbf{X},n,n} \end{bmatrix} \tag{12.20}$$

According to Equation 12.5, the element (i, j) in the covariance matrix is given by

$$\sigma_{\mathbf{x},i,j} = E\left\lfloor (\mathbf{c}_{\mathbf{X},i} - \boldsymbol{\mu}_{\mathbf{X},i})(\mathbf{c}_{\mathbf{X},j} - \boldsymbol{\mu}_{\mathbf{X},j}) \right\rfloor \tag{12.21}$$

By generalizing this equation to the elements of the feature matrix, and by considering the notation used in Equation 12.6, the covariance matrix can be expressed as

$$\Sigma_{\mathbf{X}} = \frac{1}{m}\left((\mathbf{c}_{\mathbf{X}} - \boldsymbol{\mu}_{\mathbf{X}})^T (\mathbf{c}_{\mathbf{X}} - \boldsymbol{\mu}_{\mathbf{X}}) \right) \tag{12.22}$$

Here, $\boldsymbol{\mu}_{\mathbf{X}}$ is the matrix that has columns $\boldsymbol{\mu}_{\mathbf{X},i}$. If you observe the definition of the covariance given in the previous section, you will notice that the diagonal of the covariance matrix defines the variance of a feature and that given the symmetry in the definition of the covariance, the covariance matrix is symmetric.

A third way of defining the covariance matrix is by using the definition in Equation 12.10. That is,

$$\Sigma_{\mathbf{X}} = \frac{1}{m}\left(\mathbf{c}_{\mathbf{X}}^T \mathbf{c}_{\mathbf{X}} \right) - \boldsymbol{\mu}_{\mathbf{X}}^T \boldsymbol{\mu}_{\mathbf{X}} \tag{12.23}$$

The covariance matrix gives important information about the data. For example, by observing values close to zero we can highlight independent features useful for classification. Very high or low values indicate dependent features that will not give any new information useful for distinguishing groups in the data. PCA exploits this type of observation by defining a method to transform data in a way that the covariance matrix becomes diagonal. That is, all the values but the diagonal are zero. In this case, the data has no dependences, so features can be used to form groups. Imagine that you have a feature that is not dependent on others; then by choosing a threshold you can clearly distinguish between two groups independently of the values of other features. PCA also provides information about the importance of elements in the new data. So you can distinguish between important data for classification or compression.

12.5 Data transformation

We are looking for a transformation $\mathbf{W}$ that maps each feature vector defined in the set $\mathbf{X}$ into another feature vector for the set $\mathbf{Y}$, such that the covariance matrix of the elements in $\mathbf{Y}$ is diagonal. The transformation is linear and it is defined as

$$\mathbf{c}_{\mathbf{Y}} = \mathbf{c}_{\mathbf{X}} \mathbf{W}^T \tag{12.24}$$

or more explicitly

$$\begin{bmatrix} y_{1,1} & y_{1,2} & \cdots & y_{1,n} \\ y_{2,1} & y_{2,2} & \cdots & y_{2,n} \\ \vdots & \vdots & \cdots & \vdots \\ y_{m,1} & y_{m,2} & \cdots & y_{m,n} \end{bmatrix} = \begin{bmatrix} x_{1,1} & x_{1,2} & \cdots & x_{1,n} \\ x_{2,1} & x_{2,2} & \cdots & x_{2,n} \\ \vdots & \vdots & \cdots & \vdots \\ x_{m,1} & x_{m,2} & \cdots & x_{m,n} \end{bmatrix} \begin{bmatrix} w_{1,1} & w_{2,1} & \cdots & w_{n,1} \\ w_{1,2} & w_{2,2} & \cdots & w_{n,2} \\ \vdots & \vdots & \cdots & \vdots \\ w_{1,n} & w_{2,n} & \cdots & w_{n,n} \end{bmatrix} \tag{12.25}$$

Notice that

$$\mathbf{c}_{\mathbf{Y}}^T = \mathbf{W} \mathbf{c}_{\mathbf{X}}^T \tag{12.26}$$

or more explicitly

$$
\begin{bmatrix}
y_{1,1} & y_{2,1} & \cdots & y_{m,1} \\
y_{1,2} & y_{2,2} & \cdots & y_{m,2} \\
\vdots & \vdots & \cdots & \vdots \\
y_{1,n} & y_{2,n} & \cdots & y_{m,n}
\end{bmatrix}
=
\begin{bmatrix}
w_{1,1} & w_{1,2} & .. & w_{1,n} \\
w_{2,1} & w_{2,2} & .. & w_{2,n} \\
\vdots & \vdots & .. & \vdots \\
w_{n,1} & w_{n,2} & .. & w_{n,n}
\end{bmatrix}
\begin{bmatrix}
x_{1,1} & x_{2,1} & \cdots & x_{m,1} \\
x_{1,2} & x_{2,2} & \cdots & x_{m,2} \\
\vdots & \vdots & \cdots & \vdots \\
x_{1,n} & x_{2,n} & \cdots & x_{m,n}
\end{bmatrix}
\tag{12.27}
$$

To obtain the covariance of the features in $\mathbf{Y}$ based on the features in $\mathbf{X}$, we can substitute $\mathbf{c_Y}$ and $\mathbf{c_Y^T}$ in the definition of the covariance matrix as

$$
\Sigma_{\mathbf{Y}} = \frac{1}{m} \left[\left(\mathbf{Wc_X^T} - E\left[\mathbf{Wc_X^T} \right] \right) \left(\mathbf{c_X W^T} - E\left[\mathbf{c_X W^T} \right] \right) \right]
\tag{12.28}
$$

By factorizing $\mathbf{W}$, we have:

$$
\Sigma_{\mathbf{Y}} = \frac{1}{m} \left[\mathbf{W} \left(\mathbf{c_X} - \boldsymbol{\mu}_{\mathbf{X}} \right)^T \left(\mathbf{c_X} - \boldsymbol{\mu}_{\mathbf{X}} \right) \mathbf{W}^T \right]
\tag{12.29}
$$

or

$$
\Sigma_{\mathbf{Y}} = \mathbf{W} \Sigma_{\mathbf{x}} \mathbf{W}^T
\tag{12.30}
$$

Thus, to transform feature vectors, we can use this equation to find the matrix $\mathbf{W}$ such that $\Sigma_{\mathbf{Y}}$ is diagonal. This problem is known in matrix algebra as matrix diagonalization.

12.6 Inverse transformation

In the previous section we defined a transformation from the features in $\mathbf{X}$ into a new set $\mathbf{Y}$ whose covariance matrix is diagonal. To map $\mathbf{Y}$ into $\mathbf{X}$ we should use the inverse of the transformation. However, this is greatly simplified since the inverse of the transformation is equal to its transpose. That is,

$$
\mathbf{W}^{-1} = \mathbf{W}^T
\tag{12.31}
$$

This definition can been proven by considering that

$$
\Sigma_{\mathbf{X}} = \mathbf{W}^{-1} \Sigma_{\mathbf{Y}} \left(\mathbf{W}^T \right)^{-1}
\tag{12.32}
$$

But since the covariance is symmetric, $\Sigma_{\mathbf{x}} = \Sigma_{\mathbf{x}}^T$ and

$$
\mathbf{W}^{-1} \Sigma_{\mathbf{Y}} \left(\mathbf{W}^T \right)^{-1} = \left(\mathbf{W}^{-1} \right)^T \Sigma_{\mathbf{Y}} \left(\left(\mathbf{W}^T \right)^{-1} \right)^T
\tag{12.33}
$$

which implies that

$$
\mathbf{W}^{-1} = \left(\mathbf{W}^{-1} \right)^T \text{ and } \left(\mathbf{W}^T \right)^{-1} = \left(\left(\mathbf{W}^T \right)^{-1} \right)^T
\tag{12.34}
$$

These equations can only be true if the inverse of $\mathbf{W}$ is equal to its transpose.

Thus, to obtain the features in $\mathbf{X}$ from the $\mathbf{Y}$ we have that $\mathbf{c_Y^T} = \mathbf{Wc_X^T}$ can be written as

$$
\mathbf{W}^{-1} \mathbf{c_Y^T} = \mathbf{W}^{-1} \mathbf{Wc_X^T}
\tag{12.35}
$$

That is,

$$
\mathbf{W}^T \mathbf{c_Y^T} = \mathbf{c_X^T}
\tag{12.36}
$$

This equation is important for reconstructing data in compression applications. In compression, the data c_X is approximated by using this equation by considering only the most important components of c_Y.

12.7 Eigenproblem

By considering that $W^{-1} = W^T$, we can write Equation 12.30 as

$$\Sigma_X W^T = W^T \Sigma_Y \tag{12.37}$$

We can write the right-hand side in more explicit form as

$$W^T \Sigma_Y = \begin{bmatrix} w_{1,1} & w_{2,1} & .. & w_{n,1} \\ w_{1,2} & w_{2,2} & .. & w_{n,2} \\ : & : & .. & : \\ w_{1,n} & w_{2,n} & .. & w_{n,n} \end{bmatrix} \begin{bmatrix} \lambda_1 & 0 & .. & 0 \\ 0 & \lambda_2 & .. & 0 \\ : & : & : & : \\ 0 & 0 & .. & \lambda_n \end{bmatrix}$$

$$= \lambda_1 \begin{bmatrix} w_{1,1} \\ w_{1,2} \\ : \\ w_{1,n} \end{bmatrix} + \lambda_2 \begin{bmatrix} w_{2,1} \\ w_{2,2} \\ : \\ w_{2,n} \end{bmatrix} + \cdots + \lambda_n \begin{bmatrix} w_{n,1} \\ w_{n,2} \\ : \\ w_{n,n} \end{bmatrix} \tag{12.38}$$

Here, diagonal elements of the covariance have been named λ using the notation used in matrix algebra.

Similarly, for the left-hand side we have:

$$\Sigma_X W^T = \Sigma_X \begin{bmatrix} w_{1,1} \\ w_{1,2} \\ : \\ w_{1,n} \end{bmatrix} + \Sigma_X \begin{bmatrix} w_{2,1} \\ w_{2,2} \\ : \\ w_{2,n} \end{bmatrix} + \cdots + \Sigma_X \begin{bmatrix} w_{n,1} \\ w_{n,2} \\ : \\ w_{n,n} \end{bmatrix} \tag{12.39}$$

That is,

$$\Sigma_X \begin{bmatrix} w_{1,1} \\ w_{1,2} \\ : \\ w_{1,n} \end{bmatrix} + \Sigma_X \begin{bmatrix} w_{2,1} \\ w_{2,2} \\ : \\ w_{2,n} \end{bmatrix} + \cdots + \Sigma_X \begin{bmatrix} w_{n,1} \\ w_{n,2} \\ : \\ w_{n,n} \end{bmatrix} = \lambda_1 \begin{bmatrix} w_{1,1} \\ w_{1,2} \\ : \\ w_{1,n} \end{bmatrix} + \lambda_2 \begin{bmatrix} w_{2,1} \\ w_{2,2} \\ : \\ w_{2,n} \end{bmatrix} + \cdots + \lambda_n \begin{bmatrix} w_{n,1} \\ w_{n,2} \\ : \\ w_{n,n} \end{bmatrix} \tag{12.40}$$

Thus, W can be found by solving the following equations

$$\Sigma_X w_i = \lambda_i w_i \tag{12.41}$$

for $\mathbf{w}_i$, the ith row of $\mathbf{W}$. λ_i define the eigenvalues and $\mathbf{w}_i$ define the eigenvectors. 'Eigen' is a German word meaning 'hidden', and there are alternative names such as characteristic values and characteristic vectors.

12.8 Solving the eigenproblem

In the eigenproblem formulated in the previous section, we know Σ_X and we want to determine $\mathbf{w}_i$ and λ_i. To find them, first you should notice that $\lambda_i \mathbf{w}_i = \lambda_i \mathbf{I} \mathbf{w}_i$, where $\mathbf{I}$ is the identity matrix. Thus, we can write the eigenproblem as

$$\lambda_i \mathbf{I} \mathbf{w}_i - \Sigma_X \mathbf{w}_i = 0 \tag{12.42}$$

or

$$(\lambda_i \mathbf{I} - \Sigma_X)\, \mathbf{w}_i = 0 \tag{12.43}$$

A trivial solution is obtained for $\mathbf{w}_i$ equal to zero. Other solutions exist when the determinant det is given by

$$\det(\lambda_i \mathbf{I} - \Sigma_X) = 0 \tag{12.44}$$

This is known as the characteristic equation and it is used to solve for the values of λ_i. Once the values of λ_i are known, they can be used to obtain the values of $\mathbf{w}_i$. According to the previous formulations, each λ_i is related to one in $\mathbf{w}_i$. However, several λ_i can have the same value. Thus, when a value λ_i is replaced in $(\lambda_i \mathbf{I} - \Sigma_X)\, \mathbf{w}_i = 0$, the solution should be determined by combining all the independent vectors obtained for all λ_i. According to the formulation in the previous section, once the eigenvectors $\mathbf{w}_i$ are known, the transformation $\mathbf{W}$ is simply obtained by considering $\mathbf{w}_i$ as its columns.

12.9 PCA method summary

The mathematics of PCA can be summarized in the following steps.

1. Obtain the feature matrix $\mathbf{c}_X$ from the data. Each column of the matrix defines a feature vector.
2. Compute the covariance matrix Σ_X. This matrix gives information about the linear independence between the features.
3. Obtain the eigenvalues by solving the characteristic equation $\det(\lambda_i \mathbf{I} - \Sigma_X) = 0$. These values form the diagonal covariance matrix Σ_Y. Since the matrix is diagonal, each element is the variance of the transformed data.
4. Obtain the eigenvectors by solving for $\mathbf{w}_i$ in $(\lambda_i \mathbf{I} - \Sigma_X)\, \mathbf{w}_i = 0$ for each eigenvalue. Eigenvectors should be normalized and linear independent.
5. The transformation $\mathbf{W}$ is obtained by considering the eigenvectors as their columns.
6. Obtain the transform features by computing $\mathbf{c}_Y = \mathbf{c}_X \mathbf{W}^T$. The new features are linearly independent.
7. For classification applications, select the features with large values of λ_i. Remember that λ_i measures the variance and features that have large range of values will have large variance.

For example, two classification classes can be obtained by finding the mean value of the feature with largest λ_i.

8. For compression, reduce the dimensionality of the new feature vectors by setting to zero components with low λ_i values. Features in the original data space can be obtained by $\mathbf{c}_X^T = \mathbf{W}^T \mathbf{c}_Y^T$.

12.10 Example

Code 12.1 is a Matlab implementation of PCA, illustrating the method by a simple example with two features in the matrix cx.

In the example code, the covariance matrix is called CovX and it is computed by the Matlab function cov. The code also computes the covariance by evaluating the two alternative definitions given by Equations 12.22 and 12.23. Notice that the implementation of these equations divides the matrix multiplication by $m - 1$ instead of m. In statistics, this is called an unbiased estimator and it is the estimator used by Matlab in the function cov. Thus, we use $m - 1$ to obtain the same covariance values as the Matlab function.

To solve the eigenproblem, we use the Matlab function eig. This function solves the characteristic equation $\det(\lambda_i \mathbf{I} - \Sigma_\mathbf{X}) = 0$ to obtain the eigenvalues and find the eigenvectors. In the code the results of this function are stored in the matrices $\mathbf{L}$ and $\mathbf{W}$, respectively. In general, the characteristic equation defines a polynomial of higher degree requiring elaborate numerical methods to find its solution. In our example, we have only two features, thus the characteristic equation defines the quadratic form

$$\lambda_i^2 - 1.208\lambda_i + 0.039 = 0 \tag{12.45}$$

for which the eigenvalues can be easily obtained as $\lambda_1 = 0.0331$ and $\lambda_1 = 1.175$. The eigenvectors can be obtained by substitution of these values in the eigenproblem. For example, for the first eigenvector, we have:

$$\begin{bmatrix} 0.033 - 0.543 & -0.568 \\ -0.568 & 0.033 - 0.665 \end{bmatrix} \mathbf{w}_1 = 0 \tag{12.46}$$

Thus,

$$\mathbf{w}_1 = \begin{bmatrix} -1.11s \\ s \end{bmatrix} \tag{12.47}$$

where s is an arbitrary constant. After normalizing this vector, we obtain the first eigenvector

$$\mathbf{w}_1 = \begin{bmatrix} -0.74 \\ 0.66 \end{bmatrix} \tag{12.48}$$

Similarly, the second eigenvector is obtained as

$$\mathbf{w}_2 = \begin{bmatrix} 0.66 \\ 0.74 \end{bmatrix} \tag{12.49}$$

```
%PCA

%Feature Matrix cx. Each column represents a feature and
%each row a sample data
cx= [1.4000 1.55000
     3.0000 3.2000
     0.6000 0.7000
     2.2000 2.3000
     1.8000 2.1000
     2.0000 1.6000
     1.0000 1.1000
     2.5000 2.4000
     1.5000 1.6000
     1.2000 0.8000
     2.1000 2.5000 ];
[m,n]=size(cx);

%Data Graph
figure(1);
plot(cx(:,1),cx(:,2),'k+');        hold on;   %Data
plot(((0,0]),((-1,4]),'k-');       hold on;   %X axis
plot(((-1,4]),((0,0]),'k-');                  %Y axis
axis([-1,4,-1,4]);
xlabel('Feature 1');
ylabel('Feature 2');
title('Original Data');
%Covariance Matrix
covX=cov(cx)

%Covariance Matrix using the matrix definition
meanX=mean(cx)         %mean of all elements of each row

cx1=cx(:,1)-meanX(1);    %substract mean of first row in cx
cx2=cx(:,2)-meanX(2);    %substract mean of second row in cx

Mcx=[cx1 cx2];
covX=(transpose(Mcx)*(Mcx))/(m-1) %definition of covariance

%Covariance Matrix using alternative definition
meanX=mean(cx);             %mean of all elements of each row

cx1=cx(:,1);     %substract mean of first row in cx
cx2=cx(:,2);     %substract mean of second row in cx

covX=((transpose(cx)*(cx))/(m-1) )-
((transpose(meanX)*meanX)*(m/(m-1)))

%Compute Eigenvalues and Eigenvector
[W,L]=eig(covX)      %W=Eigenvalues   L=Eigenvector
```

```
%Eigenvector Graph
figure(2);
plot(cx(:,1),cx(:,2),'k+');                        hold on;
plot(([0,W(1,1)*4]),([0,W(1,2)*4]),'k-');          hold on;
plot(([0,W(2,1)*4]),([0,W(2,2)*4]),'k-');
axis([-4,4,-4,4]);
xlabel('Feature 1');
ylabel('Feature 2');
title('Eigenvectors');

%Transform Data
cy=cx*transpose(W)

%Graph Transformed Data
figure(3);
plot(cy(:,1),cy(:,2),'k+');        hold on;
plot(([0,0]),([-1,5]),'k-');       hold on;
plot(([-1,5]),([0,0]),'k-');
axis([-1,5,-1,5]);
xlabel('Feature 1');
ylabel('Feature 2');
title('Transformed Data');

%Classification example
meanY=mean(cy);

%Graph of classification example
figure(4);
plot(([-5,5]),([meanY(2),meanY(2)]),'k:');   hold on;
plot(([0,0]),([-5,5]),'k-');                 hold on;
plot(([-1,5]),([0,0]),'k-');                 hold on;
plot(cy(:,1),cy(:,2),'k+');                  hold on;
axis([-1,5,-1,5]);
xlabel('Feature 1');
ylabel('Feature 2');
title('Classification Example');
legend('Mean',2);

%Compression example
cy(:,1)=zeros;
xr=transpose(transpose(W)*transpose(cy));

%Graph of compression example
figure(5);
plot(xr(:,1),xr(:,2),'k+');        hold on;
plot(([0,0]),([-1,4]),'k-');       hold on;
plot(([-1,4]),([0,0]),'k-');
axis([-1,4,-1,4]);
xlabel('Feature 1');
ylabel('Feature 2');
title('Compression Example');
```

Code 12.1 Matlab PCA implementation

Figure 12.1 shows the original data and the eigenvectors. The eigenvector with the largest eigenvalue defines a line that goes through the points. This is the direction of the largest variance of the data.

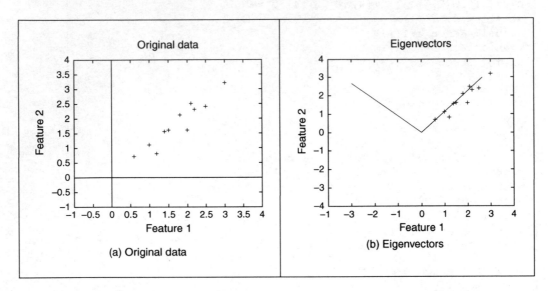

Figure 12.1   Data samples and the eigenvectors

Figure 12.2 shows the results obtained by transforming the features $c_Y = c_X W^T$. Basically, the eigenvectors become our main axes. The second feature has points more spread along the axis, and this is related to a higher value in the eigenvector. Remember that for the transformed data, the covariance matrix is diagonal, thus there is not any linear dependence between the features.

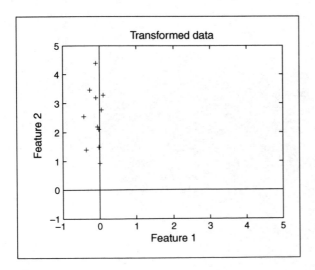

Figure 12.2   Transformed data

If we want to classify our data in two classes, we should consider the variation along the second transformed feature. Since we are using the axis with the highest eigenvalue, the classification is performed along the axis with highest variation in the data. In Figure 12.3, we divide the points by the line defined by the mean value.

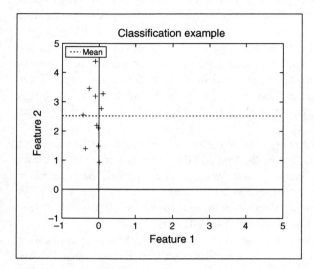

Figure 12.3 Classification via PCA

For compression, we want to eliminate the components that have less variation, so in our example, we eliminate the first feature. In the last part of the Matlab implementation, data is reconstructed by setting to zero the vales of the first feature in the matrix cy. The result is shown in Figure 12.4. Notice that the loss of one dimension in the transformed set produces

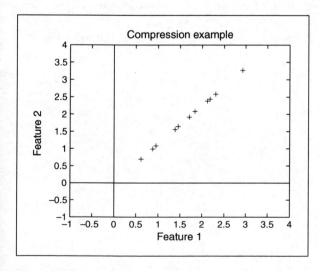

Figure 12.4 Compression via PCA

data aligned in the original space. So some variation in the data has been lost. However, the variation along the first eigenvector is maintained.

Data with two features, as shown in this example, may be useful in some applications such as reducing a stereo signal into a single channel. Other low-dimensional data such as three features can be used to reduce colour images to grey level. However, in general, PCA is applied to data with many features. In these cases, the implementation is practically the same, but it should compute the eigenvalues by solving a characteristic equation defining a polynomial of high degree.

Data with many features are generally used for image classification wherein features are related to image metrics or to pixels. For example, face classification has been done by representing pixels in an image as features. Pixels are arranged in a vector and a set of eigenfaces is obtained by PCA. For classification, a new face is compared to the others by computing a new image according to the transformation obtained by PCA. The advantage is that PCA has independent features.

Another area that has used PCA extensively is image compression. In this case, pixels with the same position are used for the vectors. That is, the first feature vector is formed by grouping all the values of the first pixel in all the images. Thus, when PCA is applied, the pixel value on each image can be obtained by reconstructing data with a reduced set of eigenvalues. As the number of eigenvalues is reduced, most information is lost. However, if you choose low eigenvalues, then the information that is lost represents low data variations.

Although classification and compression are perhaps the most important areas of application for PCA, this technique can be used to analyse any kind of data. Applications of PCA are continuously being developed in many areas of research. For example, PCA has been used in applications as diverse as compressing animation in 3D models and analysing data in spectroscopy. The difference in each application is how data is interpreted, but the fundamentals of PCA are the same.

12.11 References

Anton, H., *Elementary Linear Algebra: With Applications*, John Wiley and Sons, New York, 2005
DeGroot, M. H. and Schervish, M. J., *Probability and Statistics*, 3rd edn, Addison Wesley, Reading, MA, 2001

Index

accumulator array, 189–191, 197–206
active appearance models, 272–275
active contours (see snakes), 244–266
   parametric, 244–261
   geometric, 261–266
active pixel, 11
active shape models, 272–275
   comparison, 275
active contour without edges, 264–265
acuity, 5
adaptive Hough transform, 235
addition, 20, 72
additive operator splitting, 264
Adoculos, 15
affine
   invariance, 167, 178, 227, 281, 324
   moments, 324
   transformation 359
ageing, 11
aliasing, 45–46
   antialiasing, 198
analysis of 1st order edge operators,
   119–129
anisotropic diffusion, 96–101
antialiasing, 198
aperture problem, 172
arbitrary shape extraction, 186–193, 221–235
area description, 311
artificial neural networks, 343
aspect ratio, 14
associative cortex, 7
autocorrelation, 41, 159
averaging error, 90
averaging operator, 84–90
   direct, 84–88
   Gaussian, 88–90

backmapping, 199, 206
background
   estimation, 93–94, 167
   subtraction, 167
band-pass filter, 65, 125, 152

bandwidth, 6, 12, 34, 65, 86, 152
basis functions, 51, 59, 61, 63
Benham's disk, 8
Bhattacharyya distance, 339
binary morphology, 103–107
biometrics, 2, 101, 236, 343
blind spot, 5
blooming, 12
boundary, 282
boundary descriptors, 282–311
Bresenham's algorithm
   lines, 199
   circles, 206
brightness, 16, 34
   addition, 72
   clipping, 72
   division, 72
   inversion, 16–18, 72
   multiplication, 72
   scaling, 72
Brodatz texture images, 330
burn, 11

C implementation, 14
C++, 14
camera, 10–12
   ageing, 11
   blooming, 12
   burn, 11
   CCD, 10–12
   CCIR standard, 10
   CMOS, 10–12
   digital, 10, 14
   digital video, 14
   high resolution, 12
   hyperspectral, 12
   infrared, 12
   interlacing, 13
   lag, 11
   low-light, 12
   progressive scan, 14
   readout effects, 12
   vidicon, 10

Canny edge detection operator, 129–136
canonical analysis, 343
cartesian moments, 315–320
CCD camera, 10–12
CCIR standard, 10
CMOS camera, 10–12
central limit theorem, 90, 187, 331
centralised moments, 316
chain codes, 283–285
charge coupled device, 10–12
Chebyshev moments, 324
choroid, 4
chrominance, 7
ciliary muscles, 4
circle drawing, 206
circle finding, 203–207, 212–216
classification, 339–343
clipping, 72
closing operator, 106
    closure, 106
coding, 14, 36, 59, 63, 274
colour, 34–36
compactness, 312–314, 338
comparison
    active shape, 275
    circle extraction, 235
    corner extraction, 158, 167
    deformable shapes, 275
    edge detection, 145, 156
    filtering images, 97, 102
    Hough transform, 210, 235
    moments, 315, 324
    optical flow, 177
    statistical operators, 102
    template matching, 275
    texture, 335, 343
    thresholding, 136
complete snake implementation, 251
complementary metal oxide silicon, 10
complex magnitude, 39
complex moments, 324
complex phase, 39
computer software, 14
computer vision system, 10–14
computer interface, 12–14
computerised tomography, 2
cones, 5
    types, 5
connectivity analysis, 135, 282
continuous Fourier transform, 37–42
continuous signal, 37
continuous symmetry operator, 271

convolution, 41, 59, 86, 186–192
    duality, 41, 86
    template, 81–84, 118–119, 186
co-occurrence matrix, 335–337
co-ordinate systems, 16, 355–380
corner detection, 153–163
    chain code, 295
    comparison, 158, 167
    differencing, 154–156
    differentiation, 156–158
    Harris operator, 159–163
    improvement, 163
    Moravec operator, 159
    performance, 158, 163
correlation, 41, 159, 170, 188, 193–194
    function, 159
correlation optical flow, 167, 170
cosine transform, 58–59, 335
cross-correlation, 167, 188, 193
cubic splines, 326
curvature, 145, 152, 246, 249, 263, 326
    definition, 153–154
    primal sketch, 163
    scale space, 163
curve fitting, 165, 382
CVIP tools, 15

d.c. component, 48, 52, 64
deformable template, 242–243
delta function, 41
demonstrations, 25, 28
Deriche operator, 131
descriptors
    3D Fourier, 310
    elliptic Fourier, 301–310
    Fourier, 286–310
    real Fourier, 290–292
    region, 311–326, 337
    texture, 332–339
digital camera, 14
    video, 14
difference of Gaussian, 140, 164
differential optical flow, 171–176
digital video camera, 12, 14
dilation, 104–109
direct averaging, 84–88
discrete cosine transform, 58–59, 335
discrete Fourier transform, 47–54, 129, 324,
discrete Hartley transform, 59–61
discrete sine transform, 59
discrete symmetry operator, 268–272

distance measure, 325, 339–340
    Euclidean, 248, 339
distance transform, 266–268
drawing lines, 199
drawing circles, 206
dual snake (active contour), 258–260
duality convolution, 41, 86

Ebbinghaus illusion, 8
edge detector, 117–147
    Canny, 129–136
    comparison, 145, 156
    Deriche, 131
    first order, 117–136
    horizontal, 117
    Laplacian, 137–139, 145
    Laplacian of Gaussian, 139
    Marr-Hildreth, 139–144, 147
    Petrou, 144
    Prewitt, 121–123, 145
    Roberts cross, 120–121
    second order, 137–144
    Sobel, 123–129, 145
    Spacek, 144
    survey, 156
    Susan, 145
    surveys, 146
    vertical, 117
edge
    direction, 121–123, 126–128, 131, 139
    magnitude, 121–123
    vectorial representation, 121
eigenvalue, 161, 273, 385
eigenvector, 273, 385
ellipse finding, 207–209, 216–221
elliptic Fourier descriptors, 301–310
energy, 149, 333
energy minimisation, 243, 245, 326
entropy, 166, 333
equalisation, 75–77
erosion, 104–109
estimation of background, 93–94, 167
estimation theory, 381
Euclidean distance, 248, 339
Euler number, 315
evidence gathering, 197
example worksheets, 21, 24, 349–354
eye, 4–6

face recognition, 2, 28, 63, 260, 274
fast Fourier transform, 50, 86, 193, 295

fast Hough transform, 235
fast marching methods, 264
feature space, 339
feature extraction, 2–347(!)
feature subset selection, 343
FFT application, 102, 193, 332
fields, 13
filter
    averaging, 84–90
    band-pass, 65, 125, 152
    high-pass, 65, 121, 128, 143
    low-pass, 64, 85, 128, 143
    median, 91–93, 102, 113
    mode, 94–96
    truncated median, 94–96, 102
filtering image comparison, 97, 102
firewire, 13
first order edge detection, 117–136
fixed pattern noise, 12
flash A/D converter, 12
fexible shape extraction, 241–279
flexible shape models, 272
flow detection, 167–178
foot-of-normal description, 200
force field transform, 101–102
form factor, 196
fovea, 5
Fourier descriptors, 285–311
    3D, 310
    elliptic, 301–310
    real Fourier, 290–292
Fourier transform, 37–42
    applications, 64–66, 86, 147, 332
    display, 52, 332
    discrete, 47–54, 129, 324
    frequency scaling, 56–57, 332
    inverse, 38–39, 49, 149
    log polar, 196
    Mellin, 196
    moments, 324
    of Sobel operator, 128–129
    of Marr-Hildreth operator, 142–143
    ordering, 52
    pair, 40, 42, 47, 53
    phase congruency, 147
    pulse, 38
    reconstruction, 38–39, 49, 149, 324
    replication, 50
    reordering, 52
    rotation, 56
    separability, 51
    shift invariance, 54–55, 289–290

Fourier transform, *(Cont'd)*
 superposition, 57
 texture analysis, 332
Fourier-Mellin transform, 196
frames, 12
framegrabber, 13
frequency domain, 36
frequency scaling, 56–57, 332
frequency, 36
fuzzy Hough Transform, 235

Gabor wavelet, 61–62, 151, 334
 log-Gabor, 151–152
gait recognition, 177, 236, 271, 324
Gaussian
 averaging, 88–90
 function, 41, 53, 88–90
 noise, 90, 186, 381
 operator, 88–90
 smoothing, 97, 103, 124, 130
general form of Sobel operator, 125
generalized Hough transform, 221–228
Generic Image Library, 15
genetic algorithm, 244
geometric active contour, 261–266
GIL, 15
greedy algorithm, 246
 implementation, 248–227
greedy snake, 246–252
greyscale, 17, 34
grey level morphology, 107–112
group operations, 81–90

Hamming window, 87, 195
Hanning window, 87, 195
Harris corner detector, 159–163
Hartley transform, 59–61
high resolution camera, 12
high-pass filter, 65, 121, 128, 143
histogram, 70
 equalisation, 75–77
 normalisation, 74–75
hit or miss operator, 103–104
homogeneous co-ordinate system, 16, 357–379
homography, 360
horizontal edge detection, 117
horizontal optical flow, 172
Hotelling transform, 63, 385
Hough Transform (HT), 196–236
 adaptive, 235

antialiasing, 198
backmapping, 199, 206
circles, 205–207, 212–216
ellipses, 207–210, 217–221
fast, 235
fuzzy, 235
generalised, 221–228
invariant, 228–235
lines, 197–205
mapping, 197
noise, 198, 206
occlusion, 198, 206
polar lines, 200–202
probabilistic, 235
randomized, 235
reviews, 235
velocity, 235
human eye, 4–6
human vision, 1–9
hyperspectral camera, 12
hysteresis thresholding, 132–136

IEEE 1394, 13
illumination, 117, 164, 177, 183, 332
image coding, 14, 36, 59, 63
image filtering comparison, 97, 102
image formation, 34–36
image processing, 2–347(!)
image texture, 2, 57, 257, 330–339
inclusion operator, 108
inertia, 334
infrared camera, 12
interlacing, 14
intensity normalization, 74–75
invariance, 167, 183, 281, 332
 affine, 167, 178, 227, 281, 324
 illumination, 117, 164, 177, 183, 332
 location, 164, 169, 183, 281, 294, 317
 position, 165, 169, 183, 281, 294, 317
 projective, 167, 281
 rotation, 164, 172, 196, 281, 305, 317
 shift, 54–55, 164, 196, 289–290, 332
 start point, 285
 scale, 164, 183, 191, 196, 305, 317 332
invariant Hough transform, 228–235
inverse Fourier transform, 38–39, 49, 149
inversion of brightness, 16–18, 72
iris, 4
irregularity, 312, 337
isochronous transfer, 11

Java, 14
journals, 24–25
JPEG coding, 14, 36, 59, 167

Karhunen-Loeve transform, 63, 273, 385–398
Kass snake, 252–257
kernel methods, 343
Khoros, 14
k-nearest neighbour rule, 339–343

L_1 and L_2 norms, 339
Laboimage, 15
lag, 11
Laplacian edge detection operator, 137–139
Laplacian of Gaussian, 139
    Fourier transform, 143
Laplacian operator, 137
lateral inhibition, 6
lateral geniculate nucleus, 7
least squares criterion, 381–382
Legendre moments, 324
lens, 4
level sets, 261–266, 276
line drawing, 199
line finding, 197–205, 210–212
line terminations, 157, 246
linearity, 41, 57–58
local energy, 149–151
location invariance, 164, 169, 183, 281, 294, 317
log-polar mappings, 196
logarithmic point operator, 73
look-up table, 12, 74
low-light camera, 12
low-pass filter, 64, 85–86, 128, 143
luminance, 7

Mach bands, 5
magazines, 25
magnetic resonance, 2
Maple mathematical system, 16
Marr-Hildreth edge detection, 139–144, 147
    Fourier transform, 142–143
Mathcad, 16–21
    example worksheet, 349–352
mathematical systems, 15
    Maple, 16
    Mathcad, 15–21
    Mathematica, 15
    Matlab, 15, 21–24

Matlab mathematical system, 15, 21–24
    example worksheet, 352–354
Matusita distance, 339
medial axis, 268
median filter, 91–93, 102, 113
Mellin transform, 196
mexican hat, 139
Minkowski operator, 109–112
mode, 94
mode filter, 94–96
moments, 315–325
    affine invariant, 324
    Cartesian, 315–320
    centralised, 316
    Chebyshev, 324
    complex, 324
    centralised, 316
    Fourier, 324
    Legendre, 324
    normalised central, 318–320
    pseudo-Zernike, 324
    reconstruction, 325
    reviews, 315, 324
    statistical, 315
    Tschebichef, 324
    velocity, 324
    Zernike, 320–324
Moravec corner operator, 159
morphology,
    binary, 103–107
    grey level, 107–112
motion detection, 167–177, 185
    area, 168–171
    differencing, 171–177, 185
    optical flow, 168–177
MPEG coding, 14, 59
multiplication of brightness, 72

narrow band, 264
nearest neighbour, 339
neighbours, 282
neural
    model, 8
    networks, 8, 305
    signals, 6
    system, 6–7
noise
    Gaussian, 90, 186, 381
    Rayleigh, 91, 96
    salt and pepper, 92, 145
    speckle, 96

non-maximum suppression, 130–133
normal force, 257
norms (distance), 339
normalisation, 74–75, 250
normalized central moments, 318–320
normal distribution, 123, 187, 381
NTSC, 13
Nyquist sampling criterion, 44

occipital cortex, 7
occlusion, 192–193
open contour, 257
opening operator, 106
open CV, 15
optical flow, 167–178
    comparison, 177
    differential, 171–176
    correlation, 167, 170
    horizontal, 172
    matching, 167
    vertical, 172
optical Fourier transform, 49, 195
optimal smoothing, 125, 129
optimal thresholding, 78–81
ordering of Fourier transform, 52
orthogonality, 207, 273, 322
orthographic projection, 16, 271, 366

PAL system, 13
palette, 34
parameter space reduction, 210–218
parametric active contour, 244–261
passive pixel, 11
pattern recognition, 25, 343–345
    statistical, 80, 315, 93, 315
    structural, 286, 343–345
perimeter, 311
    descriptors, 282–311
perspective, 16, 355–357
    camera model, 355–357
Petrou operator, 144, 146
phase, 39, 55
phase congruency, 147–152
photopic vision, 5
picture elements, 2,17
pixels, 2,17
    active, 11
    passive, 11
Poincarré measure, 317
point operators, 71–75

point distribution model, 272
polar co-ordinates, 191, 172
polar HT lines, 200–202
position invariance, 164, 169, 183, 281, 294, 317
Prewitt edge detection, 121–123, 145
projective invariance, 167, 281
primal sketch,
    curvature, 163
principal components analysis, 63, 273, 385–398
probabilistic Hough transform, 235
progressive scan camera, 14
pseudo Zernike moments, 324
pulse, 37

quadratic splines, 326
quantisation, 34–37
quantum efficiency, 12

Radon transform, 196
randomised HT, 235
rarity, 166
Rayleigh noise, 91, 96
readout effects, 12
real Fourier descriptors, 290–292
reconstruction,
    Fourier transform, 38–39, 49, 149, 324
    moments, 325
rectilinearity, 315
region, 282
region descriptors, 311–325, 337
regularization, 257
remote sensing, 2
reordering Fourier transform, 52
replication, 50
research journals, 24–25
retina, 5
review
    chain codes, 283
    circle extraction, 235
    corners, 167
    deformable shapes, 236, 276
    education, 26
    edge detection, 146
    Hough transform, 235
    level set methods, 276
    moments, 315, 324
    optical flow, 177
    pattern recognition, 343
    shape analysis, 326
    shape description, 326

template matching, 235
texture, 345
thresholding, 78
Roberts cross edge detector, 120–121
rods, 5
rotation invariance, 164, 172, 196, 281, 305, 317
rotation matrix, 161, 222, 361–362
R-table, 224

saliency, 166
salt and pepper noise, 92, 145
sampling criterion, 43–46, 290
sampling, 36–37, 290–291
sawtooth operator, 70
scale invariance, 164, 183, 191, 196, 305, 317
scale invariant feature transform
    SIFT, 164–166
scale space, 97, 113, 163, 165
    curvature, 163
scaling of brightness, 72
scotopic vision, 5
second order edge operators, 137–144
separability, 51
shape descriptions, 281–328
shape extraction,
    circle, 203–207, 212–216
    ellipses, 207–209, 216–221
    lines, 197–205, 210–212
    unknown shapes, 241–266
shape reconstruction, 287, 309, 323
shift invariance, 54–55, 164, 196, 289–290, 332
SIFT operator, 164–166
sinc function, 38, 41, 54
sine transform, 59
skeletonization, 266–272
skewed symmetry, 271
smoothness constraint, 172
snake, 244–266
    3D, 257
    active contour without edges, 264–265
    dual, 258–260
    geometric active contour, 261–266
    greedy 246–252
    Kass, 252–257
    normal force, 257
    open contour, 257
    parametric active contour, 244–261
    regularization, 257
Sobel edge detection operator, 123–129
    general form, 125
    Fourier transform 128–129

Spacek operator, 144, 146
speckle noise, 96
spectrum, 5, 38
splines, 326
start point invariance, 285
statistical geometric features, 337
statistical moments, 315
statistical pattern recognition, 80, 315,
        93, 315
structural pattern recognition, 286,
        343–345
structuring element, 103, 106
subtraction of background, 167
superposition, 57
support vector machine, 343
survey, see review
Susan operator, 145
symmetry, 268–272
    continuous operator, 271
    discrete operator, 268–271
    focus, 271
    skewed, 271
synthetic computer images, 3

television
    aspect ratio, 14
    interlacing, 14
    signal, 13
template
    convolution, 81–84, 118–119, 186
    matching, 186–196
    computation, 191
    Fourier transform, 193–195
    noise, 194
    occlusion, 194
    optimality, 187
    shape, 92
    size, 84, 87–88, 92
terminations, 157, 246
textbooks, 25–27
texture, 2, 57, 257, 330–339
    definition, 330
    classification, 339–343
    description, 332–339
    segmentation, 343–345
texture mapping, 93
thinning, 129, 326
thresholding, 77–81, 132–136, 184
    hysteresis, 132–136
    optimal, 78–81
    uniform, 77–78, 136, 119

TN-image, 15
transform
    adaptive Hough transform, 235
    continuous Fourier, 37–42
    discrete cosine, 58–59, 335
    discrete Fourier, 47–54, 129, 324
    discrete Hartley, 59–61
    discrete sine, 59
    distance, 266–268
    fast Fourier transform, 50, 86, 193, 295
    fast Hough transform, 235
    force field, 101–102
    Fourier-Mellin, 196
    Gabor wavelet, 61–63, 151–152,
    generalised Hough, 221–228
    Hotelling, 63, 385
    Hough, 196–236
    inverse Fourier, 38–39, 49, 149
    Karhunen Loève, 63, 273, 385–398
    Mellin, 196
    one-dimensional Fourier, 47–49
    optical Fourier, 49, 195
    Radon, 196
    two-dimensional Fourier, 49–54
    Walsh, 63, 310, 335
    wavelet transform, 61–66, 334
transform pair, 40, 42, 47, 53
translation invariance, 164, 169, 183, 281,
    294, 317
true colour, 34
truncated median filter, 94–96, 102
Tschebichef moments, 324
two-dimensional Fourier transform, 49–54

ultrasound, 2, 96, 102, 145
    filtering, 96, 102, 145
umbra approach, 107
uniform thresholding, 77–78, 136, 119
unpredictability, 166

velocity, 168–176
    Hough transform, 235
    moments, 324
vertical edge detection, 117
vertical optical flow, 172
vidicon camera, 10
Visiquest, 14
vision, 1–9
VXL, 15

Walsh transform, 63, 310, 335
wavelets, 60–62, 277, 296, 305
wavelet transform, 61–66, 334
    Gabor, 61–63, 151–152,
weak perspective model, 369
windowing operators, 87, 195
worksheets, 21, 24, 349–354

z transform, 128
Zernike moments, 320–324
Zernike polynomials, 321
zero crossing detection, 137, 141–143, 384
zero padding, 194
Zollner illusion, 8

Lightning Source UK Ltd.
Milton Keynes UK
15 March 2011

169257UK00007B/26/P